Advances in Experimental Medicine and Biology

Volume 1029

More information about this series at http://www.springer.com/series/5584

Yasunori Sasakura
Editor

Transgenic Ascidians

Springer

Editor
Yasunori Sasakura
Shimoda Marine Research Center
University of Tsukuba
Shimoda, Shizuoka, Japan

Additional material to this book can be downloaded from http://extras.springer.com.

ISSN 0065-2598 ISSN 2214-8019 (electronic)
Advances in Experimental Medicine and Biology
ISBN 978-981-10-7544-5 ISBN 978-981-10-7545-2 (eBook)
https://doi.org/10.1007/978-981-10-7545-2

Library of Congress Control Number: 2018934428

Printed on acid-free paper

This Springer imprint is published by the registered company Springer Nature Singapore Pte Ltd. part of Springer Nature.
The registered company address is: 152 Beach Road, #21-01/04 Gateway East, Singapore 189721, Singapore

Foreword

The ascidian *Ciona* provides one of the best experimental systems in developmental biology, developmental genomics, and evolutionary developmental biology. We ascidian developmental biologists are justifiably proud of the fact that we established the *Ciona* system.

In traditional embryology, ascidians were recognized as organisms exhibiting a "mosaic-mode of embryogenesis", in contrast to the regulatory mode of embryogenesis seen, for example, in sea urchins. A detailed description of embryonic cell lineage by Edwin G. Conklin revealed that the mosaic-mode of embryogenesis is determined by determinate cleavage of the embryo and a robust pattern of differentiation and morphogenesis. Maternal factors and cellular communications that play essential roles in gastrulation, neurulation, tailbud-embryo formation, and tadpole-type larval development work together under the harmonious, but rigid control of a gene regulatory network. In addition, newly hatched larvae comprise only about 2600 cells, including 40 notochord cells and 38 muscle cells. Thus, ascidians are considered embryonically simple compared to vertebrates and other deuterostomes, such as sea urchins. In other words, ascidians represent a system to challenge the most basic question of developmental biology: How does a complex, multicellular, metazoan body arise from a single cell, the fertilized egg?

Decoding of the draft genome of *Ciona intestinalis* in 2002 identified almost all of the developmentally relevant genes in its genome, which represent the basic set of gene components in chordates prior to the two rounds of genome duplication that occurred in the vertebrate lineage. The absence of redundancy in ascidian regulatory gene functions makes it easier to ascertain developmental roles of individual genes. In association with the genome sequencing project, a cDNA sequencing project was also carried out, providing a great quantity of information on expression profiles of regulatory genes. Many whole-mount in situ hybridization studies reveal distinct gene expression profiles in embryos at the single-cell level. Furthermore, results obtained from embryological research provide straightforward insights into chordate evolution as well as the origin of vertebrates, because ascidians are the closest relatives to vertebrates.

In this research environment, it is desirable to introduce new, state-of-the-art techniques into the *Ciona* system, with the most valuable at this time being "transgenic" techniques. These include microinjection and/or electroporation of exogenous DNA to reveal mechanisms involved in gene regulatory networks, germ-line transgenesis to create various marker lines for studies of gene regulation, and TALEN-based and CRISPR/Cas9-based knockout lines to facilitate studies of gene function.

This book, edited by Prof. Yasunori Sasakura at the University of Tsukuba, constitutes a very timely discussion of recent advances in this field. In addition, most authors of this book are new or mid-career researchers. Indeed, we owe them further development of the ascidian systems. Lastly, I wish to express once again my conviction that *Ciona* is the most promising system to explore molecular and genetic mechanisms involved in animal embryogenesis at the single-cell level, the tissue and organ level, and the individual level.

Marine Genomics Unit
Okinawa Institute of Science and Technology
Graduate University,
Okinawa, Japan

Noriyuki Satoh

Preface

Undoubtedly, transgenic technologies are inevitable for studying molecular functions during development. In several model ascidian species, creation of transgenic animals has been routinely performed, and the methods are easily retrieved from many articles and from some book chapters. However, the details of the methods have not always been provided in the literature. On the one hand, microinjection is basically a simple method for introducing exogenous DNAs without the requirement of an expensive machine. At the same time, because microinjection can be a relatively difficult process, many laboratories have their own practical tips that are suitable for their study of species. These tips are usually not seen in original articles whose main purpose is describing development mechanisms. To solve the issue, this book has been purposed to gather and describe the devices that have arisen from our own great interest and enthusiasm, and that are not usually seen or known as major methods for performing transgenesis in ascidians.

I hope this book will be useful for all researchers, including beginners in tunicate research who wish to introduce a tunicate in their laboratories and also specialists of tunicates who wish to find a clue to improve their methods.

Shimoda, Japan Yasunori Sasakura

Contents

1 Introduction

Yasunori Sasakura

Abstract

The chordate ascidians, the major group of tunicates, is the best animal group for studying molecular and cellular processes underlying formation of a chordate body plan. For these studies, transgenic technologies are powerful. Transgenesis of ascidians has a long history of more than 20 years, and many practical tips have been accumulated. This book is aimed at summarizing the accumulated techniques in ascidians in addition to concrete research in which transgenic techniques have played pivotal roles. This book is useful for fast assimilation of the techniques and for learning the unique devices developed by the enthusiasm of ascidian researchers.

Keywords

Chordate · Ascidian · Vertebrate · *Ciona* · Genome · Transgenesis

1.1 The Aim of This Book

Marine invertebrate chordates, tunicates, is an important animal group for understanding the mechanisms of chordate evolution (Satoh 2003; Lemaire 2011). This is because tunicates are regarded as the sister group of vertebrates (Delsuc et al. 2006; see Chaps. 2, 7 and 8 for details). Tunicates have characteristics that define chordates, such as a notochord, a dorsal neural tube, a gill in the pharynx, and an endostyle/thyroid gland. Moreover, it has been recently suggested that tunicates have some features that were previously thought to be specific to vertebrates, such as neural crest cells and neurogenic placodes (Jeffery et al. 2004; Abitua et al. 2015). Therefore, tunicates are the key organisms for understanding how the chordate/vertebrate-specific features evolved. Among the tunicates, which include ascidians, larvaceans, and thaliaceans, ascidians have been centered on molecular development and genetic studies in this animal group. In addition to their importance from the point of view of phylogenetic position, ascidians possess advantageous features as experimental systems. The advantage of ascidians can be summarized in their simplicity. In the case of *Ciona intestinalis*, a representative ascidian species, the simplicity can be seen in three aspects:

- Fast development, which starts with gastrulation at 5 hours postfertilization (hpf), neurulation at 7 hpf. The larval stage starts within 1 day of fertilization, and metamorphosis is completed within a few days.
- Small cell numbers constituting embryonic and larval bodies. Early gastrula embryos

Y. Sasakura (✉)
Shimoda Marine Research Center, University of Tsukuba, Shimoda, Shizuoka, Japan
e-mail: sasakura@shimoda.tsukuba.ac.jp

Y. Sasakura (ed.), *Transgenic Ascidians*, Advances in Experimental Medicine and Biology 1029,
https://doi.org/10.1007/978-981-10-7545-2_1

consist of only 110 cells. Ascidian larvae possess about 3000 cells, and cell lineages of most cells have been described.

- Compact genome. The genome size of *Ciona intestinalis* is ~160 mega base pairs with ~16,000 genes (Dehal et al. 2002). The scores indicate that a gene is present in every 10 kilobase pairs of the *Ciona* genome. *Cis* elements controlling transcription of genes are usually located near the transcription start sites. The gene sets in the ascidian genomes are basically nonredundant (Satou et al. 2003; Sasakura et al. 2003), because ascidian genomes are thought not to have experienced whole genome duplication, which has occurred twice during vertebrate evolution.

Owing to these characteristics, ascidians are regarded as the best animal for studying molecular and particularly cellular processes underlying formation of the chordate body plan. For these studies, transgenic technologies are powerful and nowadays almost inevitable. In some model ascidians, the techniques have been reported more than 20 years ago, and thus have a long history (Hikosaka et al. 1992). Since then, the methods have been used in many applications, some of which are unique in ascidians (Iitsuka et al. 2014). These technological innovations are exceptional, particularly in marine invertebrates. This book is aimed at summarizing the accumulated transgenesis technologies in ascidians, in addition to concrete examples of research in which transgenic techniques have played pivotal roles. The information provided will be useful for ascidian researchers for fast assimilation of the techniques, and for biologists working with other organisms to learn the unique techniques and ingenious attempts specific to ascidians, in addition to knowledge from recent studies.

I would like to state here a potential issue that may confuse readers of this book. Recently, a few manuscripts have been published that propose new classification of *Ciona intestinalis*. In some locations such as Europe, so-called *Ciona intestinalis* populations have been suggested to be divided into type A and B (Caputi et al. 2007). The detailed comparison of morphology and some developmental features of these types further suggested that they should be divided into two species (Pennati et al. 2015). The authors claimed that *Ciona intestinalis* type A, which is the common type of *Ciona intestinalis*, should be renamed "*Ciona robusta*", and that *Ciona intestinalis* type B, which lives in more restricted areas, is *Ciona intestinalis*. Because the renaming proposal is under consideration for *Ciona* (and the ascidian) community, I decided not to rigidly determine which name, *Ciona intestinalis* or *Ciona robusta*, should be used in this book. Therefore, although some chapters simply use *Ciona intestinalis*, others mention the rename issue in the first section of the chapters and then use *Ciona robusta*. Because all authors took so much care with regard to dealing with the name, I believe that there is no confusion when reading this book; however, please keep this potential issue in your mind.

References

Abitua P, Gainous T, Kaczmarczyk A, Winchell C, Hudson C, Kamata K, Nakagawa M, Tsuda M, Kusakabe T, Levine M (2015) The pre-vertebrate origins of neurogenic placodes. Nature 524:462–465

Caputi L, Andreakis N, Mastrototaro F, Cirino P, Vassillo M, Sordino P (2007) Cryptic speciation in a model invertebrate chordate. Proc Natl Acad Sci U S A 104(22):9364–9369

Dehal P, Satou Y, Campbell RK, Chapman J, Degnan B, De Tomaso A, Davidson B, Di Gregorio A, Gelpke M, Goodstein DM, Harafuji N, Hastings KE, Ho I, Hotta K, Huang W, Kawashima T, Lemaire P, Martinez D, Meinertzhagen IA, Necula S, Nonaka M, Putnam N, Rash S, Saiga H, Satake M, Terry A, Yamada L, Wang HG, Awazu S, Azumi K, Boore J, Branno M, Chin-Bow S, DeSantis R, Doyle S, Francino P, Keys DN, Haga S, Hayashi H, Hino K, Imai KS, Inaba K, Kano S, Kobayashi K, Kobayashi M, Lee BI, Makabe KW, Manohar C, Matassi G, Medina M, Mochizuki Y, Mount S, Morishita T, Miura S, Nakayama A, Nishizaka S, Nomoto H, Ohta F, Oishi K, Rigoutsos I, Sano M, Sasaki A, Sasakura Y, Shoguchi E, Shin-I T, Spagnuolo A, Stainier D, Suzuki MM, Tassy O, Takatori N, Tokuoka M, Yagi K, Yoshizaki F, Wada S, Zhang C, Hyatt PD, Larimer F, Detter C, Doggett N, Glavina T, Hawkins T, Richardson P, Lucas S, Kohara Y, Levine M, Satoh N, Rokhsar DS (2002) The draft genome of *Ciona intestinalis*: insight into chordate and vertebrate origins. Science 298 (5601):2157–2167

Delsuc F, Brinkmann H, Chourrout D, Philippe H (2006) Tunicates and not cephalochordates are the closest living relatives of vertebrates. Nature 439(7079):965–968

Hikosaka A, Kusakabe T, Satoh N, Makabe KW (1992) Introduction and expression of recombinant genes in ascidian embryos. Develop Growth Differ 34:627–634

Iitsuka T, Mita K, Hozumi A, Hamada M, Satoh N, Sasakura Y (2014) Transposon-mediated targeted and specific knockdown of maternally expressed transcripts in the ascidian *Ciona intestinalis*. Sci Rep 4:5050

Jeffery WR, Strickler AG, Yamamoto Y (2004) Migratory neural crest-like cells form body pigmentation in a urochordate embryo. Nature 431:696–699

Lemaire P (2011) Evolutionary crossroads in developmental biology: the tunicates. Development 138(11):2143–2152

Pennati R, Ficetola G, Brunetti R, Caicci F, Gasparini F, Griggio F, Sato A, Stach T, Kaui-Strehlow S, Gissi C, Manni L (2015) Morphological differences between larvae of the *Ciona intestinalis* species complex: hints for a valid taxonomic definition of distinct species. PLoS One 10:e0122879

Satoh N (2003) The ascidian tadpole larva: comparative molecular development and genomics. Nat Rev Genet 4:285–295

Satou Y, Imai K, Levine M, Kohara Y, Rokhsar D, Satoh N (2003) A genomewide survey of developmentally relevant genes in *Ciona intestinalis*. I. Genes for bHLH transcription factors. Dev Genes Evol 213(5–6):213–221

Sasakura Y, Yamada L, Takatori N, Satou Y, Satoh N (2003) A genomewide survey of developmentally relevant genes in *Ciona intestinalis*. VII. Molecules involved in the regulation of cell polarity and actin dynamics. Dev Genes Evol 213:273–283

2 Microinjection of Exogenous Nucleic Acids into Eggs: *Ciona* Species

Kenji Kobayashi and Yutaka Satou

Abstract

Microinjection is a common technique used to deliver nucleic acids into eggs and embryos in *Ciona* species. There are three *Ciona* species that are commonly used for research—*Ciona intestinalis* type A (*C. robusta*), *C. intestinalis* type B (*C. intestinalis*), and *C. savignyi*. Here, we present the microinjection methods using eggs and embryos of *C. intestinalis* type A and *C. savignyi*; however, our methods would also be applicable to eggs and embryos of *C. intestinalis* type B. Microinjection is a classical and widely used delivery method, which involves the use of a glass micropipette, a hollow glass needle with a microscopic tip, to inject nucleic acids into eggs and embryos under a stereo microscope. The required amount of nucleic acids is much smaller for microinjection than for electroporation, another delivery method. Proteins, and other chemicals, such as fluorescent dye, can be introduced with nucleic acids using a microinjection.

Keywords

Developmental biology · Microinjection · Ascidian · *Ciona* · Nucleic acid · DNA · RNA · Morpholino oligonucleotide · Egg · Embryo

K. Kobayashi (✉) · Y. Satou
Department of Zoology, Graduate School of Science, Kyoto University, Sakyo-ku, Kyoto 606-8502, Japan
e-mail: kobaken@ascidian.zool.kyoto-u.ac.jp

2.1 Introduction

Microinjection is a common technique used to deliver nucleic acids into eggs and embryos of *Ciona* species. The diameter of eggs of *Ciona* species typically ranges from 140 to 150 μm (Fig. 2.1); therefore, microinjection can be performed under a stereo microscope. The technique is not difficult to learn; in our experience, most persons are able to perform it after several days of training. Skilled persons can inject nucleic acids into over 100 eggs within 1 h.

Among the species belonging to the genus *Ciona*, *Ciona intestinalis* and *Ciona savignyi* have been widely used for manipulation. It has been considered that *C. intestinalis* is grouped into two subspecies—type A and type B (Suzuki et al. 2005; Nydam and Harrison 2007; Caputi et al. 2007). However, recently, a report proposed that these are indeed different species and not subspecies (Brunetti et al. 2015). According to that report, *C. intestinalis* type A is *C. robusta*, and *C. intestinalis* type B is *bona fide C. intestinalis*. However, most of the previous studies on *Ciona* species did not clearly discriminate between these two species. In addition, the reproductive systems of these two species are undoubtedly isolating, but still not completely isolated (Suzuki et al. 2005; Caputi et al. 2007; Nydam and Harrison 2011; Sato et al. 2014). Therefore, we would not discriminate between these two species, and simply describe them as *C. intesti-*

Y. Sasakura (ed.), *Transgenic Ascidians*, Advances in Experimental Medicine and Biology 1029,
https://doi.org/10.1007/978-981-10-7545-2_2

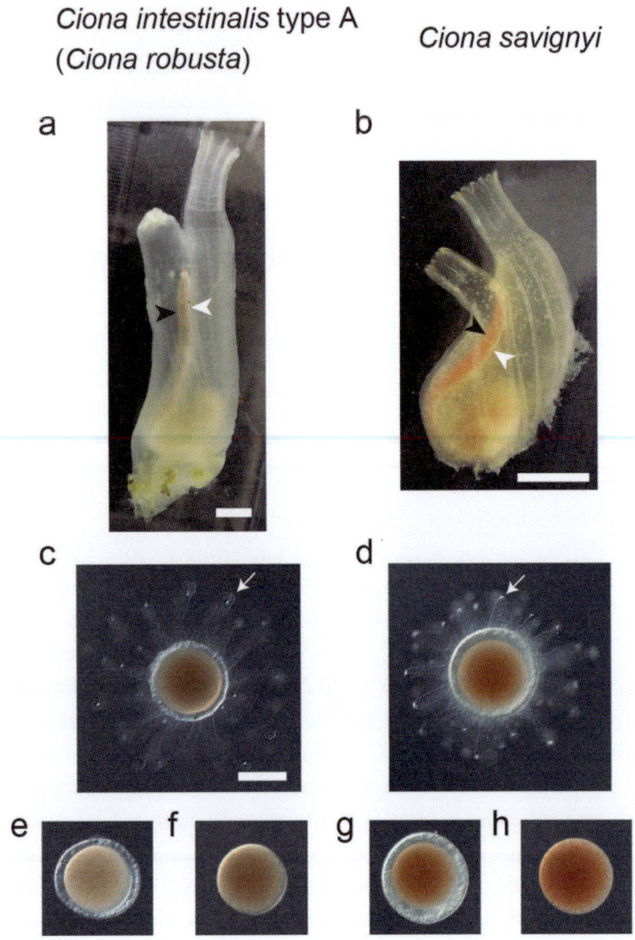

Fig. 2.1 Two *Ciona* species commonly used for experiments. Adult animals of (**a**) *Ciona intestinalis* type A (*Ciona robusta*) and (**b**) *Ciona savignyi*. *Black arrowheads* indicate oviducts, and *white arrowheads* indicate sperm ducts. Unfertilized eggs of (**c**, **e**, **f**) *C. intestinalis* type A, and (**d**, **g**, **h**) *C. savignyi*. *White arrows* indicate follicle cells. Follicle cells can easily be removed by pipetting, and such eggs with chorion (**e**, **g**) are sometimes used for microinjection. However, in most cases, the chorion is removed and used for microinjection (**f**, **h**). Scale bars, 10 mm (**a**, **b**); 100 μm (**c**)

nalis in this chapter. Animals available in our laboratory are *C. intestinalis* type A and *C. savignyi* (Fig. 2.1); therefore, the methods described here are based on these two species. However, our methods would also be applicable to eggs and embryos of *C. intestinalis* type B.

Microinjection is a classical delivery method. On the other hand, electroporation, as explained in Chap. 5, is easier to learn, and nucleic acids can be introduced into a greater number of eggs in one experiment. However, microinjection is still widely used for several reasons—the primary reason is probably that the required amount of nucleic acids is much smaller for microinjection than for electroporation. Preparation of RNAs and synthetic oligonucleotides at amounts sufficient for electroporation is often costly; additionally, simultaneous introduction of DNA, RNA, protein, and other chemicals, such as a fluorescent dye, is possible with a microinjection.

2.2 Preparation of the Microinjection

Room temperature should be adjusted to around 18 °C. Seawater and agar-coated dishes should be equilibrated to the room temperature before use.

2.2.1 Preparation of Nucleic Acids

Deoxyribonucleic acid (DNA) is typically microinjected at final concentrations of 1–20 μg/mL. While preparing the injection solution, nucleic acids are commonly mixed with a dye; therefore, the DNA stock solutions have to be prepared at a higher concentration. Injection of higher concentrations of DNA may nonspecifically inhibit development (Hikosaka et al. 1992). Plasmid DNA can be prepared from *Escherichia coli* using a conventional alkaline-lysis method. Most of the commercially available kits could be used for this purpose. Circular plasmid DNAs are sometimes linearized with a restriction enzyme, which may improve their transcriptional efficiency. It is also possible to introduce PCR products after purification. Columns with a silica membrane are convenient for this purpose. DNA is typically eluted or dissolved using 1 mM Tris-HCl (pH 8.0) and 0.1 mM EDTA (Hikosaka et al. 1994; Satou and Satoh 1996).

Ribonucleic acid (RNA) is typically microinjected at final concentrations of 0.05–2 mg/mL, which is much higher than the concentration of DNA, but that has no effect on embryogenesis. While preparing the injection solution, nucleic acids are commonly mixed with a dye; therefore, RNA stock solutions have to be prepared at a higher concentration. RNA is prepared using in vitro transcription from DNA templates using T3, T7, or SP6 promoter. For efficient translation, a cap structure analog 7mGpppG is usually incorporated as the 5'terminal or the first G of the transcript during in vitro transcription. It is convenient to use a commercially available kit, such as mMessage mMachine Transcription Kits (Thermo Fisher Scientific, Waltham, MA, USA), for this synthesis. Template DNAs need to be removed before injection following the manufacturer's protocol. RNAs are typically eluted or dissolved in nuclease-free water.

Morpholino oligonucleotides (MOs; Gene Tools, Philomath, OR, USA) are used for knocking down gene function (Satou et al. 2001), and can be injected in the same way as DNA or RNA at final concentrations of 0.25–1.5 mM; the concentration is usually determined empirically.

2.2.2 Preparation of Manipulators and a Stereo Microscope

Any stereo microscopes can be used for microinjection. Systems with zoom magnification up to ×100 are convenient. The working distance is important for setting up manipulators under the objective lens.

In our laboratory, manipulators produced by Narishige Corporation (Setagaya-ku, Tokyo, Japan) are used and shown here as a typical set (Fig. 2.2). However, the following example also works as a reference for setting up manipulators that are available from several other companies.

2.2.3 Preparation of Injection Needles

We usually prepare needles from glass capillaries with an external diameter of 1.0 mm (e.g., GD-1 or GDC-1; Narishige). Capillaries with a filament are convenient, because an aliquot (~1 μL) of nucleic acid solution is loaded at the back end of a needle with a micropipette, and then the filament leads the nucleic acid solution to the tip of the needles. For preparation of the needles, a machine called a puller is required (we use Model P-97, from Sutter Instrument, Novato, CA, USA). The puller heats a glass capillary and then pulls it to make two needles. The strength of heating and pulling power are adjustable, and greatly influence the shape of the needles. The shape is important for successful injection, but this is a matter of preference. Start with the shape of a needle that we use (Fig. 2.3a, b), and change it, if you want, by adjusting the settings of your puller (our typical setting for the puller is—heat, 550 (filament lamp test, 565); pull, 20; velocity, 90; time, 255; we adjust these values every time before use). Keep needles on a needle stand, which can be easily hand-made (Fig. 2.3c).

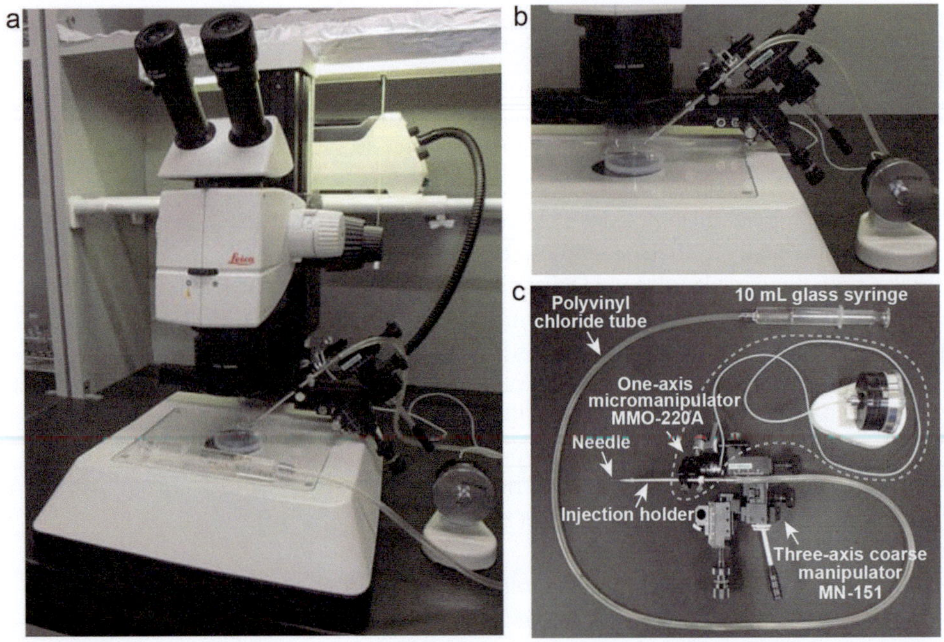

Fig. 2.2 A stereo microscope and a set of manipulators for microinjection. (**a**) A photograph of the whole microinjection system. (**b**) A high magnification view of the manipulators. (**c**) The whole set of manipulators

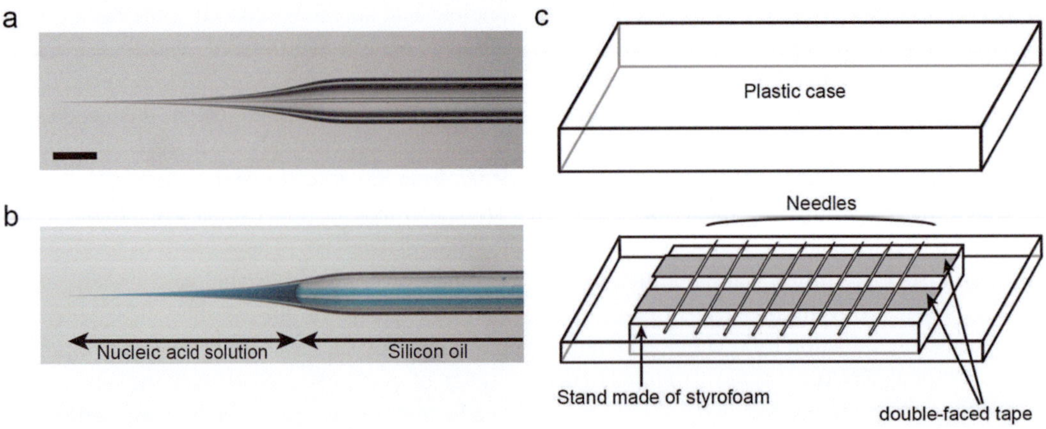

Fig. 2.3 Injection needles. (**a**) An empty glass needle with a filament. (**b**) Nucleic acid solution with green dye was loaded at the tip of the needle, and silicone oil filled the remaining part. (**c**) A needle stand used for keeping needles and filling nucleic acid solutions into needles. Scale bar, 1 mm (**a**)

Glass capillaries without a filament are also used widely. In this case, an additional type of needle is required for filling the nucleic acid solution into the tip of injection needles from the back end, which is not described here.

2.2.4 Preparation of Agar-Coated Petri Dishes

2.2.4.1 Agar-Coated Petri Dishes for Incubation of Embryos

For the incubation of dechorionated embryos, dishes need to be coated with agar, because dechorionated embryos are easily broken by sticking to the naked surface of dishes. Agar is commonly dissolved in seawater at a concentration of 1% (w/v) by heating in a microwave. To dissolve agar completely, agitate the mixture occasionally during heating. Then pour an adequate amount of 1% agar into the dishes. Keep the agar layer thin so that agar-coated dishes can hold a sufficient amount of seawater. After agar is cooled, pour seawater onto the agar gel to prevent the agar gel from drying out. Agar-coated dishes may be stored in a refrigerator. Dishes with a diameter of 6- or 9-cm would be easy to use.

2.2.4.2 Agar-Coated Petri Dishes for Microinjection

As for the above-mentioned procedure, prepare 1% agar solution using seawater. Pour the agar solution into a dish, and then float a cover glass on the agar solution in the dish (Fig. 2.4). Use tweezers to move the cover glass to an appropriate position. After the agar is cooled, remove the cover glass using tweezers. A small step would be formed on the surface of the agar in the dish, which would be useful during injection. Pour seawater onto the agar gel to prevent the agar gel from drying out.

2.2.5 Preparation of a Micro-Glass Pipette

Although it is not mandatorily required, a micro-glass pipette (Fig. 2.5) is extremely useful for transferring eggs one by one. Cut a glass tube with an external diameter of 4.0 mm and a length of 100–200 mm. Using a gas burner, heat the center of the glass tube until it is softened. Pull the glass with both hands, which will form two pipettes. Cut the tips of these pipettes with a file; the ideal pore size will be around 200 μm in diameter. Then, smooth the tip using a gas burner. A silicone tube, one end of which is plugged or tied, can be used as a bulb.

2.2.6 Preparation of Solutions

2.2.6.1 Seawater

Natural seawater should be purified using a 0.2-μm pore size filter before use. Commercially available artificial seawater may be used. We usu-

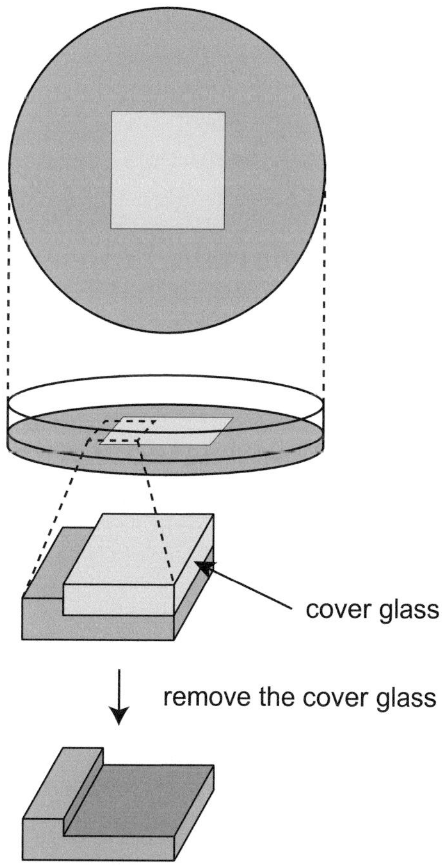

Fig. 2.4 Preparation of an agar-coated Petri dish for microinjection. By floating a cover glass before the agar is cooled, a small step can be made on the agar surface

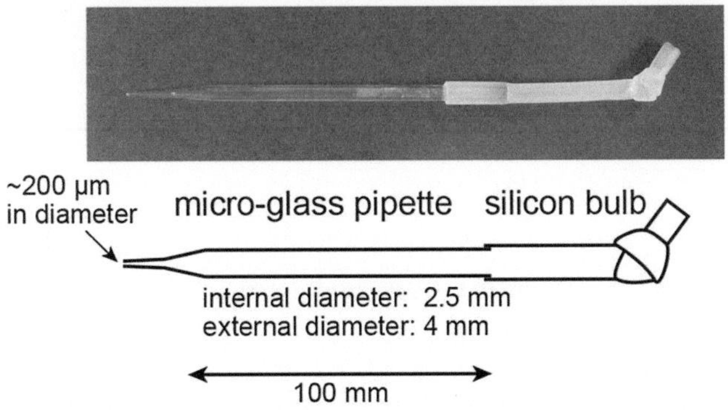

Fig. 2.5 A photograph and illustration of a micropipette, which is useful for manipulation of embryos

ally use Marine Art BR (Tomita Pharmaceutical, Naruto, Tokushima, Japan). Streptomycin can be added at the concentration of 50 µg/mL.

2.2.6.2 Dechorionation Solution

The dechorionation solution is prepared at ×2 concentration just before use.

- ×2 dechorionation solution
- 2% sodium thioglycolate
- 0.1% actinase (pronase) E
- 1 M NaOH

2.2.6.3 A Dye Indicator for Microinjection

- ×10 Fast Green FCF solution

Dissolve Fast Green FCF at a concentration of 10 mg/mL in nuclease-free water. This solution can be stored at room temperature.

2.3 Microinjection

2.3.1 Filling Needles with Nucleic Acid

Prepare an injection solution containing your nucleic acid and ×1 Fast Green FCF. You can mix multiple nucleic acids, if required. Using a micropipette, dispense up to 1 µL of the injection solution to the back end of a needle, which is placed on a needle stand. The filament within the needle guides the solution to the tip of the needle. Next, you need to fill up the rest of the needle with silicone oil using a 1- or 5-mL glass syringe with a fine metal needle (30-gauge, 2″ backfilling needle; Drummond Scientific, Broomall, PA, USA) behind the injection solution. Do not introduce air bubbles into the needle. We usually prepare five or more needles for one experiment. Keep the filled needles in a refrigerator or at room temperature.

2.3.2 Fertilization and Dechorionation

2.3.2.1 Eggs and Sperm

Pick up two healthy adult animals with a sufficient amount of eggs and sperm (ascidians are hermaphrodites), because self-fertilization rarely occurs. Because their body is transparent, you can easily find eggs (brown in *C. intestinalis* type A and orange in *C. savignyi*) and sperm along the intestine (see Fig. 2.1).

To obtain eggs and sperm, remove the tunic that covers the whole body manually using scissors. Then, carefully dissect the body to expose the oviduct and sperm duct. First, cut the oviduct and softly push out the eggs into a Petri dish filled with seawater. Second, cut the sperm duct and softly push out sperm into a different empty dish that is not filled with seawater (alternatively, you can use a Pasteur pipette to collect sperm). Repeat the same process for another ascidian. Now you will have two dishes with eggs and two dishes with sperm. Sperm can be mixed in a tube and kept in a refrigerator for several hours.

Eggs are viscous. Using a Pasteur pipette, continuously withdraw and dispense eggs into seawater to minimize their viscosity (step 1 in Fig. 2.6). Swirl the dish to collect eggs around the center. Transfer eggs with a small amount of seawater into a new dish with fresh seawater.

2.3.2.2 Fertilization Before Injection

If you inject nucleic acid solutions into unfertilized eggs, skip this step. Add an aliquot of sperm to the dish in which eggs are kept (step 2 in Fig. 2.6). Sperm from a different individual (or a mixture of sperm from multiple individuals) should be used, because *Ciona* eggs are hardly ever self-fertilized. Mix the contents with a Pasteur pipette. Incubate a mixture of eggs and sperm for 5–10 min.

If you need to remove the chorion (vitelline membrane) of the eggs before injection, proceed to the next step. If you need to inject nucleic acid solutions into the eggs with the chorion, eggs should be washed with fresh seawater at least twice, the dish should be swirled to gather the eggs around the center, and transfer them with a minimum amount of seawater to a new dish.

Activation of sperm may be helpful for synchronized fertilization. Add 25 μL 1 M Tris-HCl (pH 9.5) to 500 μL seawater, and mix well. Then, add 10 μL sperm to the mixture, and mix well again. You will see actively moving spermatozoa under the stereo microscope. To fertilize the eggs, add 50 μL of the solution containing activated spermatozoa to a 6-cm Petri dish containing eggs and seawater.

2.3.2.3 Dechorionation

If you need to inject your nucleic acid solutions into eggs with the chorion, skip this step. In most experiments, dechorionated eggs are used for injection, because they are more amenable for injection than eggs with the chorion. Injection to *C. savignyi* eggs with the chorion is much easier than that to *C. intestinalis* type A eggs with the chorion (Hikosaka et al. 1992). However, it is known that ascidian embryos lose the left–right axis without the chorion (Yoshida and Saiga 2011), and larvae developed from eggs with the chorion may metamorphose at a higher rate.

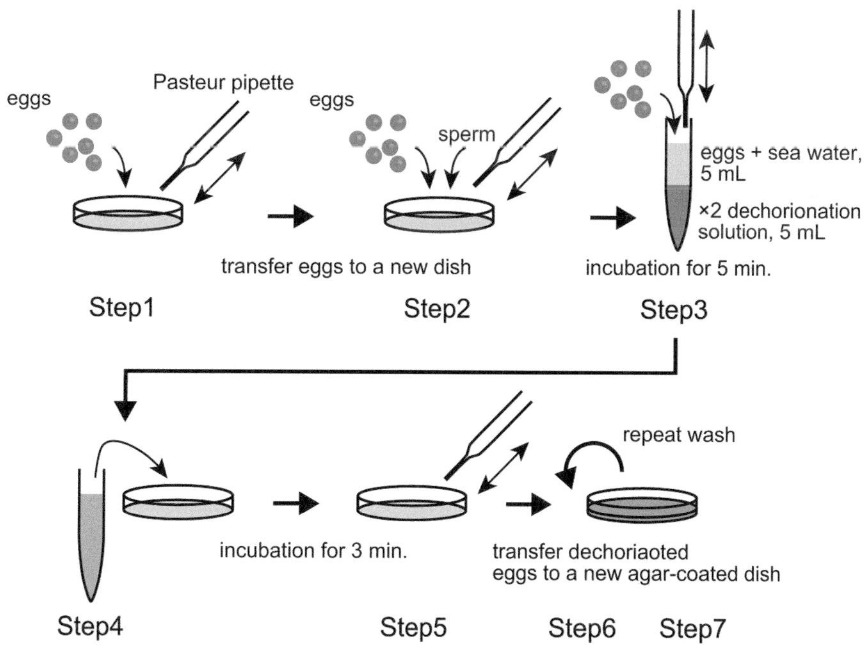

Fig. 2.6 Dechorionation procedure

Dispense 5 mL 2× dechorionation solution and 375 μL 1 M NaOH into a plastic 15-mL tube. Swirl the dish to collect eggs around the center, and transfer 5 mL seawater containing as many eggs into the plastic tube (now it becomes ×1 solution; step 3 in Fig. 2.6). Mix the content gently, and then transfer the entire content to a new 6-cm Petri dish. Incubate the reaction for at least 3 min (step 4 in Fig. 2.6). Observe the eggs under the stereo microscope to confirm that the vitelline membrane is beginning to break down. After that, mix the contents softly using a Pasteur pipette, and observe eggs under the stereo microscope (step 5 in Fig. 2.6). Immediately after chorions of almost all (~80%) the eggs are removed, swirl the dish and transfer the dechorionated eggs to an agar-coated dish filled with fresh seawater (step 6 in Fig. 2.6). After gentle mixing, swirl the content, and transfer the eggs into another agar-coated dish filled with fresh seawater (step 7 in Fig. 2.6). Repeat this wash one more time. It is important to reduce the time of the dechorionation step. Longer incubation in the dechorionation solution is harmful for normal development.

2.3.3 Microinjection Technique

2.3.3.1 Aligning Eggs

Align up to about 130 eggs along the step that was formed by a cover glass on the agar of an injection dish (step 1 in Fig. 2.7). A micro-glass needle is useful for this procedure.

2.3.3.2 Set Up the Instrument

Set a needle filled with the nucleic acid and silicone oil into an injection holder, the back end of which is connected to a polyvinyl chloride tube and a 10-mL glass syringe filled with water (see Fig. 2.2c). Take care not to introduce air bubbles into both a polyvinyl chloride tube and a 10-mL glass syringe. Then, fix the holder to the one-axis fine manipulator (micromanipulator), and fix it to the three-axis coarse manipulator.

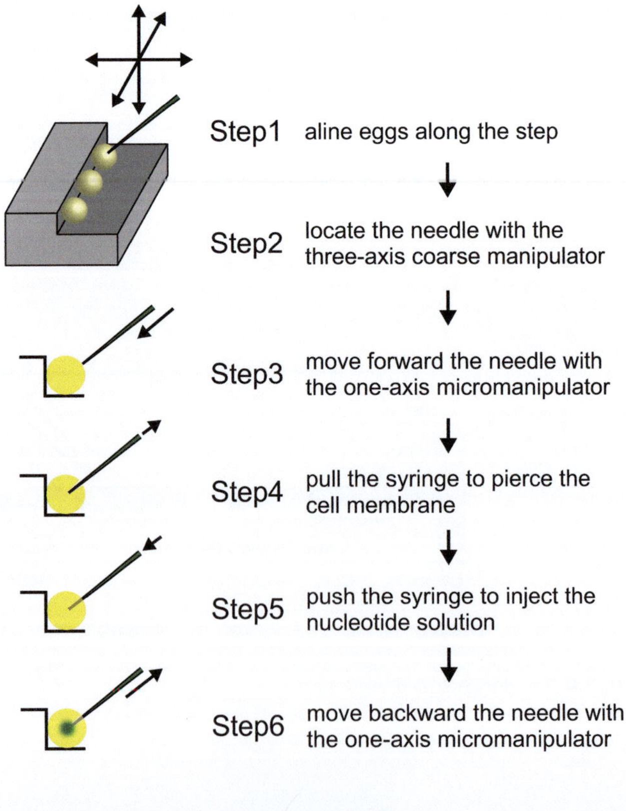

Fig. 2.7 Microinjection procedure. The step in this figure is the one created on the surface of agar, as explained in Fig. 2.4

2.3.3.3 Injection

To break the tip of the injection needle, use the three-axis coarse manipulator and the one-axis micromanipulator to gently push it against a cover glass that was kept on the dish filled with seawater. Push the 10-mL syringe, and confirm that the nucleic acid solution is being pushed out from the tip of the needle.

Using the three-axis manipulator, move the needle so that the tip of the needle can touch the surface of an egg (step 2 in Fig. 2.7). Next, use a one-axis micromanipulator to push the needle into one-third or a half-length diameter of the egg (step 3 in Fig. 2.7). The step in the agar plate is required at this stage of the procedure; without it, the eggs move forward to the other side and the needle cannot pierce the egg. Pull the syringe gently so that the needle tip can pierce the egg membrane (step 4 in Fig. 2.7). Then, push the syringe to inject the nucleic acid (step 5 in Fig. 2.7). You will see a green spot within the egg if your injection solution contains Fast Green FCF. Inject a defined amount of the solution by checking the diameter of green spots. Typically, we inject solutions until the green spot becomes one-fourth to one-third of the egg diameter. When the injection succeeds, the green spot immediately disperses into the cytoplasm of the egg. If the green spot remains in the egg permanently, the needle tip cannot piece the egg membrane; thus, the injection fails.

Pull back the needle with the one-axis micromanipulator (step 6 in Fig. 2.7). Occasionally, injected eggs are stuck at the tip of the needle. In such cases, use a micro-glass pipette to detach the eggs. Injected eggs should be transferred one by one with a micro-glass pipette to a new agar-coated dish filled with seawater.

Dechorionated (injected or uninjected) eggs should be incubated at a low density at 18 °C. If density is too high, eggs and embryos stick together.

If you use unfertilized eggs for injection, inseminate the eggs at this moment, as described above.

References

Brunetti R, Gissi C, Pennati R, Caicci F, Gasparini F, Manni L (2015) Morphological evidence that the molecularly determined *Ciona intestinalis* type A and type B are different species: *Ciona robusta* and *Ciona intestinalis*. J Zool Syst Evol Res 53(3):186–193. https://doi.org/10.1111/jzs.12101

Caputi L, Andreakis N, Mastrototaro F, Cirino P, Vassillo M, Sordino P (2007) Cryptic speciation in a model invertebrate chordate. Proc Natl Acad Sci U S A 104(22):9364–9369. https://doi.org/10.1073/pnas.0610158104

Hikosaka A, Kusakabe T, Satoh N, Makabe KW (1992) Introduction and expression of recombinant genes in ascidian embryos. Develop Growth Differ 34(6):627–634. https://doi.org/10.1111/j.1440-169X.1992.tb00031.x

Hikosaka A, Kusakabe T, Satoh N (1994) Short upstream sequences associated with the muscle-specific expression of an actin gene in ascidian embryos. Dev Biol 166(2):763–769. https://doi.org/10.1006/dbio.1994.1354

Nydam ML, Harrison RG (2007) Genealogical relationships within and among shallow-water *Ciona* species (Ascidiacea). Mar Biol 151(5):1839–1847. https://doi.org/10.1007/s00227-007-0617-0

Nydam ML, Harrison RG (2011) Introgression despite substantial divergence in a broadcast spawning marine invertebrate. Evolution 65(2):429–442. https://doi.org/10.1111/j.1558-5646.2010.01153.x

Sato A, Shimeld SM, Bishop JDD (2014) Symmetrical reproductive compatibility of two species in the *Ciona intestinalis* (Ascidiacea) species complex, a model for marine genomics and developmental biology. Zool Sci 31(6):369–274. https://doi.org/10.2108/zs130249

Satou Y, Satoh N (1996) Two cis-regulatory elements are essential for the muscle-specific expression of an actin gene in the ascidian embryo. Develop Growth Differ 38(5):565–573. https://doi.org/10.1046/j.1440-169X.1996.t01-1-00013.x

Satou Y, Imai K, Satoh N (2001) Action of morpholinos in *Ciona* embryos. Genesis 30(3):103–106. https://doi.org/10.1002/gene.1040

Suzuki MM, Nishikawa T, Bird A (2005) Genomic approaches reveal unexpected genetic divergence within *Ciona intestinalis*. J Mol Evol 61(5):627–635. https://doi.org/10.1007/s00239-005-0009-3

Yoshida K, Saiga H (2011) Repression of Rx gene on the left side of the sensory vesicle by Nodal signaling is crucial for right-sided formation of the ocellus photoreceptor in the development of *Ciona intestinalis*. Dev Biol 354(1):144–150. https://doi.org/10.1016/j.ydbio.2011.03.006

3 Practical Guide for Ascidian Microinjection: *Phallusia mammillata*

Hitoyoshi Yasuo and Alex McDougall

Abstract

Phallusia mammillata has recently emerged as a new ascidian model. Its unique characteristics, including the optical transparency of eggs and embryos and efficient translation of exogenously introduced mRNA in eggs, make the *Phallusia* system suitable for fluorescent protein (FP)-based imaging approaches. In addition, genomic and transcriptomic resources are readily available for this ascidian species, facilitating functional gene studies. Microinjection is probably the most versatile technique for introducing exogenous molecules such as plasmids, mRNAs, and proteins into ascidian eggs/embryos. However, it is not practiced widely within the community; presumably, because the system is rather laborious to set up and it requires practice. Here, we describe in as much detail as possible two microinjection methods that we use daily in the laboratory: one based on an inverted microscope and the other on a stereomicroscope. Along the stepwise description of system setup and injection procedure, we provide practical tips in the hope that this chapter might be a useful guide for introducing or improving a microinjection setup.

Keywords

Microinjection · Ascidian · *Phallusia mammillata* · Live imaging · Fluorescent protein

3.1 Introduction

3.1.1 *Phallusia mammillata* as a Model for Imaging-Based Studies from Egg to Tadpole

To take advantage of fluorescent microscopy techniques, it is standard practice to express fluorescent protein (FP)-fused constructs in living cells. In the case of ascidian models, electroporation or microinjection techniques are used to introduce either plasmid or mRNA constructs encoding FP-fusion proteins. Electroporation is not possible when the egg is covered by extracellular layers of cells, vitelline coats, egg shells, etc., and the removal of these vestments prevents correct embryonic development in many invertebrates. Remarkably, however, removing the extracellular chorion of ascidian eggs to expose

Electronic supplementary material The online version of this chapter (https://doi.org/10.1007/978-981-10-7545-2_3) contains supplementary material, which is available to authorized users.

H. Yasuo (✉) · A. McDougall
Sorbonne Universités, UPMC Univ Paris 06, CNRS, Laboratoire de Biologie du Développement de Villefranche-sur-mer, Observatoire Océanologique, Villefranche-sur-mer, France
e-mail: yasuo@obs-vlfr.fr; dougall@obs-vlfr.fr

Y. Sasakura (ed.), *Transgenic Ascidians*, Advances in Experimental Medicine and Biology 1029,
https://doi.org/10.1007/978-981-10-7545-2_3

the naked plasma membrane does not prevent their development. This has made it possible to simultaneously electroporate plasmids into hundreds of fertilized ascidian eggs, allowing the transient expression of FPs in a lineage-specific manner in the embryos of *Ciona intestinalis* (Corbo et al. 1997). Transient transgenesis is a powerful technique that has now been exploited in *Ciona* by many laboratories in the community (Christiaen et al. 2009; Matsuoka et al. 2005; Zeller et al. 2006). However, FP fluorescence, even when the FP is expressed under the control of an early acting promoter such as the *Fog* promoter, is not measurable until around the time of gastrulation (Roure et al. 2007). The technique is also dependent on the availability of promoters suitable for expressing FPs in the time and space of interest. Therefore, although it has revolutionized ascidian experimental studies, electroporation-based transient transgenesis is not useful for live-imaging events taking place from fertilization up to gastrulation.

To monitor FP fluorescence in unfertilized eggs and early embryos, mRNA or protein microinjection is necessary. As protein production is time-consuming and more challenging than making mRNA, we usually microinject mRNA. Unfortunately, microinjected mRNAs are not translated efficiently in unfertilized eggs or early embryos of some species of ascidian. Fortunately though, species such as *Phallusia mammillata* and *Ascidiella aspersa* do translate mRNAs efficiently in the unfertilized egg and during cleavage divisions (*Ascidiella*: Levasseur and McDougall 2000; *Phallusia*: Negishi and Yasuo 2015; Prodon et al. 2010). Interestingly, both these species have exceptionally transparent eggs and embryos, making them additionally useful for fluorescence microscopy (Fig. 3.1). We have been developing fluorescent GFP-based markers for use in *Phallusia* eggs/embryos for cell biology studies for many years, e.g., markers for the cytoskeleton (actin/microtubules), chromosomes/kinetochores, nucleus, centrosomes, and plasma membrane (Fig. 3.1) (McDougall et al. 2015). Here, we describe in detail two microinjection techniques we use to introduce exogenous molecules into *Phallusia mammillata* eggs: an inverted microscope setup and a stereomicroscope setup.

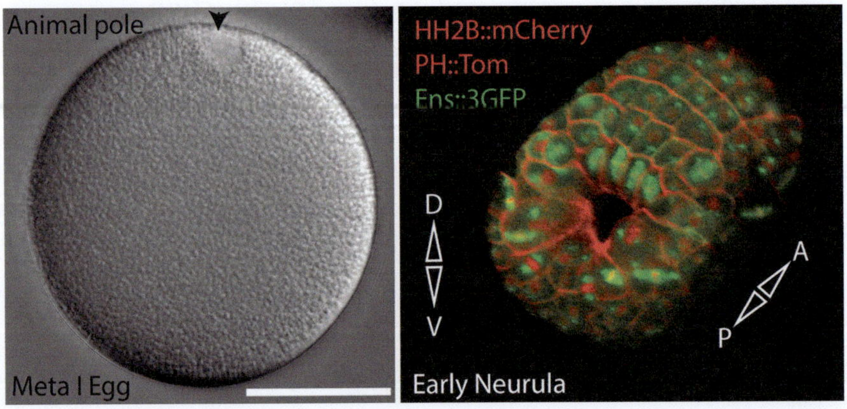

Fig. 3.1 Unfertilized *Phallusia* egg and early neurula. A differential interference contrast image of an unfertilized *Phallusia* egg at metaphase I. The animal pole containing the meiotic spindle appears as a clear zone (*arrowhead*). Early neurula embryo showing fluorescence from three mRNAs that were co-injected into the egg. PH::Tom (Tom = tdTomato) labels the plasma membrane (*red*), histone H2B::mCherry labels the chromatin (*red*) and Ens::3GFP labels the microtubules (*green*). The image is a projected z-stack from a confocal time-lapse that has been rendered in 3D using the VTK module in ICY software. *D* dorsal, *V* ventral, *A* anterior, *P* posterior. Scale bar = 50 μm

3.1.2 General Information About *Phallusia* and Transparent Ascidians

For those interested in *Phallusia* as an experimental model, there are now resources available including the genome sequence (Aniseed), several RNAseq transcriptomic datasets (Aniseed) and an arrayed cDNA library (searchable in Aniseed) (Brozovic et al. 2016). *Phallusia* is found exclusively in Europe. It is possible to obtain eggs and embryos from *Phallusia* from February through October and, if kept in the laboratory and fed, they will produce eggs during the winter, making it possible to work with them all year round (although springtime and fall is best). We have grown *Phallusia* at Villefranche-sur-Mer and the life cycle is about 6 months from the fertilized egg to the production of eggs capable of being fertilized. For those researchers wishing to use a transparent ascidian such as *Phallusia* outside of Europe, one can find *Ascidiella* on the east coast of the USA (e.g., Woods Hole), and in northern Japan (e.g., Asamushi Marine Station). Finally, note that other species of *Phallusia* such as *P. nigra* also produce transparent eggs.

3.2 Part 1: Inverted-Microscope Microinjection Setup (the McDougall Lab Setup)

3.2.1 Materials

3.2.1.1 Gelatin/Formaldehyde Coating

Gelatin crosslinked with formaldehyde produces a thin transparent film of gelatin on the glass or plastic surface when dry. Treatment of glassware and plastic dishes with 1% gelatin/formaldehyde (GF) solution (in water) makes them nonsticky for *Phallusia* eggs/embryos. To coat glass or plastic surfaces, apply a thin layer of GF solution, dry and rinse in water (Sardet et al. 2011).

3.2.1.2 Dechorionated *Phallusia* Eggs

Add 1 ml of 10× trypsin solution to 9 ml of seawater/10 mM TAPS (pH 8.2) containing chorionated *Phallusia* eggs in a GF-coated plastic petri dish (5 cm in diameter). Agitate (rotation, seesaw, etc.) gently for about 1½ to 2 h at 18–22 °C. When most eggs are "dechorionated" and have settled at the bottom of the dish, wash them several times with seawater. Transfer dechorionated eggs to a GF-coated dish.

- TAPS buffer stock solution: 500 mM TAPS, pH 8.2. Store at room temperature.
- 10× trypsin solution: 1% trypsin (Sigma, T-9201) in seawater/10 mM TAPS.

3.2.1.3 Injection Mix

We routinely microinject a number of reagents including proteins, synthetic mRNAs, plasmids, morpholino oligonucleotides, and fluorescent dyes into unfertilized eggs, fertilized eggs, and blastomeres. Generally, these reagents are diluted in water or injection buffer (180 mM KCl, 10 mM PIPES, 100 μM EGTA, pH 7.1). We dilute all mRNAs for microinjection in Molecular Biology Grade Water (5 Prime, Hamburg, Germany). We dilute dyes such as Fura-2 in injection buffer. To estimate injection volume, we measure the displacement of the cytoplasm upon microinjection (typically one tenth to one fifth of the diameter of an egg, which is roughly equivalent to 0.1% to 1% of the egg volume).

3.2.1.4 Injection Chamber, Glass Pieces, and Associated Reagents

Injection chambers are custom-made of Perspex (Fig. 3.2). VALAB (Vaseline/Lanolin/Beeswax 1:1:1) is used to attach small pieces of glass to coverslips when creating a microinjection wedge. Dow Corning Vacuum grease is used to fix coverslips to a Perspex mounting chamber.

3.2.1.5 Injection Needles, Puller, and High-Pressure Injection Box

We use glass capillaries without a filament (GC-100 T10, Harvard Apparatus) and pull needles with a Narishige horizontal puller (PN-30). Both needles are used following a single two-stage pull. Perhaps the most important factor in microinjection is that the needles are appropriate. For

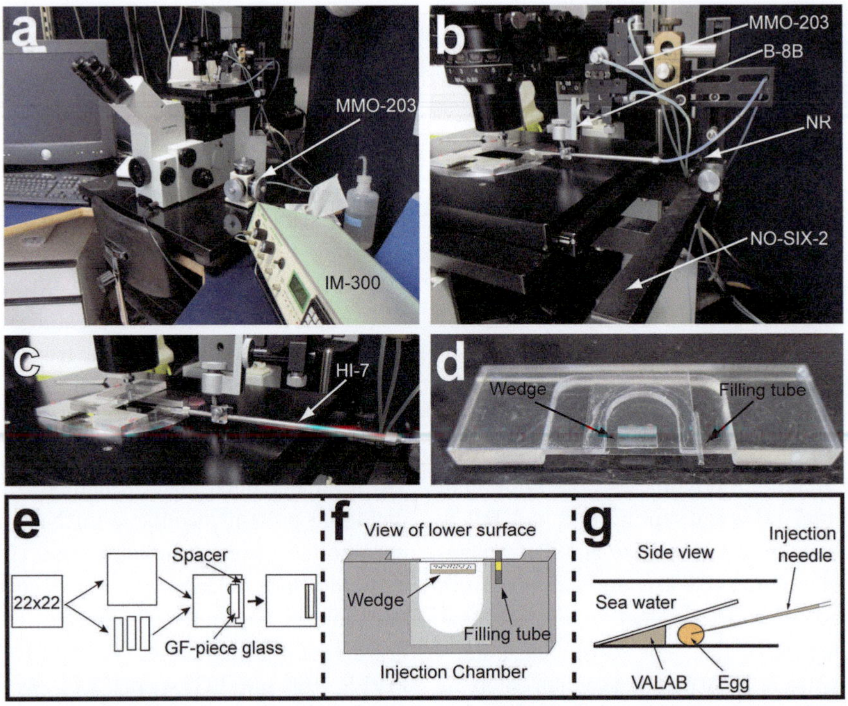

Fig. 3.2 Inverted microscope setup. (**a**) Olympus inverted microscope (IX70), with the three-way hydraulic micromanipulator highlighted (MMO-203). The high-pressure injection box is also visible (IM-300). The stage control is left mounted, allowing the use of the right hand for the micromanipulator and the left for the stage. (**b**) Close-up of the injection setup, showing the stage side mounting (NO-SIX-2), the mounting adaptor with rotation mechanism (NR), the ball joint for holding the injection needle, and the hydraulic three-way micromanipulator (MMO-203). (**c**) Close-up of the stage, injection needle holder (HI-7), and the injection chamber. (**d**) View of the lower surface of an injection chamber with a wedge and a filling tube attached. Note: as the microscope is inverted, both the wedge and the filling tube are on the side that face the objective lens (the lower side). (**e**) Schematic showing the preparation of a wedge. Briefly, a 22 × 22 no. 1.5 coverslip is coated with gelatin and cut into small pieces. A spacer is laid upon a GF coverslip. Then some VALAB is scraped onto a GF piece of glass, which is attached parallel to the spacer. Next, the coverslip is moved to a heating block to melt the VALAB, and the Spacer removed once the melted VALAB has solidified. (**f**) Schematic of the injection chamber with wedge and filling tube viewed from the lower surface. (**g**) Side view of an egg in a wedge being injected. The VALAB holds the GF piece of glass of the coverslip, but note that the angle of the wedge is sufficiently acute so that the egg does not touch the VALAB. If the eggs touch the VALAB they may die, so avoid this. If it proves difficult to get the perfect wedge at first, you could try using no. 1 or no. 0 coverslips to make the spacer form a more acute angle. Note: the position of the injection needle close to the center of the egg

Phallusia microinjection, success is improved with long thin needles (the shanks should be as narrow as possible). We pull using the following settings: second pull set to 10, heater 65.1, sub-magnet 13.8, main magnet 78.3 (however, heater and magnet settings alter as the platinum loop becomes older; thus, the shape of the needles should be inspected regularly). We use an IM-300 Narishige high-pressure microinjection box connected to a compressor providing around 100 psi air pressure to the injection box. Injection needles are tip-filled.

3.2.1.6 Injection Setup

Our injection setup is shown in Fig. 3.2 and comprises the following equipment.

- Inverted microscope (Olympus IX70) using a ×10 objective and transmitted light.
- Three-axis hydraulic micromanipulator (Narishige MMO-203)
- Stage-mounting equipment (Narishige NO-SIX-2)
- Needle holder (Narishige HI-7)

3.2.2 Method

3.2.2.1 Prepare Glass Wedge Setup for Microinjection

1. Glass coverslips (22 × 22, no. 1.5) are dipped in GF solution. Coat and air dry about 20 coverslips. Once dry, wash with tap water and air dry. Once dried, store for later use.
2. Using one of these gelatin coverslips, cut small pieces (about 10 mm long by 4 mm wide) with a diamond knife. Store for later use. For convenience, we call these GF pieces.
3. Prepare similar sized glass pieces (15 mm long, 4 mm wide) from virgin coverslips and store (we call these spacers).
4. To prepare the wedge, place a gelatin coverslip on a clean surface (a black plastic mat is ideal).
5. Using forceps, add a spacer to one side of the coverslip so that it lies parallel to one edge, slightly overhanging.
6. Using forceps, scrape some VALAB onto one GF piece of coverslip to create two small feet on one of the long edges (Fig. 3.2).
7. Place the GF piece parallel to the spacer so that the VALAB feet of the GF piece of coverslip attach to the coverslip next to the spacer (but avoid touching the spacer with the VALAB; Fig. 3.2).
8. Tap the GF piece of glass with the forceps so that it lies on the spacer at an angle.
9. Place immediately on a heated block to melt the VALAB, which flows along the edge of the GF piece of glass. Remove from the heated block and the VALAB solidifies, becoming opaque (this takes about 30 s).
10. Now, carefully remove the spacer leaving the small GF piece of coverslip glass attached to the coverslip. This creates a wedge-shaped chamber coated in gelatin to which the eggs can be added (See Fig. 3.2).

3.2.2.2 Fill the Filling Tube with Injection Material

1. Place the filling tube on the edge of a 5-ml petri dish and add 1 μl mineral oil.
2. Add 1 μl mRNA (or other injection material).
3. Add 1 μl mineral oil, forming an oil sandwich with mRNA in the middle.

Note: the filling tube containing mRNA can be kept for several weeks at 4 °C and reused.

3.2.2.3 Attach Glass Wedge to the Perspex Injection Chamber, Fill with Eggs, and Attach Filling Tube

1. Apply Dow Corning High Vacuum Grease to the upper and lower sides of the injection chamber where the coverslips will be attached.
2. Attach the wedge to the lower surface of the injection chamber.
3. Using a binocular microscope and a mouth pipette, add eggs to the wedge. With practice, up to about 100 eggs can be loaded into a wedge.
4. Once the eggs have been added to the wedge, a second coverslip is attached to the upper surface of the injection chamber, with the grease forming a sandwich.
5. Fill the resultant space between the wedge and coverslip with seawater (approximately 400 μl).
6. Attach the filling tube to the lower surface of the injection chamber parallel to the coverslip with the attached wedge so that the filling tube protrudes a little in front of the seawater opening (Fig. 3.2).
7. Move the injection chamber with the eggs to the inverted microscope for injection.

3.2.2.4 Microinjection

1. Lay the injection chamber with the wedge containing the eggs on the stage of an inverted microscope for injection. Focus on the filling tube using a ×10 objective lens. Note: the injection chamber needs to be raised about 1 cm above the surface of the stage, as the injection is horizontal.
2. Place a pre-pulled needle into the injection holder, clip into place, and advance the needle toward the filling tube by eye. As the viewer first focuses on the filling tube, both needle and tube are visible in the field of view using a ×10 objective.

3. View the needle and filling tube using oculars and with the three-way hydraulic micromanipulator, gently break the tip of the needle against the side of the filling tube, and move the needle into the filling tube to fill with mRNA, etc.
4. Go to the fill program on the IM-300 injection box and fill the needle. We use two programs: a fill program and an injection program. Both should be programmed using the instruction manual before starting. Fill the needle so that the tip of the needle and the meniscus both remain visible in the field of view with a ×10 objective.
5. Stabilize the movement of the meniscus with the balance pressure.
6. Once the meniscus is stable, activate the inject program, remove the needle from the filling tube, and advance it toward the eggs in the adjacent wedge.
7. Insert the needle into an egg such that the tip of the needle is about half way into the egg (Fig. 3.2).
8. Apply a short pulse of suction pressure using the fill button (100-ms pulse) to break the plasma membrane.
9. Inject by tapping the foot pedal, activating the injection program: our injection program consists of one 100-ms pulse each time the foot pedal is tapped. Adjust the injection pressure knob to select the appropriate pressure (typically this is between 10–20 psi [pounds per square-inch]). Note: the cytoplasm is displaced by the high-pressure rapid injection, allowing estimation of the injection size.
10. Remove the needle from the egg. Caution: the most delicate part of injecting *Phallusia* eggs is removing the needle. There is no single method for this, as it depends on the batch of eggs, some being more fragile than others. However, a good rule of thumb is to remove the needle slowly until the egg starts to move together with the needle, using the three-way hydraulic micromanipulator. At this point, the needle should be removed very rapidly using the stage control.
11. If the injected eggs are to be recovered and moved to a petri dish they can be pushed out of the wedge so that they lie at the edge of the wedge and can easily be recovered later using a mouth pipette or similar device on a binocular microscope.

With practice, about one egg per minute can be injected using this technique (or fewer eggs if the eggs are particularly fragile). If injecting unfertilized eggs proves difficult at first, fertilized eggs at the pronucleus stage are a little easier to inject, but there is a time constraint, because once they enter mitosis, it is not advisable to inject them, as this often perturbs cell division.

3.3 Part 2: Stereomicroscope Microinjection Setup (the Yasuo Lab's Setup)

3.3.1 Materials

3.3.1.1 Injection Mix

We routinely inject morpholino oligonucleotides, plasmids, in vitro synthesized mRNAs, and proteins into ascidian eggs/blastomeres. Solutions containing these molecules are mixed with Fast Green (see below), which allows the injected solution to be visualized in eggs/blastomeres and its volume to be estimated. When it is important to trace descendants of injected blastomeres, the injection mix is supplemented with fluorescent dextrans (see below; e.g., Haupaix et al. 2013).

- Fast Green FCF (Sigma–Aldrich): 1 mg/ml in distilled water (×2 to ×4 stock). Keep the working aliquot at room temperature, storing the remaining aliquots at −20 °C.
- Fluorescent dextrans, 10-kDa molecular weight (Texas-Red coupled, fluorescein-coupled, rhodamine-coupled, etc.; Molecular Probes): 2 mM in distilled water (×2 stock). Keep aliquots at −20 °C.
- Mineral oil (Sigma–Aldrich)
- Artificial seawater (ASW): 420 mM NaCl, 9 mM KCl, 10 mM $CaCl_2$, 24.5 mM $MgCl_2$,

25.5 mM $MgSO_4$, 2.15 mM $NaHCO_3$, and 10 mM HEPES buffer, pH 8.0. Sterilize with a 0.2-μm filter and add 0.05 g/l of kanamycin sulfate.

3.3.1.2 Injection Setup

The stereomicroscope injection setup is as shown in Fig. 3.3a, b using the following materials:

- Stereomicroscope (Leica S8APO with Leica TL BFDF, a brightfield–darkfield transmitted light base)
- Three-axis hydraulic micromanipulator (Narishige MMO-203)
- Three-axis coarse manipulator (Narishige M-152 or Harvard apparatus MM-33)
- Magnetic stand (Narishige GJ-1)
- Needle holder (Narishige HI-7)
- Iron plate

3.3.1.3 Tubing

We use the following materials to connect a glass syringe to the Narishige HI-7 needle holder via a Teflon tube (Fig. 3.3c). The Bio-Rad Tefzel tube that we use has an inner diameter (ID) of 0.5 mm, whereas the HI-7 needle holder is compatible with a tube of 1 mm ID. It is thus necessary to enlarge one end of the tube using the tip of a fine forceps. The other end of the tube is fitted with a set of Delrin nut/ferrule/lock ring (Bio Rad 1.6 mm OD Post-Pump Fittings 750-0554) and can be connected to the male luer of glass syringe via a 1/4 28 female to male luer (Bio-Rad Luer to BioLogic System Fittings Kit, 732-0113; Fig. 3.3d). The entire system is filled with mineral oil.

- Teflon tube (Bio-Rad 1.6 mm OD Tefzel Tubing, 750-0602).
- Tube fittings (Bio-Rad Luer to BioLogic System Fittings Kit, 732-0113; Bio-Rad 1.6 mm OD Post-Pump Fittings, 750-0554).
- 2-ml glass syringe with a male Luer–Lock connection fitting.

3.3.1.4 Injection Needles

We use glass capillaries with filament and pull needles with a Narishige needle puller (PN-31). We use both needles resulting from a single pull. Needle pulling is one of the critical steps for a successful injection: the efficiency of the injection can change dramatically depending on the shape of the needles. The current setting of our PN-31 is: heater 84.4, sub-magnet 32.7, and main magnet 103.3. However, the optimal setting should be obtained for each machine and regularly revised. Needles are cut so that the straight capillary part (=pre-tapering part) is about 3 cm long. The blunt end should be rounded using the flame of a lighter so that the inner silicone gasket of the needle holder is not damaged.

- Glass capillaries (Harvard Apparatus Borosilicate Thin Wall with Filament, 1.0 mm OD, 0.78 mm ID, 100 mm L, 30-0038).
- Needle puller (Narishige PN-31)
- Glass cutter
- 5-ml glass syringe fitted with a metal 25-gauge hub needle. This syringe is used to fill injection needles with mineral oil after they are loaded with the injection mix.
- Needle carrier: we make this carrier using a plastic petri dish measuring 140 × 20 mm. Stick a polystyrene block of 15 mm × 15 mm × 150 mm on the petri dish lid with double-sided tape. Stick a strip of double-sided tape on the surface of the polystyrene block. Needles are stuck to the tape while they are loaded with injection mix and mineral oil and are held on the block until used for injection.

3.3.2 Injection Chamber

We make the injection chamber with 1.5% agarose in ASW using a mold (see Fig. 3.3e for how to make the mold and injection chamber. Note: refer also to Gregory and Veeman (2013) for a 3D printing-based method of making a mold.

- Coverslip (#1).
- Plastic block (about 15 mm × 15 mm × 5 mm).
- Double face tape.
- Plastic petri dishes (5 cm diameter).
- 1.5% agarose in ASW. Note: do not over-boil the agarose/ASW in a microwave so as not to change the salt concentration of ASW.

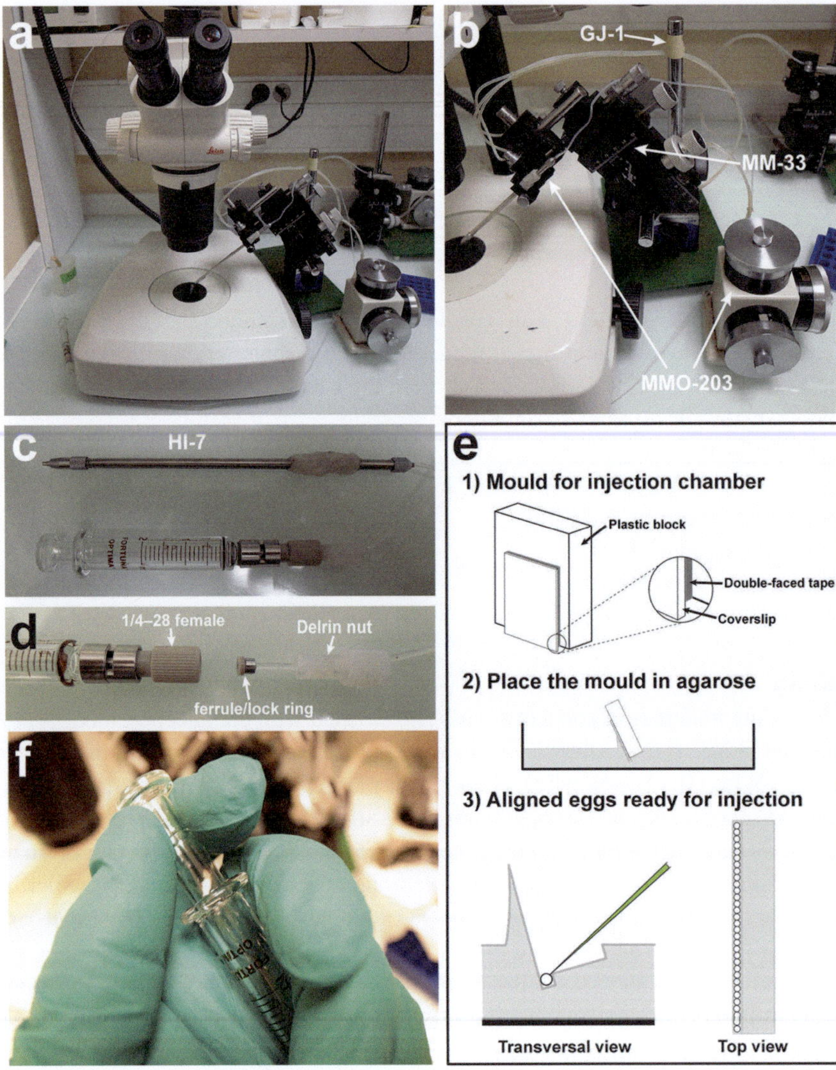

Fig. 3.3 Stereomicroscope microinjection setup. (**a**) Overview of our stereomicroscope microinjection setup. (**b**) Micromanipulator setup. (**c**) The assembly of the needle holder, Teflon tube and glass syringe. (**d**) Close-up of the connection of a Teflon tube to a glass syringe with male luer lock using Bio-Rad fittings kit components. (**e**) Schematics showing (*1*) how to make a mold for the injection chamber, (*2*) how to place the mound in melted 1.5% agarose/ASW, and (*3*) aligned eggs in the injection chamber. To obtain a perfect mold, the coverslip has to stick out from the edge of the plastic block, ideally by about 120 μm. This is not an easy task and requires much trial and error. Gregory and Veeman (2013) recently reported their successful application of a 3D–printing technology to make a microwell mold suitable for ascidian microinjection (Gregory and Veeman 2013). Their method represents a sophisticated alternative to mold making. To make the injection chamber using our mold, pour melted 1.5% agarose/ASW using a plastic pipette into a petri dish 5 cm in diameter and float the petri dish on ice-cold water. Before the agarose hardens, place the mold slightly tilted, as shown in the drawing, using forceps: make sure that the edge of the coverslip is not in direct contact with the bottom of the petri dish. Keep the mold in place until the agarose hardens and then let the petri dish float on ice-cold water for a few minutes. Take the mold gently out of the agarose. Cover the agarose with ASW and keep injection chambers at 4 °C. We use an injection chamber for several times over 1 week before disposing it. To align eggs into the groove of the injection chamber, first place dechorionated unfertilized eggs on the platform next to the groove and, using a gentle water flow from the micropipette, displace the eggs toward the groove and let them fall into it. (**f**) Image showing how to hold the injection syringe. This holding position allows both slight pulling and pushing of the plunger

- Micropipette: we use a hand-made micropipette to handle eggs/embryos. A glass tube (OD 5 mm, ID 3 mm) is pulled under the flame, so that one end becomes tapered to around 1 mm in diameter. To the large end, attach a rubber tube with the other end stapled (Sardet et al. 2011). Soak newly made micropipettes in water overnight; this treatment renders glass pipettes less sticky.

3.3.3 Method

3.3.3.1 Preparation of Injection Needles

1. Place the needle carrier with the needles vertical on the bench with the cut end of needles pointing upward. We normally prepare two needles per one injection chamber.
2. With a Gilson P2 pipette (or equivalent), place a tiny drop of injection mix (0.25 to 0.5 μl) onto the cut end of the needle. The injection mix should quickly go down along the inner filament toward the needle tip.
3. Tap the needle gently with a pencil to remove most of the air bubbles present in the tapered part of the needle.
4. Using the 5-ml glass syringe filled with mineral oil, fill up the needles with mineral oil. It is acceptable to leave small air bubbles at the interface between the injection mix and mineral oil, but make sure that there are no air bubbles in the straight capillary part of the needle up to the cut end.
5. Needles are now ready to be fitted into the needle holder.

3.3.3.2 Microinjection

1. Make sure that there are no air bubbles in the mineral oil in the assembly of the needle holder/tubing/syringe. Note: a small bubble in the syringe can be tolerated.
2. Insert a needle into the needle holder, making sure to avoid air bubbles. Note: push a little oil out of the needle holder using the syringe while pushing the needle into the holder; it gets a little messy with mineral oil spilling onto fingers.
3. Align dechorionated-unfertilized eggs into the groove of the injection chamber using a micropipette (Fig. 3.3e).
4. Cut a coverslip into a strip about 5 mm wide. Mark a spot (or line) on one side of the strip so that the mark is under the seawater. Stick the strip into the agarose of the injection chamber so that it leans on the wall of the petri dish with the marked surface facing toward you.
5. Set the injection chamber under the stereomicroscope and position and adjust the manipulator setup so that the needle tip is in the seawater around the centre of the injection chamber. Displace the injection chamber so that the needle tip comes very close to the marked surface of the coverslip strip. Under the stereomicroscope, focus on the needle tip and bring the needle tip “very slowly” toward the marked surface by controlling the oil hydraulic micromanipulator. During this process, maintain a gentle pressure on the glass syringe. When the needle tip makes contact with the coverslip, it breaks and the green injection solution oozes out of the tip. It is important to keep the opening of the needle tip as small as possible: the opening should not be seen. It is now ready to inject. Note: wear a powder-free nitrile glove on the hand handling the glass syringe.
6. Position the needle tip against the middle of the first egg.
7. Push the needle into the egg with the oil hydraulic micromanipulator so that its tip is placed around the center of the egg.
8. Pull back the syringe gently to break the egg membrane (nothing can be seen happening; this step takes trial and error). See Fig. 3.3f for how to hold the syringe.
9. Push the syringe gently to dispense the injection mix into the egg. A colored sphere of injected solution forms around the needle tip, but it should quickly disperse. Injection volume can be controlled based on the diameter of the sphere before it disperses: we aim to control the diameter of the sphere to about one quarter of the egg’s diameter. Note: if the

injected solution remains as a distinct sphere, the egg membrane was not broken at step 8. If this is the case, repeat the step.

10. Pull the needle gently out from the egg. Position the needle at the second egg and repeat steps 7 to 10. Note: during injection, keep the eggs in focus.

3.3.3.3 Troubleshooting

1. In our experience, using the stereomicroscope injection setup, *Phallusia* eggs are more fragile than *Ciona* eggs, making it more challenging to inject them (e.g., around 20 injected eggs can be obtained from one injection chamber of lined-up *Phallusia* eggs compared with about 50 with *Ciona* eggs). It is clear when *Phallusia* eggs are killed by injection as they become opaque (Movie 3.1).
2. The plunger of the injection syringe has to move "very smoothly." Tiny dusts between the plunger and barrel could prevent its smooth movement, which significantly hinders successful injection. Make sure that the plunger and inner wall of the barrel are clean and move smoothly (always use powder-free gloves to handle the syringe).
3. Make sure that there are no air bubbles in the mineral oil in the assembly of the needle/needle holder/Teflon tubing.
4. When the needle tip starts to become clogged, there are three options: (a) stab the needle tip into the agarose of the injection chamber to remove cellular debris from the needle tip, (b) repeat step 5 to reopen the needle tip or (c) change to a new needle.

References

Brozovic M, Martin C, Dantec C, Dauga D, Mendez M, Simion P, Percher M, Laporte B, Scornavacca C, Di Gregorio A, Fujiwara S, Gineste M, Lowe EK, Piette J, Racioppi C, Ristoratore F, Sasakura Y, Takatori N, Brown TC, Delsuc F, Douzery E, Gissi C, McDougall A, Nishida H, Sawada H, Swalla BJ, Yasuo H, Lemaire P (2016) ANISEED 2015: a digital framework for the comparative developmental biology of ascidians. Nucleic Acids Res 44:D808–D818

Christiaen L, Wagner E, Shi W, Levine M (2009) Electroporation of transgenic DNAs in the sea squirt Ciona. Cold Spring Harb Protoc. https://doi.org/10.1101/pdb.prot5345

Corbo JC, Levine M, Zeller RW (1997) Characterization of a notochord-specific enhancer from the Brachyury promoter region of the ascidian, Ciona intestinalis. Development 124:589–602

Gregory C, Veeman M (2013) 3D-printed microwell arrays for Ciona microinjection and timelapse imaging. PLoS One 8:e82307

Haupaix N, Stolfi A, Sirour C, Picco V, Levine M, Christiaen L, Yasuo H (2013) p120RasGAP mediates ephrin/Eph-dependent attenuation of FGF/ERK signals during cell fate specification in ascidian embryos. Development 140:4347–4352

Levasseur M, McDougall A (2000) Sperm-induced calcium oscillations at fertilisation in ascidians are controlled by cyclin B1-dependent kinase activity. Development 127:631–641

Matsuoka T, Awazu S, Shoguchi E, Satoh N, Sasakura Y (2005) Germline transgenesis of the ascidian Ciona intestinalis by electroporation. Genes 41:67–72

McDougall A, Chenevert J, Pruliere G, Costache V, Hebras C, Salez G, Dumollard R (2015) Centrosomes and spindles in ascidian embryos and eggs. Methods Cell Biol 129:317–339

Negishi T, Yasuo H (2015) Distinct modes of mitotic spindle orientation align cells in the dorsal midline of ascidian embryos. Dev Biol 408:66–78

Prodon F, Chenevert J, Hébras C, Dumollard R, Faure E, Gonzalez-Garcia J, Nishida H, Sardet C, McDougall A (2010) Dual mechanism controls asymmetric spindle position in ascidian germ cell precursors. Development 137:2011–2021

Roure A, Rothbächer U, Robin F, Kalmar E, Ferone G, Lamy C, Missero C, Mueller F, Lemaire P (2007) A multicassette Gateway vector set for high throughput and comparative analyses in ciona and vertebrate embryos. PLoS One 2:e916

Sardet C, McDougall A, Yasuo H, Chenevert J, Pruliere G, Dumollard R, Hudson C, Hebras C, Le Nguyen N, Paix A (2011) Embryological methods in ascidians: the Villefranche-sur-Mer protocols. Methods Mol Biol 770:365–400

Zeller RW, Virata MJ, Cone AC (2006) Predictable mosaic transgene expression in ascidian embryos produced with a simple electroporation device. Dev Dyn 235:1921–1932

Check for updates

4 Microinjection of Exogenous DNA into Eggs of *Halocynthia roretzi*

Gaku Kumano

Abstract

Exogenous gene expression assays during development, including reporters under the control of 5′ upstream enhancer regions of genes, constitute a powerful technique for understanding the mechanisms of tissue-specific gene expression regulation and determining the characteristics, behaviors, and functions of cells that express these genes. The simple marine chordate *Halocynthia roretzi* has been used for these transgenic analyses for a long time and is an excellent model system for such studies, especially in comparative analyses with other ascidians. In this study, I describe simple methods for microinjecting *H. roretzi* eggs with exogenous DNA, such as a promoter construct consisting of a 5′ upstream region and a reporter gene, which are prerequisites for transgenic analyses. I also describe basic knowledge regarding this ascidian species, providing reasons why it is an ideal subject for developmental biology studies.

Keywords

Halocynthia roretzi · Microinjection · Ooplasmic segregation · Micromanipulator · Fertilized eggs · Enhancer regions

G. Kumano (✉)
Asamushi Research Center for Marine Biology, Graduate School of Life Sciences, Tohoku University, Asamushi, Aomori, Japan
e-mail: kumano@m.tohoku.ac.jp

4.1 Generalities Regarding *H. roretzi*

4.1.1 Biology of *H. roretzi*

H. roretzi is an edible ascidian consumed in Japan and Korea. This species belongs to the order Stolidobranchia (Fig. 4.1a, class Ascidiacea), different than the order containing the ascidians *Ciona* and *Phallusia*, and characterized by the presence of paired gonads located on each side of the body, the presence of folded pharyngeal baskets, and the absence of an abdomen (Satoh 1994). *H. roretzi* is hermaphroditic like other ascidian species, shows self-incompatibility during fertilization, and spawns eggs and sperm almost simultaneously every day for approximately a month when the weather becomes colder as winter approaches in the wild. It starts releasing eggs and sperm by light stimulation after sunrise, and this process can be reproduced in the laboratory by setting a light and temperature control system. There are three types of *H. roretzi* according to differences in spawning time during the day and during the year (Fig. 4.1b, Numakunai and Hoshino 1973, 1974). The aquaculture of one of these three types, called noon-type *Halocynthia*, has been successful for several decades in the northern parts of Japan, mainly along the Sanriku coast of the Tohoku area, and on the east and south coasts of Korea. Aquaculture of the morning-type *Halocynthia* has recently

Y. Sasakura (ed.), *Transgenic Ascidians*, Advances in Experimental Medicine and Biology 1029,
https://doi.org/10.1007/978-981-10-7545-2_4

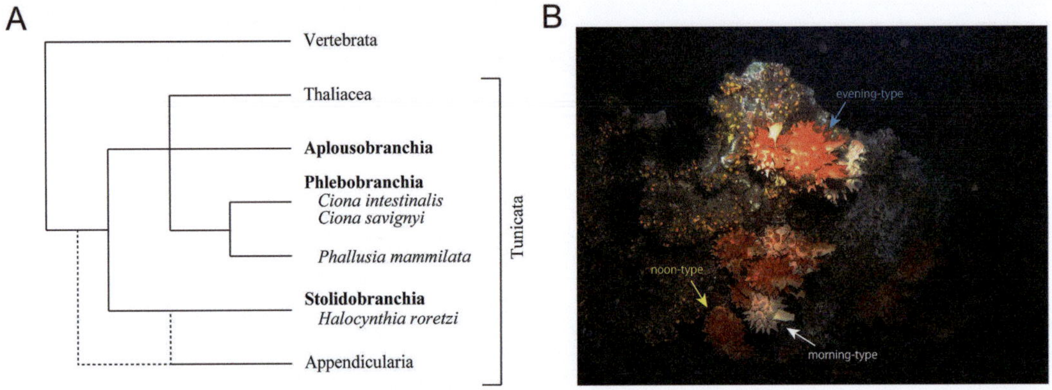

Fig. 4.1 (**a**) Cladogram of the major solitary ascidian species in the context of the tunicate phylogeny. The orders that belong to the class Ascidiacea are indicated in bold letters. The relative position of Thaliacea, Aplousobranchia, and Phlebobranchia, and the position of Appendicularia (*dotted lines*) are controversial (Tsagkogeorga et al. 2009). (**b**) Occurrence of the three types of *H. roretzi* in Mutsu Bay, located at the northern end of the main island of Japan. Photo taken and kindly provided by Dr. Takaharu Numakunai. *H. roretzi* has been grouped into three types depending on the time of day at which eggs and sperm start to spawn in the wild: the morning type (*white arrow*), noon type (*yellow arrow*), or evening type (*blue arrow*). Mutsu Bay is the only place known so far where the three types reside. The three types also differ in the time of year at which their spawning seasons start. The evening type is the earliest among them and starts spawning by the end of October in the Mutsu Bay area. It is followed by the morning type by mid-November, and the noon type by the end of December. The noon-type *H. roretzi* has been found in a broad region along the coast of the main island of Japan, in addition to the east and south coasts of Korea. In contrast, the morning type has been reported only in the northern parts of Japan, including the Aomori and Hokkaido Prefectures, and on the east coast of Korea. Interestingly, the evening type has been found only in Mutsu Bay. In appearance, the morning type can be easily distinguished from the other two types by its whiter colored tunic mantle with longer and thinner spikes. However, the noon and evening types are sometimes difficult to distinguish from each other in appearance

begun in the Mutsu Bay area in Japan. Consequently, *Halocynthia* eggs and adults can be easily obtained in large amounts for scientific use, and the species has been extensively studied for a long time, particularly in the field of developmental biology.

4.1.2 Benefits of Using *H. roretzi*

Halocynthia roretzi shares several fascinating biological features with other ascidian species, such as *Ciona* and *Phallusia*. These features include gamete spawning induced by light stimulation (Castle 1896; Conklin 1905; Rose 1939; Hŭus 1939; Berrill 1947; Castello et al. 1957; Hirai and Tsubata 1956), self and nonself-recognition during fertilization (Morgan 1923, 1938; Rosati and De Santis 1978; Fuke 1983), invariant cell lineages during embryogenesis (Conklin 1905; Nishida 1987), establishment of a chordate body plan with relatively small numbers of cells and cell types (Kowalevsky 1861, 1871; Monroy 1979; Yamada and Nishida 1999; Satoh 2003), metamorphosis (Willey 1893a, 1893b; Cloney 1961, 1987; Numakunai et al. 1964; Karaiskou et al. 2015), and cellulose synthesis (Herzog and Gonell 1924; Hunt 1970). Research into *H. roretzi* is important in that it is the only well-studied solitary ascidian species that belongs to the order Stolidobranchia, and that it is possible to investigate evolutionary aspects of biological processes within the class Ascidiacea, which belongs to the closest extant relative group of the vertebrates, the Tunicata (Delsuc et al. 2006; Vienne and Pontarotti 2006). *Halocynthia* is phylogenetically far away (several hundred million years) from the most well-studied ascidian species *Ciona intestinalis/robusta* (Lemaire and Piette 2015); however, larval morphologies and cell lineages between these two species appear to be almost identical, thus making them ideal subjects for studies of developmental system drift (Hudson and Yasuo 2008; Stolfi et al. 2014; Lemaire and Piette 2015).

Eggs of *H. roretzi* are relatively large (approximately 280 μm in diameter) compared with those of other ascidian species such as *C. intestinalis* and *Phallusia mammillata* (approximately 140 and 120 μm in diameter respectively) (Lemaire et al. 2008). This feature makes *Halocynthia* more suitable for the application of micromanipulation techniques in developmental biology analyses. Although microinjection into eggs has been performed regularly on several ascidian species, *H. roretzi* is the only one that has been extensively used for other micromanipulation techniques, such as blastomere isolation and cytoplasm transplantation (Nishida 1997, 2005). Consequently, molecular and cellular knowledge about embryogenesis gained from different experimental approaches has been accumulated for *Halocynthia* (Nishida 1994; Kim et al. 2000; Kobayashi et al. 2003; Negishi et al. 2007; Kumano and Nishida 2009).

4.1.3 Developmental Table

In nature, *H. roretzi* spawns eggs and sperm in early winter, and the embryos develop at cold temperatures. Figure 4.2 shows the developmental timetables of *H. roretzi* under laboratory conditions. Fertilized eggs of ascidian species go through several shape changes and undergo two rounds of ooplasmic segregation before the first cell division (Fig. 4.2a, Conklin 1905; Hirai 1941; Sawada and Osanai 1981; Jeffery and Meier 1983; Sardet et al. 1989; Nishida 1994). The egg shape changes result from cortical actin dynamics that segregate maternal factors responsible for development to the vegetal pole (the first phase of ooplasmic segregation, Fig. 4.2a) (Kumano and Nishida 2007; Prodon et al. 2007; Sardet et al. 2007; Lemaire 2009; Makabe and Nishida 2012). A subset of these vegetally localized factors is further transported by the action of sperm asters to the future posterior pole (the second phase of ooplasmic segregation, Fig. 4.2a) (Kumano and Nishida 2007; Prodon et al. 2007; Sardet et al. 2007; Lemaire 2009; Makabe and Nishida 2012). These segregation processes can be easily observed in *H. roretzi* eggs using stereo microscopes, as the segregated cytoplasm containing maternal factors is transparent against a yellowish background. The distinction among the different stages before the first cell division (Fig. 4.2a) is important because, at a certain stage, eggs can be microinjected more easily (see below).

After the first cell division, *Halocynthia* embryos undergo several rounds of cell cleavage until the 110-cell stage, when they start to gastrulate, and further develop into neurula and tailbud-stage embryos (Fig. 4.2b). They finally hatch to become larvae in approximately 35 hours after fertilization at 13 °C (Fig. 4.2b). In the sea, hatched larvae are thought to swim for a few days before they settle on the sea bottom to undergo metamorphosis. After metamorphosis, a few years pass before they become sexually mature.

4.2 Microinjecting *H. roretzi* Eggs

4.2.1 Materials

4.2.1.1 Microinjection Setup

Microinjection of *H. roretzi* eggs is performed in most cases after fertilization, with a rare exception that it has been done before fertilization (see below, Nishida and Sawada 2001). An Olympus stereo microscope SZX12 (Olympus corporation, Tokyo, Japan) with a right-hand-mounted Narishige Joystick Micromanipulator MN-151 (Narishige Group, Tokyo, Japan) is used (Fig. 4.3a). A Narishige Injection Holder HI-7 (Narishige Group, Tokyo, Japan) with an injection needle inserted into its front end is set on the micromanipulator at an angle of 30–45° with

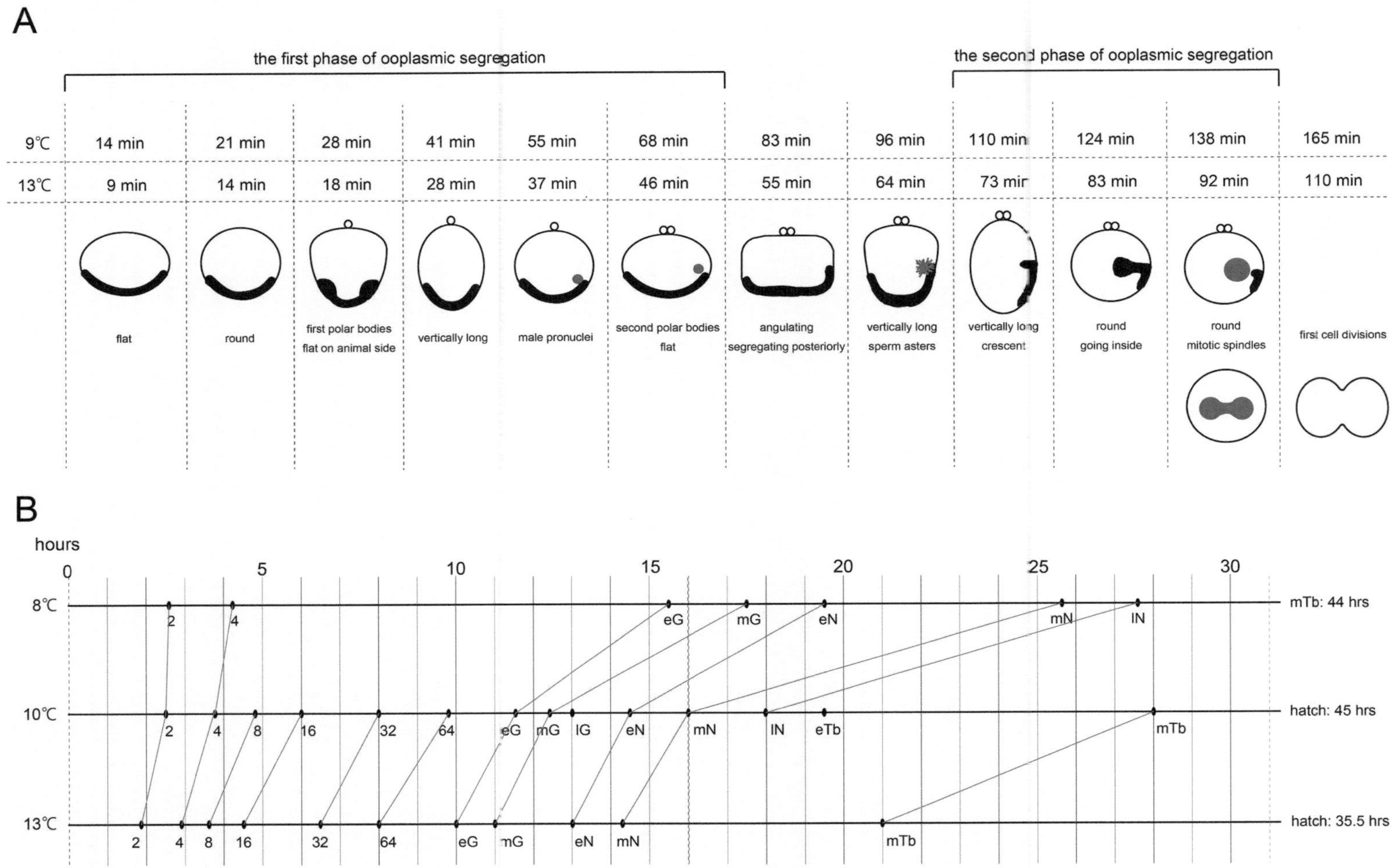

Fig. 4.2 Developmental timetables of *H. roretzi*. Data obtained and kindly provided by Dr. Hiroki Nishida (Osaka University). (**a**) Schematic diagrams of egg shape change and ooplasmic segregation processes after fertilization. Egg shape is expressed as either “flat,” “round,” “vertically long,” or “angulating” in this figure. *Black* regions indicate segregated cytoplasm. *Gray* represents pronuclei, sperm asters or mitotic spindles. Above each of the diagrams, minutes after fertilization at 9 and 13 °C are indicated. Eggs are shown with their animal poles at the top. Eggs from 37 min to 92 min at 13 °C are oriented with their future posterior pole to the right. The egg at the bottom for 92 min at 13 °C is viewed posteriorly. (**b**) Developmental timing at 8, 10, and 13 °C. The numbers at the top indicate hours after fertilization: *2* two-cell stage, *4* four-cell stage, *8* eight-cell stage, *16* 16-cell stage, *32* 32-cell stage, *64* 64-cell stage, *eG* early gastrula stage, *mG* mid gastrula stage, *lG* late gastrula stage, *eN* early neurula stage, *mN* mid neurula stage, *lN* late neurula stage, *eTb* early tailbud stage, *mTb* mid tailbud stage

regard to the microscopic stage, and the back end of HI-7 is connected with a plastic tube to a 5-ml glass syringe (Fig. 4.3a). The inside of the injection holder, the plastic tube, and the syringe are filled with tap water such that the pushing or pulling force created by the syringe is directly propagated to the needle. The injection needle is prepared by pulling a Narishige Glass Capillary containing a filament GDC-1 (Narishige Group, Tokyo, Japan) with a needle puller (e.g., Micropipette Puller P-97; Sutter Instrument, Novato, CA, USA). When loaded onto the needle, the DNA solution used for injection is transported to the tip of the needle by the capillarity exerted by the filament inside the Narishige Glass Capillary. The remainder of the needle is filled with silicone oil (Shin-Etsu Chemical Co., Tokyo, Japan; Fig. 4.3b). Loaded needles should be placed on ice in a plastic case until ready for microinjection. Fertilized eggs are injected in Millipore-filtered seawater containing 50 μg/ml of streptomycin and kanamycin (MFSW), on a new 24 x 24 -mm micro cover glass placed on the bottom of a flipped 35 mm petri dish lid (Fig. 4.3c).

Injection of DNA for transgenesis in *H. roretzi* has been only performed with fertilized eggs. Injected DNA should be purified from bacteria using a commercially supplied endotoxin-free plasmid prep kit. A final concentration of up to 20 ng/μl of plasmid DNA can be injected into one quarter to one fifth of the egg diameter without perturbing normal develop-

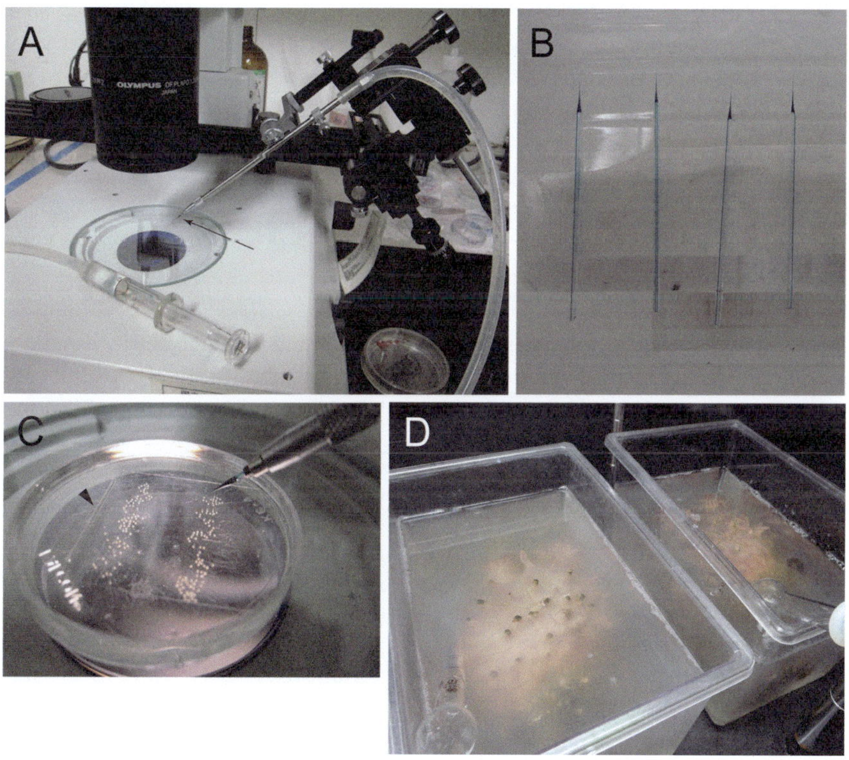

Fig. 4.3 Materials used for microinjecting *H. roretzi* eggs. (**a**) Microinjection apparatus, including an Olympus stereo microscope, a Narishige micromanipulator (to the right of the microscope), and injection holder (on the micromanipulator) with an injection needle inserted at the front end (*left down*, *arrow*), and a plastic tube attached at the back end (*right top*), through which a syringe (on the microscopic stage) is connected to the injection holder. (**b**) Injection needles loaded with DNA/Fast Green solution on their tips (*top*) followed by silicone oil. (**c**) Fertilized eggs lined up for microinjection on a micro cover glass in MFSW placed on the lid of a petri dish. The *arrowhead* points to a piece of a glass capillary used for breaking the tip of the needle. (**d**) *Halocynthia* adults after artificially induced spawning in containers. Sea water is densely cloudy with spawned sperm. Unfertilized eggs stay at the bottom and are not shown in the image

ment (Matsumoto et al. 2007). The injection solution consists of DNA, 3 mg/ml of Fast Green and autoclaved distilled water. Fast Green is used to roughly measure the injected volume and to distinguish injected eggs from un-injected ones. Because the dye also emits red fluorescence, nonfluorescent Fast Red can be used as an alternative dye at the same concentration when red fluorescent reporters such as mCherry are used in promoter constructs. The Fast Green stock solution (10 mg/ml) should be kept after filtration through a 0.22-μm pore size filter, and the DNA/Fast Green solution should be centrifuged before injection to remove lumps and avoid needle clogging.

4.2.1.2 Preparing Eggs and Sperm

Although the spawning season for *H. roretzi* only lasts approximately a month in the wild, it can be extended up to seven or 8 months in the laboratory. *H. roretzi* mature adults should be purchased from local fishermen or collected from the sea bottom by diving, before they start spawning in early winter. They are then kept in tanks with cold sea water at 8 °C or lower temperatures under constant light. These conditions prevent *H. roretzi* from spawning for several months. To obtain eggs and sperm, ascidian adults can be artificially induced to spawn by keeping them in the dark for at least 6 h before exposing them to light in sea water at a higher temperature of 11–13 °C. By placing one individual per plastic container during the light and temperature stimulation, it is possible to obtain unfertilized eggs and sperm from the same individual in a container (Fig. 4.3d), because the ascidian species show strict self-incompatibility and do not self-fertilize, as mentioned above. Fertilized eggs for microinjection can be obtained when desired by mixing eggs and sperm from different containers. Alternatively, sperm for fertilization can be prepared from undiluted and preserved dry sperm (see below). The same individuals can be used for artificial spawning several times during the season. However, a minimum interval of 2 weeks between the trials is required.

4.2.2 Methods

The procedures described here are performed at 8–13 °C.

4.2.2.1 Microinjecting Fertilized Eggs

1. Fertilize eggs by mixing eggs and sperm from different individuals.
   - Follow the protocol below with the use of dry sperm in case the spawned sperm does not work. It may become infertile in several hours after it is spawned or when it is spawned from old adults (kept without food for several months).
     (1) Collect undiluted dry sperm from the cuts made to the sperm ducts of the dissected gonads using a Pasteur pipette and put it into an Eppendorf tube. Keep the tube on ice while collecting. Dry sperm can be used for fertilization for approximately a week when kept at 4 °C.
     (2) Dilute a few microliters of dry sperm with sea water in a 35-mm petri dish and wait until the sea water becomes a little cloudy.
     (3) Activate sperm by adding one drop of 0.025 M of NaOH with a Pasteur pipette to the sea water and let it sit for 1 min.
     (4) Fertilize eggs in a 60-mm petri dish gradually over time by adding one drop, two drops, and then five drops of the activated sperm solution at approximately 3-min intervals.
2. Confirm fertilization by observing the elevation of the vitelline membranes (Fig. 4.4a, b) approximately 20 min after insemination.
3. Remove follicle cells from the vitelline membranes by incubating the fertilized eggs in 0.05% actinase E (Kaken Pharmaceutical Co, Tokyo, Japan) in MFSW for 5–10 min. Check whether the follicle cells easily fall off when scratching them with a tungsten needle. Once they do, immediately wash the eggs five times with MFSW (Fig. 4.4c).
   - The follicle removal step makes the egg cell more visible through the vitelline

membrane (Fig. 4.4b, c) and makes the vitelline membrane sticky such that the fertilized eggs can be fixed on the surface of a micro cover glass without moving when a pushing force from the needle is applied during injection.

– A prolonged incubation time with the actinase E solution softens the vitelline membrane and makes it difficult for the injection needle to break through the membrane and reach the egg.

4. Transfer and line up the eggs for injection on a new micro cover glass that is placed on the petri dish lid filled with MFSW, as described above (Fig. 4.3c).
5. Adjust the positions of the needle holder using the micromanipulator and the petri dish such that the eggs can be injected around the center of the microscopic field.

– Eggs are the easiest to inject without breaking them during the second phase of ooplasmic segregation. However, during the first phase of ooplasmic segregation, and before the initiation of the second phase, eggs can also be injected successfully (Takatori et al. 2010).

– Once the mitotic spindles appear it is better to stop injecting; otherwise, the chances of asymmetric inheritance of the injected substances in the first cell division increase.

6. Break the tip of the injection needle against a piece of a glass capillary laid on the bottom of the petri dish lid near the micro cover glass (arrowhead in Fig. 4.3c). The glass capillary piece is aligned perpendicular to the axis of the injection holder so that it rolls away on the bottom of the petri dish lid when it is hit by the needle, caus-

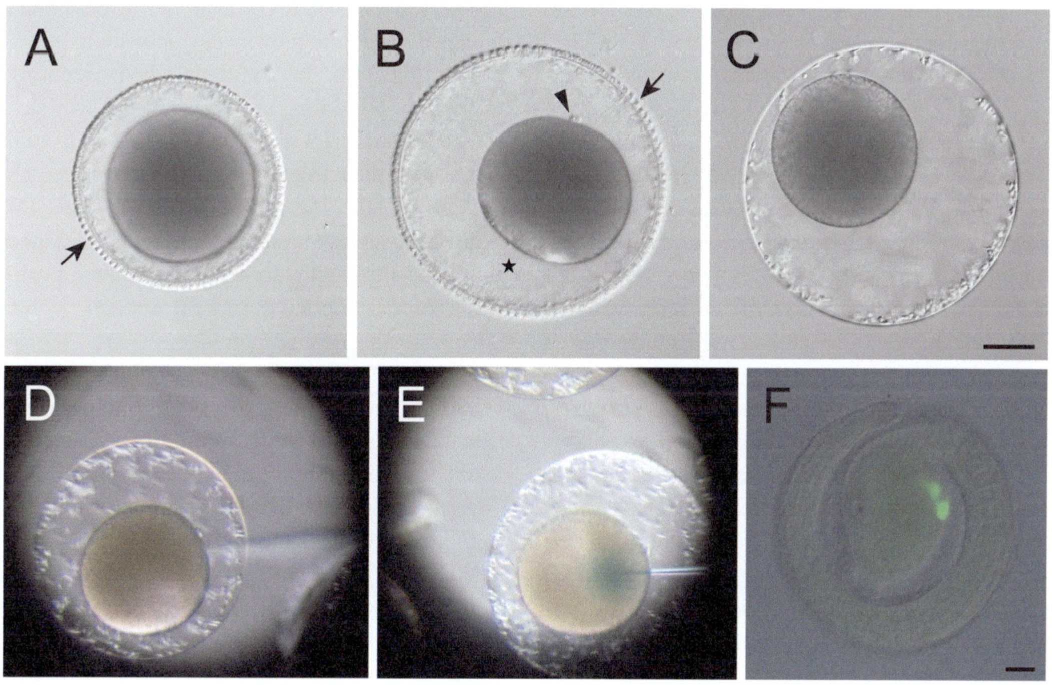

Fig. 4.4 *Halocynthia roretzi* eggs and embryo. (**a**) An unfertilized egg. (**b**) A fertilized egg approximately 50 min after insemination. *Arrows* in (**a**) and (**b**) indicate a layer of follicle cells on the vitelline membrane. The *arrowhead* in (**b**) points to the polar bodies. Transparent cytoplasm is localized to the vegetal region (*star*), opposite the polar bodies. (**c**) A fertilized egg after follicle removal. Scale bar: 50 μm for **a**, **b**, and **c**. (**d**) An egg with the tip of an injection needle (out of focus) on its surface. (**e**) An egg injected with DNA/Fast Green. Images in **d** and **e** were taken through the ocular lens. (**f**) A tailbud-stage embryo in which an enhanced green fluorescent protein reporter is expressed in the mesenchymal cells of the trunk under the control of a 1,054-bp 5′ upstream region of the *Twist-r* gene. Scale bar: 50 μm

ing as small an impact as possible on the needle.
7. Advance the injection needle toward the egg using the operation screws of the micromanipulator, which control three-dimensional large movements, break through the vitelline membrane, and place the tip of the needle on the egg surface (Fig. 4.4d).
8. Insert the needle into the center of the egg using the joystick of the micromanipulator, which controls fine movement, and remain still for 10 s.
9. Apply negative pressure by slightly pulling the syringe to break the plasma membrane, and inject DNA into the center of the egg until the size of the injected dye becomes one quarter to one fifth of the egg's diameter (Fig. 4.4e).
10. Wait another 10 s before removing the injection needle very rapidly to avoid breaking the egg using the operation screws of the micromanipulator.
11. Once enough eggs have been injected, transfer the eggs stained with Fast Green to a used petri dish filled with MFSW and allow them to develop until the stages of interest. With the follicles removed, eggs do not adhere to the bottom of a used petri dish.

4.2.2.2 Microinjecting Unfertilized Eggs

Microinjection can be performed in unfertilized eggs of *H. roretzi*, although microinjecting fertilized eggs satisfies most experimental purposes, including DNA injection for transgenesis, and is the easier way to introduce exogenous substances into the embryo. Unfertilized eggs have been microinjected for the purpose of depleting maternal mRNA with antisense oligodeoxynucleotides complementary to subset regions of the mRNA, using the protocol described below (Nishida and Sawada 2001).

1. Prepare chorion removal solution by adding six drops of 2.5 M of NaOH with a Pasteur pipette to approximately 7 ml of MFSW, containing 0.05% actinase E and 1% sodium thioglycolate.
2. Remove the vitelline membranes by incubating unfertilized eggs in the chorion removal solution for approximately 10 min. Check whether test cells come out of the peri-vitelline space through breaks on the vitelline membrane or the membrane breaks when touching it even gently with a tungsten needle. Once they do or it does, immediately but gently wash the eggs five times with MFSW. As devitellinated eggs sink faster than those with the vitelline membranes broken in MFSW when washing, try to discard the latter by removing the supernatant before everything comes to sit on the bottom in each wash.
3. Transfer the devitellinated eggs to an agar-coated petri dish (1% agar in MFSW) filled with MFSW.
4. Prepare another agar-coated petri dish for injection, which is filled with artificial seawater containing half the normal Ca^{2+} concentration and is coated by agar dissolved in the same seawater to prevent egg activation during microinjection. In addition, when preparing the dish, build a square bank on the agar surface by floating a micro cover glass (0.17 mm thick) on the hardening agar solution.
5. Transfer and line up devitellinated eggs for injection along one side of the banks described above so that they do not move when a pushing force is applied during injection.
6. Follow the above protocol 6 to 10 in Sect. 4.2.2.1 for injection, with the exception that the glass capillary to break the tip of the needle is laid on the agar and the needle does not go through the vitelline membranes.
7. Transfer the injected and inactivated eggs to another agar-coated petri dish filled with MFSW. The eggs can be kept overnight at 11 °C before fertilization (Nishida and Sawada 2001).
8. For fertilization, dilute dry sperm with egg seawater, the supernatant from the overnight suspension of eggs with vitelline membranes in MFSW (a ratio of 1 to 1), and let the diluted sperm solution sit for several minutes. After the incubation, activate sperm and fertilize the

eggs according to the above protocol 1–4) in Sect. 4.2.2.1.

9. Transfer the fertilized eggs undergoing cell shape changes to another agar-coated petri dish filled with MFSW and allow them to develop.

4.3 Examples of DNA Microinjection in *H. roretzi*

The introduction of transgenes into *H. roretzi* embryos has been successfully performed for the purposes of identifying enhancer regions responsible for tissue-specific gene expression, visualizing certain types of cells, and evolutionarily analyzing enhancer activities between ascidian species.

Examples of identified tissue-specific enhancers include those from *Brachyury* for the notochord (Takahashi et al. 1999; Matsumoto et al. 2007), *tyrosinase* for pigment cells in the brain (Wada et al. 2002; Toyoda et al. 2004), *Otx* for the brain (Oda-Ishii and Saiga 2003), *synaptotagmin* for motor neurons (Katsuyama et al. 2002; Matsumoto et al. 2008), *Tbx*, *muscle actin* and *myosin heavy chain* for the muscle (Mitani et al. 2001; Kusakabe et al. 1995; Araki and Satoh 1996), *EpiC* for the epidermis (Ishida and Satoh 1999), and *Twist-r* (Kumano et al. 2014) for the mesenchyme (unpublished by G. Kumano, Fig. 4.4f). A 3.4-kb-long 5′ upstream region of the *synaptotagmin* gene fused to green fluorescent protein was used to identify the number, position, and axonal projection pattern of the motor neurons (Okada et al. 2002). Finally, interspecific enhancer analyses have been performed between *H. roretzi* and *C. intestinalis/robusta*, using *Brachyury* (Takahashi et al. 1999), *Otx* (Oda-Ishii et al. 2005), and *synaptotagmin* (Matsumoto et al. 2008) in attempts to unravel a mystery regarding how different ascidian species might produce highly similar embryos, despite the extensive divergence in their genome sequences (Berna and Alvarez-Valin 2014).

Acknowledgments I would like to thank Dr. Takaharu Numakunai for providing the image in Fig. 4.1b, Dr. Hiroki Nishida for providing the data on the developmental timetable in Fig. 4.2, and Ms. Kaori Miyaoku and Mr. Tao Zheng for their help in preparing the images in Fig. 4.4a–f.

References

Araki I, Satoh N (1996) Cis-regulatory elements conserved in the proximal promoter region of an ascidian embryonic muscle myosin heavy-chain gene. Dev Genes Evol 206:54–63

Berna L, Alvarez-Valin F (2014) Evolutionary genomics of fast evolving tunicates. Genome Biol Evol 6:1724–1738. https://doi.org/10.1093/gbe/evu122

Berrill NJ (1947) The development and growth of *Ciona intestinalis*. J Mar Biol Assoc 26:616–625

Castle WE (1896) The early embryology of *Ciona intestinalis* Flemming (L.) Bull Mus Comp Zool, Harvard 27:201–280

Cloney RA (1961) Observations on the mechanism of tail resorption in ascidians. Am Zool 1:67–87

Cloney RA (1987) Ascidian metamorphosis: review and analysis. In: Chia FS, Rice ME (eds) Settlement and metamorphosis of marine invertebrate larvae. Elsevier, Amsterdam, pp 255–282

Conklin EG (1905) The organization and cell lineage of the ascidian egg. J Acad Nat Sci Philadelphia 13:1–119

Costello DP et al. (1957) Methods for obtaining and handling marine eggs and embryos. Marine Biological Laboratory, Woods Hole

Delsuc F, Brinkmann H, Chourrout D, Philippe H (2006) Tunicates and not cephalochordates are the closest living relatives of vertebrates. Nature 439:965–968

Fuke MT (1983) Self and non-self recognition between gametes of the ascidian, *Halocynthia roretzi*. Rouxs Arch Dev Biol 192:347–352

Herzog RO, Gonell HW (1924) Ueber den feinbau der kunstseide. Kolloid-Zeitschrift 35:201–202

Hirai E (1941) The early development of *Cynthia roretzi*. Sci Rep Tohoku Imp Univ Biol 16:217–232

Hirai E, Tsubata B (1956) On the spawning of an ascidian, *Halocynthia roretzi*. Bull Mar Biol Stat Asamushi, Tohoku Univ 8:1–4

Hudson C, Yasuo H (2008) Similarity and diversity in mechanisms of muscle fate induction between ascidian species. Biol Cell 100:265–277. https://doi.org/10.1042/BC20070144

Hunt S (1970) Polysaccharide-protein complexes in invertebrates. Academic Press, London, p 149

Hŭus J (1939) The effect of light on the spawning in ascidians. Avhandlinger utgitt av Det Norske Videnskaps-Akademi l Oslo I. Mat-Naturv Klasse 4:5–49

Ishida K, Satoh N (1999) Genomic organization and the 5′ upstream sequences associated with the specific spatio-temporal expression of HrEpiC, an epidermis-specific gene of the ascidian *Halocynthia roretzi*. Cell Mol Biol 45:523–536

Jeffery WR, Meier S (1983) A yellow crescent cytoskeletal domain in ascidian eggs and its role in early development. Dev Biol 96:125–143

Karaiskou A, Swalla BJ, Sasakura Y, Chambon JP (2015) Metamorphosis in solitary ascidians. Genesis 53:34–47

Katsuyama Y, Matsumoto J, Okada T, Ohtsuka Y, Chen L, Okado H, Okamura Y (2002) Regulation of synaptotagmin gene expression during ascidian embryogenesis. Dev Biol 244:293–304

Kim GJ, Yamada A, Nishida H (2000) An FGF signal from endoderm and localized factors in the posterior-vegetal egg cytoplasm pattern the mesodermal tissues in the ascidian embryo. Development 127:2853–2862

Kobayashi K, Sawada K, Yamamoto H, Wada S, Saiga H, Nishida H (2003) Maternal macho-1 is an intrinsic factor that makes cell response to the same FGF signal differ between mesenchyme and notochord induction in ascidian embryos. Development 130:5179–5190

Kowalevsky A (1861) Entwicklungsgeschichte der einfachen Ascidien. Mem l'Acad St Petersbourg Ser 7(10):1–19

Kowalevsky A (1871) Weitere studjen uber dje Entwicklung der einfachen Ascidien. Arch Mikr Anat 7:101–130

Kumano G, Nishida H (2007) Ascidian embryonic development: an emerging model system for the study of cell fate specification in chordates. Dev Dyn 236:1732–1747

Kumano G, Nishida H (2009) Patterning of an ascidian embryo along the anterior-posterior axis through spatial regulation of competence and induction ability by maternally localized PEM. Dev Biol 331:78–88. https://doi.org/10.1016/j.ydbio.2009.04.024

Kumano G, Negoro N, Nishida H (2014) Transcription factor Tbx6 plays a central role in fate determination between mesenchyme and muscle in embryos of the ascidian, *Halocynthia roretzi*. Develop Growth Differ 56:310–322. https://doi.org/10.1111/dgd.12133

Kusakabe T, Hikosaka A, Satoh N (1995) Coexpression and promoter function in two muscle actin gene complexes of different structural organization in the ascidian *Halocynthia roretzi*. Dev Biol 169:461–472

Lemaire P (2009) Unfolding a chordate developmental program, one cell at a time: invariant cell lineages, short-range inductions and evolutionary plasticity in ascidians. Dev Biol 332:48–60

Lemaire P, Piette J (2015) Tunicates: exploring the sea shores and roaming the open ocean. A tribute to Thomas Huxley. Open Biol 5:150053. https://doi.org/10.1098/rsob.150053

Lemaire P, Smith WC, Nishida H (2008) Ascidians and the plasticity of the chordate developmental program. Curr Biol 18:R620–R631. https://doi.org/10.1016/j.cub.2008.05.039

Makabe KW, Nishida H (2012) Cytoplasmic localization and reorganization in ascidian eggs: role of postplasmic/PEM RNAs in axis formation and fate determination. Wiley Interdiscip Rev Dev Biol 1:501–518

Matsumoto J, Kumano G, Nishida H (2007) Direct activation by Ets and Zic is required for initial expression of the Brachyury gene in the ascidian notochord. Dev Biol 306:870–882

Matsumoto J, Katsuyama Y, Ohtsuka Y, Lemaire P, Okamura Y (2008) Functional analysis of synaptotagmin gene regulatory regions in two distantly related ascidian species. Develop Growth Differ 50:543–552. https://doi.org/10.1111/j.1440-169X.2008.01049.x.

Mitani Y, Takahashi H, Satoh N (2001) Regulation of the muscle-specific expression and function of an ascidian T-box gene, as-T2. Development 128:3717–3728

Monroy A (1979) Introductory remarks on the segregation of cell lines in the embryo. In: Le Douarin N (ed) In cell lineage, stem cells and cell determination. North-Holland Biomedical Press, Amsterdam, pp 3–13

Morgan TH (1923) Removal of the block to self-fertilization in the ascidian *Ciona*. Proc Nat Acad Sci USA 9:170–171

Morgan TH (1938) The genetic and the physiological problems of self-sterility in *Ciona*. I. Data on self- and cross-fertilization. J Exp Zool 78:271–318

Negishi T, Takada T, Kawai N, Nishida H (2007) Localized PEM mRNA and protein are involved in cleavage-plane orientation and unequal cell divisions in ascidians. Curr Biol 17:1014–1025

Nishida H (1987) Cell lineage analysis in ascidian embryos by intracellular injection of a tracer enzyme. III. Up to the tissue restricted stage. Dev Biol 121:526–541

Nishida H (1994) Localization of determinants for formation of the anterior-posterior axis in eggs of the ascidian *Halocynthia roretzi*. Development 120:3093–3104

Nishida H (1997) Cell lineage and timing of fate restriction, determination and gene expression in ascidian embryos. Semin Cell Dev Biol 8:359–365

Nishida H (2005) Specification of embryonic axis and mosaic development in ascidians. Dev Dyn 233:1177–1193

Nishida H, Sawada K (2001) Macho-1 encodes a localized mRNA in ascidian eggs that specifies muscle fate during embryogenesis. Nature 409:724–729

Numakunai T, Hoshino Z (1973) Biology of the ascidian, *Halocynthia roretzi* (Drasche), in Mutsu Bay. I. Differences of spawning time and external features. Bull Mar Biol Stat Asamushi, Tohoku Univ 14:191–196

Numakunai T, Hoshino Z (1974) Biology of the ascidian, *Halocynthia roretzi* (Drasche), in Mutsu Bay II. One of the three types which has the spawning season and the different time from two others. Bull Mar Biol Stat Asamushi, Tohoku Univ, 15:23–27

Numakunai T, Ishikawa M, Hirai E (1964) Changes of structure stainable with modified Gomori's aldehyde-fuchsin method in the tadpole larvae of the ascidian, Halocynthia roretzi (V. Drasche), relating to tail resorption. Bull Mar Biol Stat Asamushi Tohoku Univ 12:161–172

Oda-Ishii I, Saiga H (2003) Genomic organization and promoter and transcription regulatory regions for the expression in the anterior brain (sensory vesicle) of Hroth, the otx homologue of the ascidian, *Halocynthia roretzi*. Dev Dyn 227:104–113

Oda-Ishii I, Bertrand V, Matsuo I, Lemaire P, Saiga H (2005) Making very similar embryos with divergent genomes: conservation of regulatory mechanisms of Otx between the ascidians *Halocynthia roretzi* and *Ciona intestinalis*. Development 132:1663–1674

Okada T, Katsuyama Y, Ono F, Okamura Y (2002) The development of three identified motor neurons in the larva of an ascidian, *Halocynthia roretzi*. Dev Biol 244:278–292

Prodon F, Yamada L, Shirae-Kurabayashi M, Nakamura Y, Sasakura Y (2007) *Postplasmic/PEM* RNAs: a class of localized maternal mRNAs with multiple roles in cell polarity and development in ascidian embryos. Dev Dyn 236:1698–1715

Rosati F, De Santis R (1978) Studies on fertilization in ascidians. I. Self-sterility and specific recognition between gametes of *Ciona intestinalis*. Exp Cell Res 121:111–119

Rose MS (1939) Embryonic induction in ascidia. Biol Bull 77:216–232

Sardet C, Speksnidjer J, Inoue S, Jaffe L (1989) Fertilization and ooplasmic movements in the ascidian eggs. Development 105:237–249

Sardet C, Paix A, Prodon F, Dru P, Chenevert J (2007) From oocyte to 16-cell stage: cytoplasmic and cortical reorganizations that pattern the ascidian embryo. Dev Dyn 236:1716–1731

Satoh N (1994) Developmental biology of ascidians. Cambridge University Press, New York

Satoh N (2003) The ascidian tadpole larva: comparative molecular development and genomics. Nat Rev Genet 4:285–295

Sawada T, Osanai K (1981) The cortical contraction related to the ooplasmic segregation in *Ciona intestinalis* eggs. Wilhelm Roux's Arch Dev Biol 190:208–214

Stolfi A, Lowe E, Racioppi C, Ristoratore F (2014) Divergent mechanisms regulate conserved cardiopharyngeal development and gene expression in distantly related ascidians. elife 3:e03728. https://doi.org/10.7554/eLife.03728

Takahashi H, Mitani Y, Satoh G, Satoh N (1999) Evolutionary alterations of the minimal promoter for notochord-specific Brachyury expression in ascidian embryos. Development 126:3725–3734

Takatori N, Kumano G, Saiga H, Nishida H (2010) Segregation of germ layer fates by nuclear migration-dependent localization of Not mRNA. Dev Cell 19:589–598. https://doi.org/10.1016/j.devcel.2010.09.003

Toyoda R, Kasai A, Sato S, Wada S, Saiga H, Ikeo K, Gojobori T, Numakunai T, Yamamoto H (2004) Pigment cell lineage-specific expression activity of the ascidian tyrosinase-related gene. Gene 332:61–69

Tsagkogeorga G, Turon X, Hopcroft RR, Tilak M-K, Feldstein T, Shenkar N, Loya Y, Huchon D, Douzery EJ, Delsuc F (2009) An updated 18S rRNA phylogeny of tunicates based on mixture and secondary structure models. BMC Evol Biol 9:187. https://doi.org/10.1186/1471-2148-9-187

Vienne A, Pontarotti P (2006) Metaphylogeny of 82 gene families sheds a new light on chordate evolution. Int J Biol Sci 2:32–37

Wada S, Toyoda R, Yamamoto H, Saiga H (2002) Ascidian otx gene Hroth activates transcription of the brain-specific gene HrTRP. Dev Dyn 225:46–53

Willey A (1893a) Studies on the Protochordata. I. On the origin of the branchial stigmata, praeoral lobe, endostyle, arterial cavities etc. in *Ciona intestinalis* Linn, with remarks on *Clavelina lepadiformis*. Q J Microbiol Sci 34:317–360

Willey A (1893b) Studies on the Protochordata. II. The development of the neuro-hypophyseal system in Ciona intestinalis and Clavelina lepadiformis, with an account of the origin of the sense organ in Ascidia mentula. Q J Microbiol Sci 35:295–334

Yamada A, Nishida H (1999) Distinct parameters are involved in controlling the number of rounds of cell division in each tissue during ascidian embryogenesis. J Exp Zool 284:379–391

Electroporation in Ascidians: History, Theory and Protocols

5

Robert W. Zeller

Abstract

Embryonic development depends on the orchestration of hundreds of regulatory and structural genes to initiate expression at the proper time, in the correct spatial domain(s), and in the amounts required for cells and tissues to become specified, determined, and ultimately to differentiate into a multicellular embryo. One of the key approaches to studying embryonic development is the generation of transgenic animals in which recombinant DNA molecules are transiently or stably introduced into embryos to alter gene expression, to manipulate gene function or to serve as reporters for specific cell types or subcellular compartments. In some model systems, such as the mouse, well-defined approaches for generating transgenic animals have been developed. In other systems, particularly non-model systems, a key challenge is to find a way of introducing molecules or other reagents into cells that produces large numbers of embryos with a minimal effect on normal development. A variety of methods have been developed, including the use of viral vectors, microinjection, and electroporation. Here, I describe how electroporation was adapted to generate transgenic embryos in the ascidian, a nontraditional invertebrate chordate model that is particularly well-suited for studying gene regulatory activity during development. I present a review of the electroporation process, describe how electroporation was first implemented in the ascidian, and provide a series of protocols describing the electroporation process, as implemented in our laboratory.

Keywords

Electroporation · *Ciona* · *Ascidiella* · *Ascidia* · *Molgula* · *Phallusia* · *Boltenia* · Ascidian · Transgenic embryo · Exponentially decaying pulse · Square wave pulse

5.1 Methods of Transferring Molecules into Ascidian Eggs

The two techniques regularly used to introduce molecules into ascidian eggs are both physical methods: microinjection and electroporation. Transgenic embryos were first produced in ascidians by using microinjection (Hikosaka et al. 1992). Microinjection is effective and can be easily implemented in most laboratories; details on the microinjection procedure can be found in other chapters of this volume and in several recent publications (Christiaen et al. 2009a, b;

R. W. Zeller (✉)
Center for Applied and Experimental Genomics, Department of Biology, San Diego State University, San Diego, CA, USA
e-mail: rzeller@mail.sdsu.edu

Y. Sasakura (ed.), *Transgenic Ascidians*, Advances in Experimental Medicine and Biology 1029,
https://doi.org/10.1007/978-981-10-7545-2_5

Sardet et al. 2011; McDougall et al. 2014; Kari et al. 2016). An advantage of microinjection is that very small quantities of reagents are required and this is one reason why it has been used extensively to introduce lineage tracers (Nishida and Satoh 1983; Nishida and Satoh 1985; Nishida 1987), dyes (McDougall and Levasseur 1998), mRNAs (Prodon et al. 2010), and morpholino oligonucleotides (Satou et al. 2001) into unfertilized or fertilized eggs and blastomeres. Microinjection is typically performed on unfertilized eggs. In some species such as *Phallusia mammillata* unfertilized eggs can be loaded with exogenous mRNAs, where they are subsequently translated before fertilization (Prodon et al. 2010). In other species, such as *Ciona robusta*, translation of injected mRNAs is very inefficient in the early embryo. For some species where electroporation has not been successful, such as *Halocynthia* (Stolfi and Christiaen 2012), microinjection is the only way to introduce exogenous molecules into the egg. The main disadvantage of microinjection is that the overall throughput is low; several hundred eggs can be injected in a day. Because of this, experiments must be carefully designed to ensure that enough eggs can be injected to provide biological and technical replicates.

Electroporation is the second physical method of generating transgenic embryos. Unlike microinjection, electroporation can generate large numbers of embryos and thus provide sufficient biological material to conduct functional screens and other types of experiments that utilize large numbers of constructs. A single round of electroporation can generate hundreds to thousands of embryos for 15–20 different transgenes from a single fertilization (Fig. 5.1). Because of this throughput, it is possible to perform studies requiring hundreds of electroporations, for example, to study *cis*-regulation (Brown et al. 2007); these studies would be difficult and expensive to perform in other systems. In addition, the overall process is straightforward and takes about 30 min from start to completion. Electroporation requires significantly larger amounts of reagents than microinjection making it less suited to introducing expensive reagents, such as morpholino oligonucleotides, although as discussed below, we have scaled the reaction volume down to 20 µl, thus opening up the possibility of using electroporation to introduce additional types of molecules beyond plasmid DNAs. Morpholino oligonucleotides have been successfully electroporated into both zebrafish (Cerda et al. 2006) and *Xenopus* embryos (Falk et al. 2007), but to date, this has not been reported in ascidians.

The standard ascidian electroporation experiment introduces super coiled plasmid DNAs into fertilized eggs. Gene expression is often mosaic, meaning that not all of the cells that should express the transgene do (Fig. 5.2). The level of mosaic expression in ascidians depends on the electroporation settings and the amount of DNA introduced (Zeller et al. 2006). Using our custom-built electroporators we can minimize the level of mosaic expression (Zeller et al. 2006). This phenomenon of mosaic expression of endogenously introduced DNA has been described for several other species including *C. elegans* (Stinchcomb et al. 1985) and sea urchins (Franks et al. 1988; Hough-Evans et al. 1988; Livant et al. 1991). In *C. elegans*, supercoiled plasmid DNA forms extrachromosomal arrays that are heritable (Stinchcomb et al. 1985). In sea urchins, when linearized DNA, together with linearized genomic "carrier" DNA, is injected into eggs, it is also expressed in a

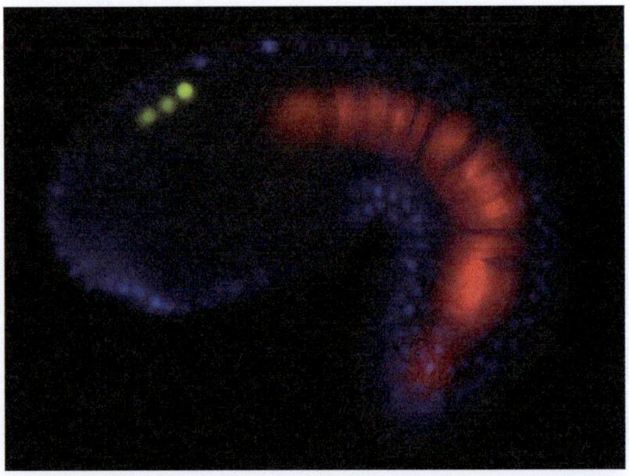

Fig. 5.1 A triple transgenic *Ciona robusta* embryo. Three different transgenes were mixed together and simultaneously electroporated into fertilized eggs. The transgenic embryo expresses: (*1*) red fluorescent protein in the notochord driven by a Brachyury promoter, (*2*) histone cyan fluorescent protein in the epidermis driven by an EpiB promoter and histone yellow fluorescent protein in the pigment cell lineage driven by a tyrosinase promoter. Photo credit: Angela Cone and Robert Zeller

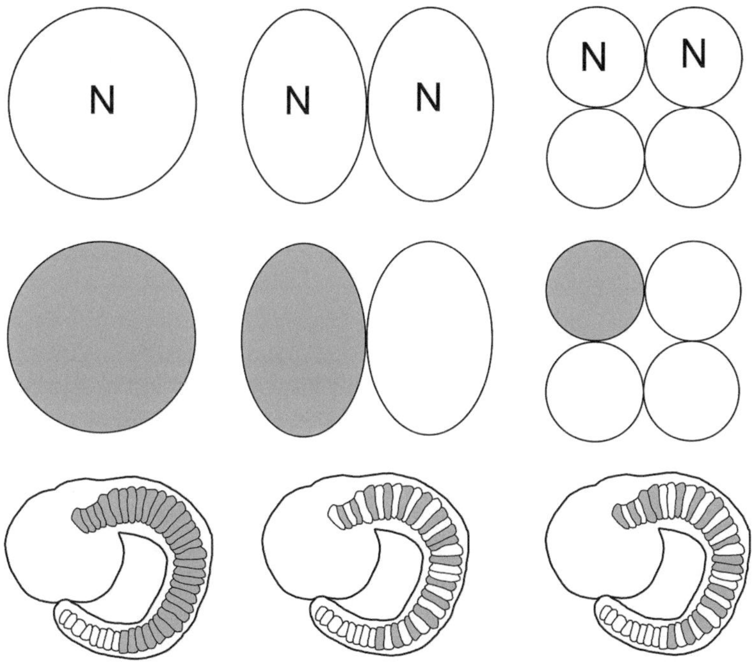

Fig. 5.2 Mosaic transgene expression in ascidian embryos and larvae. The *top row*, from *left to right*, depicts an egg, a two-cell and a four-cell embryo. The "N" designation indicates anterior notochord fate. *Middle row*: *colored in cell* indicates that the electroporated transgene has become stabilized in that cell, most likely forming an extrachromosomal array. The *bottom row* shows the resulting expression pattern in the larva. When the transgene array is inherited by the egg and all its progeny, then the entire anterior notochord expresses the transgene (*left column*). When one of the cells of the two cell embryo inherits the array, then half of the notochord cells express the transgene (*middle column*). In the four-cell embryo, there are still two cells with an anterior notochord fate and if only one of these cells contains the array, half of the notochord cells express the transgene. If the array is inherited in one of the two cells that do not have a notochord fate, the transgene is not expressed

mosaic manner (Hough-Evans et al. 1988) and can integrate into the genome (Flytzanis et al. 1985). Supercoiled DNA is not replicated or expressed in sea urchins (Flytzanis et al. 1985). The level of mosaicism can be reduced by performing multiple injections into sea urchin eggs, if each injection is performed several minutes apart (Livant et al. 1991). This reduction in mosaicism is not observed if the injections are repeated immediately one after another or if a single large-volume injection is made. To account for these observations it was hypothesized that multiple injections produce multiple DNA aggregates that increase the likelihood that the DNA will be stably inherited by all of the cells in the embryo. The absence of mosaic expression is consistent with the DNA integrating before first cleavage, or if integration occurs independently within all of the cells of the early embryo.

In ascidians the electroporated supercoiled DNA is hypothesized to form extrachromosomal arrays (Zeller et al. 2006), as observed in *C. elegans* (Stinchcomb et al. 1985). An analysis of transgene reporters for two different cell types, muscle and notochord, is consistent with the hypothesis that electroporated DNA becomes stabilized, i.e., capable of replication and inheritance during mitosis at some point during the first few cell cycles (Zeller et al. 2006). These DNAs are probably forming extrachromosomal arrays, which is consistent with the observation that occasionally expression can be partially lost from a given cell lineage (Zeller et al. 2006). These observations suggest that if the electroporated DNA becomes stabilized before first cleavage, then all of the cells of the embryo might inherit the DNA. If the DNA is stabilized in one of the cells of the two-cell embryo, then a 50% mosaic embryo is produced. Although mosaic expression may at first seem undesirable, it can also be useful as an internal control for functional experiments. For example, in 50% mosaic embryos, half of the cell lineage is affected by a transgene expressing a dominant-negative molecule, whereas the other half of the lineage is not, thus serving as an internal control. We have also found that co-electroporated DNAs are inherited/expressed together (Zeller et al. 2006) and that although the absolute amount of expression may vary from embryo to embryo, the ratio of the expression levels of the mixed transgenes is relatively constant within an embryo (Chen et al. 2011). Mosaic expression can also be advantageous in some situations. For example, it may be easier to determine the shape of a cell when a transgene reporter is only expressed in some cells. Mosaic expression may also be useful for examining whether or not a gene acts in a cell autonomous manner.

We have unpublished observations suggesting that electroporated linearized DNAs can be randomly integrated into the ascidian genome and may also recombine with the genome. These observations are based on a series of recombination vectors that we have been developing to perform genomic knock-ins. These vectors contain a GFP coding region flanked by one or more exons of a given gene, but no known regulatory elements. These vectors are transcriptionally silent when electroporated as a supercoiled plasmid. However, if the DNA is linearized before electroporation, we observe two aspects of expression that hint at random integration and recombination. First, we frequently observe expression in the cell types in which the gene is normally expressed, suggesting that recombination might have occurred. For example, if the exons flanking the GFP coding sequence are for tyrosinase, a gene normally expressed in the pigment cells, then we can find embryos in which the pigments cells express GFP. Second, we observe random GFP expression in the embryo, suggesting that random genomic integration might have occurred and that the linearized vector is acting like a gene trap. In the case of our tyrosinase example, GFP expression would be found outside of the pigment cell lineage. Future experiments need to be done to completely understand these observations. At the present time, we caution against electroporating linearized DNA, as this may lead to unexpected gene expression.

5.2 The Electroporation Process

Electroporation is a process whereby a DC electrical pulse causes cells to take up exogenous molecules present in the surrounding media. It was first used in the 1980s to transfer DNA into cells (Neumann et al. 1982), and at the time, it was hypothesized that the electrical discharge might produce pores within the cell membrane, which allowed exogenous molecules to enter the cell. There is little experimental evidence to support the formation of large pores; however, it has been shown that if the electric field strength experienced by the cell exceeds a certain threshold, then the cell becomes transiently permeabilized (Escoffre et al. 2009); greatly exceeding this threshold irreversibly permeabilizes the cell, thus causing death. The sensitivity of cells to field strength depends on cell size, with larger cells being more sensitive than smaller cells (Escoffre et al. 2009); thus, low voltage electrical pulses are required for large cells, such as the eggs of marine invertebrates. Within the electroporation cuvette, cells are randomly oriented relative to the electrodes. For small molecules, the surface of the cell facing the anode (+) is permeabilized to a greater extent than the surface facing the cathode (−) and this has been demonstrated by propidium iodide experiments (Golzio et al. 2002). These dye experiments suggest that small molecules might be able to enter the cell at either side facing the electrodes. For large molecules such as DNA, the cell surface nearest the cathode interacts with DNA, and it is hypothesized that the DNA might essentially be electrophoresed during the length of the pulse (Phez et al. 2005). As DNA is negatively charged, it does not enter the side of the cell facing the anode. Instead, the DNA on the cathode (−) side is electrophoresed towards the anode (+) side. For gene expression to occur, the DNA must migrate from the cytoplasm to the nucleus and this occurs more frequently in dividing cells (Escoffre et al. 2009). Interestingly, DNA added after the pulse has occurred is not transfected, suggesting that DNA entrance into the cell might require the interaction of the DNA and membrane during the electroporation pulse (Escoffre et al. 2009).

Electroporation can be used to introduce a variety of molecules into cells including DNA, RNA, proteins, and charged dyes such as Lucifer Yellow. The pulse that is produced generally takes one of two forms: an exponentially decaying pulse or a square wave pulse.

- An exponentially decaying pulse is typically generated by creating a resistor-capacitor (RC) circuit. A capacitor (*C*) is used to store an electrical charge, which is then discharged across the cuvette, which has a resistance (*R*) that depends on the contents of the electroporation medium. In this mode, the discharged pulse is initially at the maximal level of the charged voltage and rapidly decreases to zero in an exponential fashion. The time constant of the discharged pulse, *T*, can be derived from Eq. 5.1:

$$T = RC \tag{5.1}$$

where *R* is the resistance in Ohms and *C* is the capacitance in Farads. The time constant *T*, measured in seconds, is the time required for the stored pulse to decrease to approximately 37% of the initial voltage.

- In square-wave electroporation, the voltage of the discharged pulse is constant and reached quickly after the pulse is initiated. At the conclusion of the pulse, the voltage rapidly decreases to zero – the shape of the pulse thus has a square shape.

One of the main differences between exponentially decaying pulse and square wave pulses is that multiple square wave pulses can be applied to improve transfection efficiency. Both types of pulses have been successfully used to introduce DNA transgenes in ascidian embryos.

5.3 A History of Electroporation in Ascidians

Adaption of electroporation for use in *Ciona* occurred in the mid-1990s. Around this time, a number of publications reported the use of sperm-mediated gene transfer (SMGT) in both the mouse (Lavitrano et al. 1989) and the sea urchin (Arezzo 1989) to generate transgenic embryos. Although not readily repeatable by others at the time (Brinster et al. 1989), there is now evidence that it might work for some species (Smith and Spadafora 2005). A second series of published reports suggested that electroporation of sperm from a variety of different species was also capable of being used to generate transgenic embryos (Muller et al. 1992; Symonds et al. 1994). When electroporated, sperm were hypothesized to take up exogenous DNA and transgenic progeny resulted from using this sperm to fertilize recipient eggs. As a graduate student in the laboratory of Dr. Eric Davidson at Caltech, the author tried unsuccessfully to repeat both processes for the sea urchin, *Strongylocentrotus purpuratus*, but these failed experiments were important for initiating the notion that electroporation could be used to introduce foreign DNA into the eggs of marine invertebrates.

In the early to mid-1990s, a series of papers reported the use of electroporation to generate transgenic fish of several species (Inoue et al. 1990; Powers et al. 1992) and transgenic mollusks (Powers et al. 1995). As a postdoc in Mike Levine's laboratory at UCSD, the author and Joe Corbo, an MD/PhD student, successfully developed a reporter gene for the *Ciona robusta Brachyury* gene and generated transgenic embryos via microinjection (Corbo et al. 1997). A key step in developing the electroporation protocol was the availability of this functioning transgene reporter; if the electroporation process was successful, we knew that this reporter gene would be expressed. We reasoned that the chorion and follicle cells needed to be removed to minimize the number of barriers that the introduced DNA needed to cross before being able to enter the egg. To dechorionate a large number of eggs, we adapted the sodium thioglycolate/pronase E method first used to dechorionate *C. savignyi* eggs (Mita-Miyazawa et al. 1985). Initial attempts to generate transgenic ascidian embryos by electroporation resulted in significant heating and arcing in the electroporation cuvette, which was most likely due to the high concentration of ions in sea water. To reduce the ionic strength of sea water, we adapted the mannitol-based cell fusion media used by Hiroki Nishida to fuse egg fragments of the ascidian *Halocynthia roretzi* (Nishida 1992). This proved to be an important addition to the protocol and allowed us to successfully generate transgenic embryos by electroporation (Corbo et al. 1997).

This original electroporation protocol used the BioRad Gene Pulser with a capacitance extender, which produced an exponentially decaying pulse. Since that time, ascidian laboratories have adapted the protocol to other commercial exponential decay and square wave machines. In our experience, the custom electroporators we developed provide superior results to commercial machines (Zeller et al. 2006). In our custom machines, we incorporate low-value timing resistors in parallel with the cuvette. These resistors essentially fix the pulse parameter conditions, minimizing variation and additionally diverting the discharged current away from the cuvette and through the timing resistor. We have recently tested a simple square wave electroporator (Bullmann et al. 2015) using the optimal settings derived by the Irvine laboratory (Vierra and Irvine 2012) and found that we can also generate transgenic embryos with this simple machine. Both our custom electroporator (Zeller et al. 2006) and the simple square wave machine (Bullmann et al. 2015) can run using battery power, making them ideal for portability and for use in a variety of laboratory/field conditions, without needing to adapt the machines to mains power. Most ascidian laboratories successfully use commercial electroporators; the custom machines provide an economical alternative, especially to those laboratories that may not have access to commercial machines.

5.4 Suggestions for Adapting Electroporation to Other Species

Our laboratory has successful adapted the electroporation protocol to a variety of species, including *C. savignyi, C. intestinalis, C. robusta, Ascidiella aspersa, Styela plicata, Ascidia ceratodes*, and *Ascidia zara*. Other laboratories have reported success in *Molgula occidentalis* (Stolfi et al. 2014), *Boltenia villosa* (Di Gregorio and Levine 2002), *Phallusia mammillata* (Roure et al. 2014) and in some of the species listed above. In our experience, the most difficult element of adapting the procedure to other species is devising an effective way to remove the follicle cells and chorions. The electroporation process causes the eggs to swell and we have found that if the chorions are not removed then the eggs fail to cleave. In addition, the presence of the chorion is an impediment for introducing DNA into the egg. The electroporation field strength, essentially the charge going across the cuvette, needs to be adjusted for each new species. In general, eggs larger than *Ciona* require a lower voltage and eggs smaller than *Ciona* require a higher voltage for successful electroporation. The following formula (Multiporator manual, www.eppendorf.com) can be used to estimate the voltage required for initial electroporation conditions (Eq. 5.2):

$$E_c = V_c / \left(0.75 X d_{\text{cell}}\right) \tag{5.2}$$

where E_c is the critical field strength in volts per centimeter, V_c is the membrane permeation voltage (1 at 20 °C) and d_{cell} is the cell diameter in centimeters.

In our experience with ascidian electroporations, higher capacitance values improve transfection, but decrease viability (Zeller et al. 2006); thus, a compromise of electroporation settings must be reached that will provide adequate transgene expression with acceptable embryo survivability. In trying to adapt the protocol to new species, we suggest starting with an initial setting of around 1000 μF for an exponential decay machine with a voltage value derived from the above formula. If the selected conditions produce about 50% survival, then this should be adequate for introducing molecules into the eggs. If a transgene is not available, one can electroporate Lucifer Yellow dye at a final working concentration of 0.5 M in the electroporation solution. After electroporation, remove the eggs from the media and wash several times with clean sea water to remove exogenous dye. If the electroporation has been successful, the dye can be visualized inside the embryo on a fluorescent microscope. Once electroporation has been successful, the parameters can be altered to improve survivability, while maintaining acceptable levels of transgenesis.

5.5 Suggestions for Scaling Electroporation Volumes

The original electroporation protocol used cuvettes with 4-mm electrode gaps that hold a volume of 800 µl. Using our custom electroporators, we have been able to scale the total volume of the electroporation down to 20–25 µl when using cuvettes with a 1-mm gap. This requires an adjustment of the voltage applied to the cuvette (see Eq. 5.2 above) and the use of pipettors to accurately measure volumes. The voltage reduction approximately scales with the decrease in cuvette gap width. Transitioning from a cuvette with a 4-mm gap to one with a 1-mm gap requires a four-fold reduction in voltage (i.e., from 50 to 12.5 V across the cuvette). This lower voltage should serve as a starting point, as the final voltage level needs to be empirically determined. As mentioned above, a solution of 0.77 M mannitol is mixed with sea water/eggs to avoid heating and electrical arcing during electroporation. Although other ascidian laboratories have adjusted the amount and concentration of mannitol used, in our laboratory, we maintain the original concentration and ratio: 5 volumes of 0.77 M mannitol to 3 volumes of sea water/eggs. The small volumes obtained by this modified protocol are useful to introduce reagents that may be expensive to produce (such as RNA, siRNA or proteins).

For routine experiments, we typically perform electroporations in a total volume of 200 µl (125 µl of 0.77 M mannitol and 75 µl of sea water/eggs) using cuvettes with a 4-mm gap. As previously reported (Zeller et al. 2006), our custom electroporators employ low-value timing resistors that produce much more consistent exponentially decaying pulses than the commercial machine we tested. This allows the use of smaller volumes without significant changes to the electroporation parameters. For a 200-µl reaction, only 6–8 µg of plasmid DNA is needed; thus, a miniprep of DNA can last for several different electroporations. Before electroporation, we ethanol precipitate the DNA and resuspend in the required volume of 0.77 M mannitol. The reaction volume in a 4 mm cuvette can be reliably scaled down to 100 µl; smaller volumes require a cuvette with a narrower gap to be used and DNA precipitation.

5.6 Suggestions for Creating Transgenes

Our laboratory has standardized the design and generation of transgenes. In general, our transgenes are translational fusions with the endogenous genes. One of the benefits of this approach is that the transgene contains the endogenous 5′ UTR sequence. Because many of the genes in ascidians are transpliced (Matsumoto et al. 2010), it is not readily known where the actual start of transcription occurs; this can be problematic when using basal promoters to drive expression. We have examined many of the published basal promoters shown to work in *Ciona* and found that several are internal to the transcript. The *Brachyury* basal promoter (Erives et al. 1998) seems to be an exception and is likely to be an actual basal promoter. We have also found that incorporating an intron into the construct improves expression, as has been reported in *C. elegans* (Okkema et al. 1993). Lastly, as reported by numerous laboratories, the use of comparative genomics to identify putative *cis*-regulatory elements can be informative in defining relevant regions of regulatory DNA (Johnson et al. 2004).

5.7 Required Reagents, Supplies, and Equipment

1. Gelatin (e.g., Knox® gelatin from grocery store).
2. Formaldehyde, 37% solution (e.g., Sigma F1635).
3. Gelatin coated dishes, 35 mm, 60 mm or 100 mm in diameter (see below for preparation). One 35-mm and four 60-mm dishes for dechorionation and one 60-mm dish per electroporation are needed.
4. Penicillin/streptomycin stock (we use tissue-culture grade 10,000 U penicillin and 10,000 μg streptomycin per milliliter for convenience and treat this as a ×1000 stock; e.g., Sigma P4333).
5. Sodium thioglycolate (e.g., Sigma T0632).
6. Protease type XIV (Actinase E or Pronase E) from *Streptomyces griseus* (e.g., Sigma P5147). Freeze in 100-μl aliquots at a concentration of 2.5 mg per aliquot in sea water. We usually buy 100-mg vials, add 4 ml of sea water to resuspend, and freeze in 40, 100-μl aliquots.
7. 15-ml and 50-ml conical tubes.
8. 0.5-ml and 1.5-ml microcentrifuge tubes.
9. Electroporator—square wave or exponential decay, either commercial or custom-built.
10. Electroporation cuvettes: gaps 1, 2 or 4 mm wide, depending on needs.
11. Incubator for culturing embryos. We use plates that are cooled by a recirculating chiller; an air incubator at the correct temperature also works. For *Ciona*, we carry out all activities at 18 °C.
12. Filtered sea water for performing fertilizations (artificial sea water works too). Note that this sea water is filtered for particulates only at this point by passing water though a filter-paper-lined funnel.
13. Sterile filtered (0.22-μm bottle-top filter) sea water for culturing embryos.
14. Disposable glass Pasteur pipettes, 22 cm long.
15. 10 M NaOH to pH thioglycolate solution.
16. 0.5 M ethylenediaminetetraacetic acid (EDTA), pH 8.0.

5.8 Procedures

Over the last few years, a number of reviews and protocol papers have been published describing how electroporation is performed in various ascidian laboratories. One should refer to these publications for additional information related to electroporation (Zeller 2004; Matsuoka et al. 2005; Zeller et al. 2006; Christiaen et al. 2009a, b; Kari et al. 2016).

5.8.1 Gelatin-Coated Dishes

1. Make a ×5 stock solution of gelatin/formaldehyde to coat dishes:

   Add 250 μg of gelatin to 40 ml of water in a 50-ml conical tube and microwave for a few seconds; mix to dissolve. Add 250 μl of 37% formaldehyde. Bring final volume to 50 ml with water.
2. Make a ×1 working stock of gelatin/formaldehyde by diluting the ×5 stock with water. Fifty milliliters of ×1 stock is sufficient to coat a case of dishes.
3. Lay out the bottoms of plastic 35-, 60- or 100-mm dishes, pour gelatin solution into the first dish, then pour from dish to dish until all are coated.
4. After coating, invert dishes to dry COMPLETELY.
5. Rinse dishes ×2 with sea water immediately before use. Dishes may be stored dry for up to a year. We typically stripe the coated side of a dish with a mark to indicate that it has been coated. Dishes may also be rinsed in distilled water after use, dried, and reused several times (Sardet et al. 2011).

5.8.2 DNA Preparation

Our laboratory uses DNA produced from minipreps for all of our electroporations. We have found that miniprep kits from Omega Biotek (omegabiotek.com) work well and produce large amounts of DNA from a miniprep format. The E.Z.N.A.® Plasmid Mini Kit I produces 30 μg from a 5-ml culture and we typically obtain yields of 100 μg from the E.Z.N.A.® Plasmid Mini Kit II when starting with 15–20 ml of overnight culture. The amount of DNA per electroporation depends on the overall volume that is placed into the cuvette. We try not to exceed 125 ng/μl of DNA (100 μg in a volume of 800 μl), as higher concentrations produce more abnormal development. The appropriate amount of DNA is ethanol-precipitated and then resuspended in the required volume of 0.77 M mannitol before electroporation. Four of the larger scale minipreps yield DNA equivalent to a typical DNA midiprep; coupled with the low quantity of DNA required with our reduced volume electroporations, we no longer perform midipreps, saving significant time and cost.

5.8.3 Fertilization and Dechorionation

The decision should be made whether to fertilize and then dechorionate, or to dechorionate before fertilization. Both methods work well, but we typically fertilize, then dechorionate. Either way, eggs must be fertilized before electroporation; otherwise, unfertilized eggs will be activated, making it impossible to fertilize them.

5.8.3.1 Fertilization of *Ciona robusta/intestinalis/savignyi*

Skip this step if dechorionating first. Before starting, make a 1% solution of sodium thioglycolate in sea water (100 mg per 10 ml filtered sea water). The solution should remain clear. If it is not clear, the sodium thioglycolate is no longer useable and does not dechorionate eggs well. For this reason, we typically buy the smallest amount of sodium thioglycolate, aliquot, and use each aliquot until finished. This minimizes oxidation of the thioglycolate.

Note 1 In our experience, it is not necessary to activate *C. robusta* or *C. intestinalis* sperm before fertilization. For some species, such as *Ascidiella aspersa*, this is required. To activate sperm, add a drop of concentrated sperm to 10 ml of seawater that has been supplemented with 50 μl of 1 M Tris pH 9.5. Check sperm for motility under the microscope before use.

Note 2 In our experience, it is important to hold Pasteur pipettes vertically during the following steps to reduce the probability of egg lysis. Although the pipettes may also be coated with gelatin, we find that keeping the pipettes vertical minimizes egg lysis sufficiently.

1. Use a pair of scissors to make an incision along the excurrent siphon to expose the gonoducts. Be careful not to nick the sperm duct. Use two or more animals as they are not self-fertile.
2. For C. *robusta/C. intestinalis*, nick the oviduct and squeeze the eggs carefully into a bowl containing 100–200 ml of filtered sea water. For *C. savignyi*, carefully remove eggs from the oviduct with a Pasteur pipette and put eggs into small volume filtered sea water. In our experience, *C. savignyi* eggs do not fertilize easily unless removed with the pipette.
3. Nick the sperm duct and use a Pasteur pipette to collect "dry" sperm into a 1.5-ml microcentrifuge tube. Sperm may be kept at 4 °C for several days.
4. Add 1–2 drops of sperm from each animal into a finger bowl containing eggs from both animals in 100–200 ml of sea water. Mix and allow fertilization to occur for 2 min, longer for *C. savignyi*.
5. Remove sperm by pouring eggs into a washing basket (a plastic cup with a 100-μm mesh bottom) and transfer basket quickly to 2–4 finger bowls of sea water. This rapidly removes sperm.
6. Using a Pasteur pipette, transfer fertilized eggs to a 35-mm gelatin-coated dish. From this point on, all eggs should be placed in gelatin-coated dishes to avoid sticking. In addition, all sea water at this point is filtered

with a 0.22 μm filter and supplemented with 0.1 mM EDTA.

7. Add 28 μl of 10 M NaOH to the tube of 1% sodium thioglycolate. This brings the pH within the correct range to ensure protease activity.
8. Remove most of the sea water from the eggs, add a few milliliters of the thioglycolate solution, gently mix, then allow the eggs to settle. Repeat this until about 3 ml of the thioglycolate solution is left.
9. Remove most of the solution, add the remaining thioglycolate solution (a 35-mm dish holds about 5 ml) and add one tube (100 μl) of protease. Mix gently.
10. Gently pipette the egg mixture every few minutes. The follicle cells surrounding the egg are removed first, followed by the chorions. As this process occurs, the solution turns yellow.
11. When about 80% of the eggs are dechorionated, begin the washing procedure. It may be easier to monitor this process under a stereo microscope; with experience, you can tell by the change in egg coloration when the reaction has proceeded long enough.
12. Gently swirl the dish to concentrate fertilized eggs into the middle of the dish. Using a Pasteur pipette, transfer eggs to a dish of clean filtered sea water. Repeat until you have placed eggs in the fourth wash dish. The entire process from fertilization to dechorionation typically takes about 12–15 min to complete. The eggs in the fourth wash dish are now ready to undergo electroporation.

5.8.3.2 Dechorionation Before Fertilization

If eggs are in limited supply, or if the dechorionation process takes a long time, such as for *Ascidia ceratodes* and *A. zara*, you can dechorionate the eggs, wash them, and then use them throughout the day for electroporation. To do so, start at step 6 above and proceed to step 12. Before electroporation, the eggs need to be fertilized. For *C. intestinalis*, it is important to make sure that the sperm are activated and that fertilization proceeds for at least 10 min. After fertilization, briefly the wash eggs several times into clean dishes of sea water to remove excess sperm.

5.8.4 Electroporation

For each electroporation, prepare a solution of DNA in mannitol in a microcentrifuge tube and have a cuvette and gelatin-coated dish of filtered sea water available. As mentioned above, our routine electroporations are performed in a volume of 200 μl. DNA is first ethanol-precipitated and resuspended in 125 μl of 0.77 M mannitol. To make measuring sea water/eggs easier, mark the Pasteur pipettes to indicate a volume of 75 μl (a piece of tape placed from the edge of the laboratory bench at the correct distance makes marking pipettes quick and easy). The electroporation process itself should take about a minute per sample.

1. Using a Pasteur pipette, measure 75 μl of sea water/eggs and transfer to the tube of DNA/mannitol.
2. Gently pipette up and down 2–3 times to mix.
3. Gently transfer the solution to the cuvette and place the cuvette in a holder.
4. Electroporate the sample, remove the solution, and gently dispense into a gelatin coated dish of sea water. For small volumes, you can add some sea water from the dish into the cuvette to help resuspend the eggs.
5. Repeats steps 1–4 for each sample.
6. Let eggs sit in the dish for about 5 min to allow the cell membranes to recover. Gently pipette or swirl the dish to spread eggs across the bottom. It is important to allow the eggs to rest and then to disperse them, as they stick to the coated dish if they are not dislodged after the 5-min rest.

5.8.5 Embryo Culturing

For all embryo culturing, particularly for dechorionated eggs/embryos, we culture in dishes containing 0.22 μm filtered sea water supplemented with 100 U penicillin and 100 μg streptomycin per milliliter and 0.1 mM EDTA. Low concentrations of EDTA in sea water were shown to improve fertilization in sea urchins (Borei and Bjorklund 1953) and development in ascidians (Crowther and Whittaker 1983); however, the exact mechanism of action is not known. We use a refrigerated circulating bath to cool a series of cold plates to 18 °C to keep embryos at that temperature. An air incubator can also be used.

When the embryos are dechorionated they become "sticky" about 16–18 h after fertilization (when grown at 18 °C) and remain sticky while the test is being produced by the epidermal cells (Sato and Morisawa 1999). When they reach about 22 h of development, they are much less sticky. Between 16 and 22 h, if the embryos contact one another, they stick together and are often unable to be separated, so avoid moving the dish during this time. The propensity of embryos to stick to one another can be minimized by spreading the eggs out at low density (using a 100-mm rather than a 60-mm dish for example). Bovine serum albumin (BSA) at a final concentration of 0.1 mM can also be added to sea water to help reduce sticking. Lastly, tricaine (MS-222) can be used to anesthetize embryos. We add 2–3 mg of tricaine powder to a 60-mm dish. We have found that this must be added by 15 h of development or it is not very effective at preventing the larvae from swimming together. Tricaine can cause developmental abnormalities; thus, we minimize its use by not adding it earlier than 15 h of development.

References

Arezzo F (1989) Sea urchin sperm as a vector of foreign genetic information. Cell Biol Int Rep 13(4):391–404

Borei HG, Bjorklund U (1953) Effect of versene on sea urchin eggs. Exp Cell Res 5(1):216–219

Brinster RL, Sandgren EP et al (1989) No simple solution for making transgenic mice. Cell 59(2):239–241

Brown CD, Johnson DS et al (2007) Functional architecture and evolution of transcriptional elements that drive gene coexpression. Science 317(5844):1557–1560

Bullmann T, Arendt T et al (2015) A transportable, inexpensive electroporator for in utero electroporation. Dev Growth Differ 57(5):369–377

Cerda GA, Thomas JE et al (2006) Electroporation of DNA, RNA, and morpholinos into zebrafish embryos. Methods 39(3):207–211

Chen JS, Pedro MS et al (2011) miR-124 function during Ciona intestinalis neuronal development includes extensive interaction with the Notch signaling pathway. Development 138(22):4943–4953

Christiaen L, Wagner E, et al (2009a) Electroporation of transgenic DNAs in the sea squirt Ciona. Cold Spring Harb Protoc 2009(12):pdb prot5345

Christiaen L, Wagner E et al (2009b) Microinjection of morpholino oligos and RNAs in sea squirt (Ciona) embryos. Cold Spring Harb Protoc 2009(12):pdb prot5347

Corbo JC, Levine M et al (1997) Characterization of a notochord-specific enhancer from the Brachyury promoter region of the ascidian, Ciona Intestinalis. Development 124(3):589–602

Crowther RJ, Whittaker JR (1983) Developmental autonomy of muscle fine structure in muscle lineage cells of ascidian embryos. Dev Biol 96(1):1–10

Di Gregorio AD, Levine M (2002) Analyzing gene regulation in ascidian embryos: new tools for new perspectives. Differentiation 70(4–5):132–139

Erives A, Corbo JC et al (1998) Lineage-specific regulation of the Ciona snail gene in the embryonic mesoderm and neuroectoderm. Dev Biol 194(2):213–225

Escoffre JM, Portet T et al (2009) What is (still not) known of the mechanism by which electroporation mediates gene transfer and expression in cells and tissues. Mol Biotechnol 41(3):286–295

Falk J, Drinjakovic J et al (2007) Electroporation of cDNA/Morpholinos to targeted areas of embryonic CNS in Xenopus. BMC Dev Biol 7:107

Flytzanis CN, McMahon AP et al (1985) Persistence and integration of cloned DNA in postembryonic sea urchins. Dev Biol 108(2):431–442

Franks RR, Hough-Evans BR et al (1988) Direct introduction of cloned DNA into the sea urchin zygote nucleus, and fate of injected DNA. Development 102(2):287–299

Golzio M, Teissie J et al (2002) Direct visualization at the single-cell level of electrically mediated gene delivery. Proc Natl Acad Sci U S A 99(3):1292–1297

Hikosaka A, Kusakabe T et al (1992) Introduction and expression of recombinant genes in ascidian embryos. Develop Growth Differ 34(6):627–634

Hough-Evans BR, Britten RJ et al (1988) Mosaic incorporation and regulated expression of an exogenous gene in the sea urchin embryo. Dev Biol 129(1):198–208

Inoue K, Yamashita S et al (1990) Electroporation as a new technique for producing transgenic fish. Cell Differ Dev 29(2):123–128

Johnson DS, Davidson B et al (2004) Noncoding regulatory sequences of Ciona exhibit strong correspondence between evolutionary constraint and functional importance. Genome Res 14(12):2448–2456

Kari W, Zeng F et al (2016) Embryo microinjection and electroporation in the Chordate Ciona intestinalis. J Vis Exp 2016(116)
Lavitrano M, Camaioni A et al (1989) Sperm cells as vectors for introducing foreign DNA into eggs: genetic transformation of mice. Cell 57(5):717–723
Livant DL, Hough-Evans BR et al (1991) Differential stability of expression of similarly specified endogenous and exogenous genes in the sea urchin embryo. Development 113(2):385–398
Matsumoto J, Dewar K et al (2010) High-throughput sequence analysis of Ciona intestinalis SL trans-spliced mRNAs: alternative expression modes and gene function correlates. Genome Res 20(5):636–645
Matsuoka T, Awazu S et al (2005) Germline transgenesis of the ascidian Ciona intestinalis by electroporation. Genesis 41(2):67–72
McDougall A, Levasseur M (1998) Sperm-triggered calcium oscillations during meiosis in ascidian oocytes first pause, restart, then stop: correlations with cell cycle kinase activity. Development 125(22):4451–4459
McDougall A, Lee KW et al (2014) Microinjection and 4D fluorescence imaging in the eggs and embryos of the ascidian Phallusia mammillata. Methods Mol Biol 1128:175–185
Mita-Miyazawa I, Ikegami S et al (1985) Histospecific acetylcholinesterase development in the presumptive muscle cells isolated from 16-cell-stage ascidian embryos with respect to the number of DNA replications. J Embryol Exp Morphol 87:1–12
Muller F, Ivics Z et al (1992) Introducing foreign genes into fish eggs with electroporated sperm as a carrier. Mol Mar Biol Biotechnol 1(4–5):276–281
Neumann E, Schaefer-Ridder M et al (1982) Gene transfer into mouse lyoma cells by electroporation in high electric fields. EMBO J 1(7):841–845
Nishida H (1987) Cell lineage analysis in ascidian embryos by intracellular injection of a tracer enzyme. III. Up to the tissue restricted stage. Dev Biol 121(2):526–541
Nishida H (1992) Regionality of egg cytoplasm that promotes muscle differentiation in embryo of the ascidian, Halocynthia roretzi. Development 116(3):521–529
Nishida H, Satoh N (1983) Cell lineage analysis in ascidian embryos by intracellular injection of a tracer enzyme. I. Up to the eight-cell stage. Dev Biol 99(2):382–394
Nishida H, Satoh N (1985) Cell lineage analysis in ascidian embryos by intracellular injection of a tracer enzyme. II. The 16- and 32-cell stages. Dev Biol 110(2):440–454
Okkema PG, Harrison SW et al (1993) Sequence requirements for myosin gene expression and regulation in Caenorhabditis elegans. Genetics 135(2):385–404
Phez E, Faurie C et al (2005) New insights in the visualization of membrane permeabilization and DNA/membrane interaction of cells submitted to electric pulses. Biochim Biophys Acta 1724(3):248–254
Powers DA, Hereford L et al (1992) Electroporation: a method for transferring genes into the gametes of zebrafish (Brachydanio rerio), channel catfish (Ictalurus punctatus), and common carp (Cyprinus carpio). Mol Mar Biol Biotechnol 1(4–5):301–308
Powers DA, Kirby VL et al (1995) Electroporation as an effective means of introducing DNA into abalone (Haliotis rufescens) embryos. Mol Mar Biol Biotechnol 4(4):369–375
Prodon F, Chenevert J et al (2010) Dual mechanism controls asymmetric spindle position in ascidian germ cell precursors. Development 137(12):2011–2021
Roure A, Lemaire P et al (2014) An otx/nodal regulatory signature for posterior neural development in ascidians. PLoS Genet 10(8):e1004548
Sardet C, McDougall A et al (2011) Embryological methods in ascidians: the Villefranche-sur-Mer protocols. Methods Mol Biol 770:365–400
Sato Y, Morisawa M (1999) Loss of test cells leads to the formation of new tunic surface cells and abnormal metamorphosis in larvae of Ciona intestinalis (Chordata, ascidiacea). Dev Genes Evol 209(10):592–600
Satou Y, Imai KS et al (2001) Action of morpholinos in Ciona embryos. Genesis 30(3):103–106
Smith K, Spadafora C (2005) Sperm-mediated gene transfer: applications and implications. BioEssays 27(5):551–562
Stinchcomb DT, Shaw JE et al (1985) Extrachromosomal DNA transformation of Caenorhabditis elegans. Mol Cell Biol 5(12):3484–3496
Stolfi A, Christiaen L (2012) Genetic and genomic toolbox of the chordate Ciona intestinalis. Genetics 192(1):55–66
Stolfi A, Lowe EK et al (2014) Divergent mechanisms regulate conserved cardiopharyngeal development and gene expression in distantly related ascidians. elife 3:e03728
Symonds JE, Walker SP et al (1994) Electroporation of salmon sperm with plasmid DNA: evidence of enhanced sperm/DNA association. Aquaculture 119(4):313–327
Vierra DA, Irvine SQ (2012) Optimized conditions for transgenesis of the ascidian Ciona using square wave electroporation. Dev Genes Evol 222(1):55–61
Zeller RW (2004) Generation and use of transgenic ascidians. In: Ettensohn C, Wray G, Wessels G (eds) Experimental analysis of the development of sea urchins and other non-vertebrate deuterostomes. Academic
Zeller RW, Virata MJ et al (2006) Predictable mosaic transgene expression in ascidian embryos produced with a simple electroporation device. Dev Dyn 235(7):1921–1932

6 The Use of *cis*-Regulatory DNAs as Molecular Tools

Kotaro Shimai and Takehiro G. Kusakabe

Abstract

Ascidians possess relatively small and compact genomes. This feature enables us to easily isolate *cis*-regulatory DNAs of genes of interest. Particularly, *cis*-regulatory DNAs of genes showing tissue- or cell-type-specific expression are routinely used for the artificial induction of gene expression. This strategy helps us to label cells, tissues, and organs of interest, and to investigate gene functions through overexpression, ectopic expression, and the disruption of functions by dominant-negative forms. Thus, *cis*-regulatory DNAs provide a powerful tool for tissue-specific genetic manipulation in studies of ascidian development and physiology. This chapter summarizes the types of *cis*-regulatory DNAs as a genetic manipulation tool, describes the methods used for isolating *cis*-regulatory DNAs, and provide reported examples of the use of *cis*-regulatory DNAs as molecular tools for investigating gene functions.

K. Shimai
Institute for Integrative Neurobiology, Konan University, Kobe, Japan
e-mail: shimai@center.konan-u.ac.jp

T. G. Kusakabe (✉)
Institute for Integrative Neurobiology & Department of Biology, Faculty of Science and Engineering, Konan University, Kobe, Japan
e-mail: tgk@center.konan-u.ac.jp

Keywords

cis-regulatory DNAs · Transcription · Basal promoter · Enhancer · Overexpression · Knockdown · Cell lineage · Ca^{2+} imaging

6.1 A Brief History of the Use of *cis*-Regulatory DNAs in Ascidians

The introduction of foreign DNAs into ascidians was first reported in 1992. Hikosaka et al. (1992) microinjected plasmid DNAs into fertilized eggs of *Ciona savignyi* and *Halocynthia roretzi* and then observed the expression of exogenous genes driven by cloned *cis*-regulatory DNAs in embryos and larvae. In *Ciona*, they expressed the bacterial β galactosidase gene (*lacZ*) in various tissues of larvae by the chicken β-actin promoter and the Rous sarcoma virus enhancer. In *Halocynthia*, the bacterial chloramphenicol acetyltransferase (*CAT*) gene was specifically expressed in muscle cells of tailbud embryos by the promoter of an *H. roretzi* muscle actin gene, *HrMA4a* (Hikosaka et al. 1992; Kusakabe et al. 1992).

In 1997, Corbo et al. (1997) reported the electroporation of exogenous DNAs into *Ciona intestinalis* embryos. Electroporation is technically easier than microinjection and hundreds of synchronously developing embryos can be transformed immediately with electroporation.

Y. Sasakura (ed.), *Transgenic Ascidians*, Advances in Experimental Medicine and Biology 1029,
https://doi.org/10.1007/978-981-10-7545-2_6

Because of these advantages, the electroporation method has now become the most commonly used method of introducing exogenous DNAs into *Ciona* embryos. A number of *cis*-regulatory DNAs from various sources have now been used to artificially drive gene expression for developmental, physiological, and evolutionary studies (Di Gregorio and Levine 2002; Kusakabe 2005; Wang and Christiaen 2012). *Cis*-regulatory DNAs were also used to demonstrate *trans*-splicing in ascidian embryos (Vandenberghe et al. 2001).

Relatively short 5′ flanking sequences, usually ≤5 kb, are generally sufficient to recapitulate the endogenous gene expression patterns of most tissue-specific genes in ascidians (Kusakabe 2005). The small genome size of ascidians allows us to easily clone *cis*-regulatory DNAs by polymerase chain reaction (PCR). This advantage was further facilitated by the completion of genome sequencing and detailed gene expression profiles in *Ciona* (Dehal et al. 2002; Satou et al. 2005; Sasakura et al. 2012a). In this chapter, we summarize the types of *cis*-regulatory DNAs used in the biology of ascidians, and we present examples of and tips for the practical applications of *cis*-regulatory DNAs as molecular tools. Outlines of expression vector construction for the use of *cis*-regulatory DNAs are shown in Fig. 6.1.

6.2 Types of *cis*-Regulatory DNAs

The *cis*-regulatory DNAs that have been used as molecular tools in ascidians can be categorized into three groups. One group is the *cis*-regulatory DNAs that direct transcription in particular cell types or tissues. The second group comprises the *cis*-regulatory DNAs that widely drive gene expression in various cell types and tissues. The remaining group is the basal or minimal promoters that have been used as an effective component of various gene expression vectors.

6.2.1 Tissue-, Cell Type-, and Lineage-Restricted *cis*-Regulatory DNAs

The *cis*-regulatory DNAs that direct transcription in particular cell types or tissues have been the best characterized and used for various purposes. The most common are those of the genes whose expression is restricted to particular cell types or tissues of embryos, larvae or juveniles, such as the notochord (Corbo et al. 1997; Takahashi et al. 1999b; Di Gregorio and Levine 1999), tail muscle (Hikosaka et al. 1992; Kusakabe et al. 1995, 2004; Vandenberghe et al. 2001), epidermis (Ueki and Satoh 1995; Sasakura et al. 2005, 2010), heart (Davidson et al. 2005), intestine (Nakazawa et al. 2013), endostyle (Sasakura et al. 2003), and nervous systems (see below).

The ascidian central nervous system consists of numerous cell types, including various types of neurons and sensory cells, pigment cells, and glial cells (Meinertzhagen et al. 2004; Horie et al. 2009; Sasakura et al. 2012b; Kusakabe 2017). *Cis*-regulatory DNAs that specifically drive transcription in particular cell types in the nervous system have been identified (Yoshida et al. 2004; Shimeld et al. 2005; Horie et al. 2008, 2010; Nishino et al. 2010; Takamura et al. 2010; Kusakabe et al. 2012; Razy-Krajka et al. 2012). *Cis*-regulatory DNAs for genes specifically but widely expressed in the nervous system are also useful. For example, *cis*-regulatory DNAs for pan-neuronal genes, such as synaptotagmin and β-tubulin, have been identified and used in developmental and neurobiological studies (Katsuyama et al. 2002, 2005; Okada et al. 2002; Kusakabe et al. 2004; Imai and Meinertzhagen 2007a, b). *Ci-Nut1*, a gene encoding G-protein-coupled receptor (GPCR), is widely but specifically expressed in the neural tube and has been used as a tool to direct transcription in the neural tube (Shimai et al. 2010; Sasakura et al. 2010; Iitsuka et al. 2014).

A number of genes encoding developmental regulatory factors, such as transcription factors

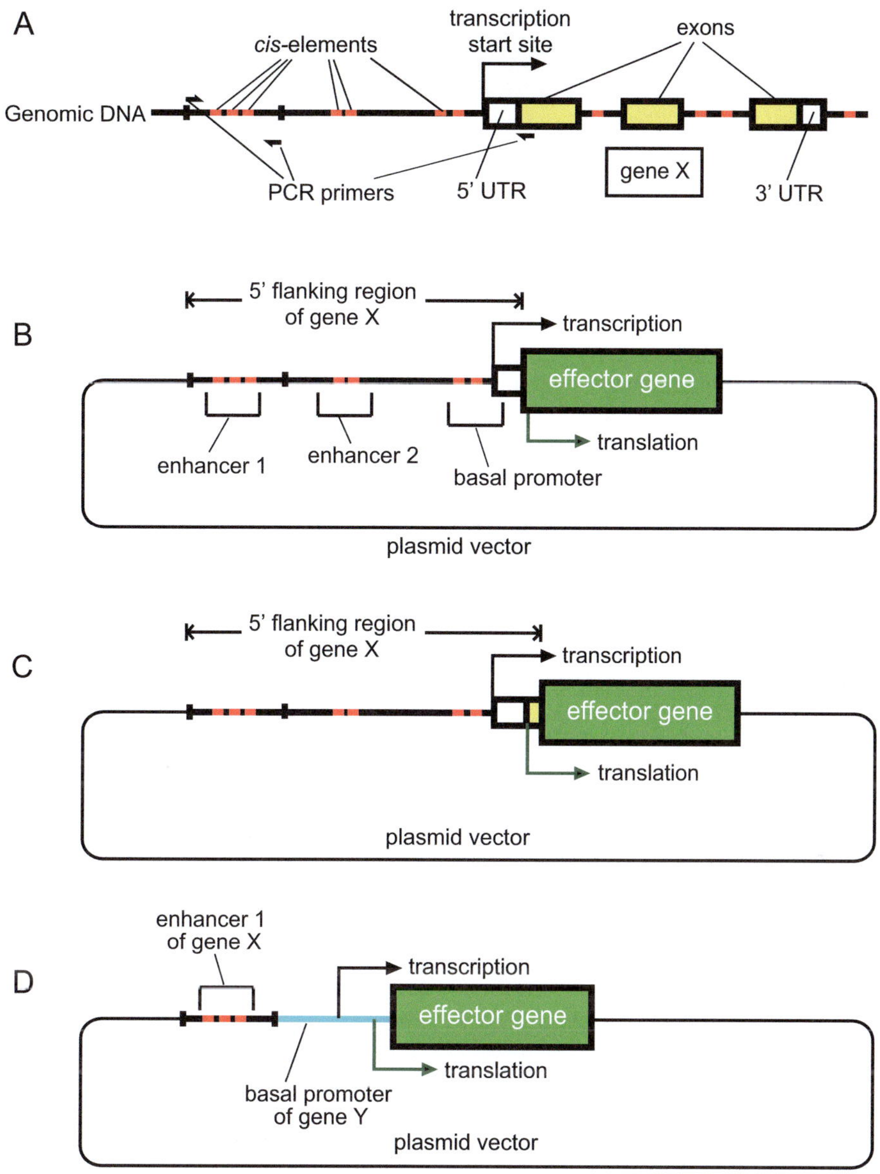

Fig. 6.1 Outlines of the construction of DNA vectors for the use of *cis*-regulatory DNAs in studies of ascidians. (**a**) Schematic diagram showing typical locations and PCR amplification of *cis*-regulatory DNAs. (**b**) A *cis*-regulatory upstream region including a near entire 5′ UTR, but not including the first Met codon, is placed upstream of the coding sequence with a short linker sequence. (**c**) A 5′ flanking region including the start codon followed by a short coding sequence of the driver gene is inserted in-frame into a restriction site upstream of the coding sequence of the effector gene. (**d**) The use of a basal promoter as an adaptor. A basal promoter is connected to the effector coding sequence in an expression vector, and an enhancer of interest is placed upstream of the basal promoter

and signaling proteins, are expressed in particular lineages of cells during embryogenesis (Imai et al. 2004, 2006). *Cis*-regulatory DNAs for these genes have been used to drive exogenous genes in specific cells of interest. For example, *cis*-regulatory DNAs for the transforming growth factor β (TGFβ) family *bmp2b* gene and the transcription factor *FoxD* gene were used to drive gene expression in animal and vegetal blastomeres respectively (Christiaen et al. 2009). In another example, the *cis*-regulatory DNA for the *Dmrt* gene, a homologue of *Drosophila doublesex*, was used to drive gene expression in a-lineage neural plate cells, whereas that for the *FoxB* gene was used to label A-lineage neural plate cells (Wagner and Levine 2012; Navarrete and Levine 2016; Oonuma et al. 2016). The *friend of GATA* gene (*FOG*) is specifically expressed in animal blastomeres of *C. intestinalis* embryos, and its *cis*-regulatory region was used to drive a pan-animal gene expression (Rothbächer et al. 2007).

6.2.2 *Cis*-Regulatory DNAs for Ubiquitous Gene Expression

Cis-regulatory DNAs that ubiquitously drive gene expression are also useful for functional studies in vivo. For example, the role of a transcription factor gene whose normal expression is restricted to a certain set of cells can be assessed by expressing it using a ubiquitous promoter in cells where the gene is not normally expressed. A ubiquitously active *cis*-regulatory DNA can also be used to ubiquitously express an endonuclease for genome editing experiments (Treen et al. 2014; Sasaki et al. 2014; Stolfi et al. 2014; Kawai et al. 2015; Gandhi et al. 2017). As candidate ubiquitous *cis*-regulatory DNAs, flanking DNA sequences of several genes, from both ascidians and non-ascidian species, have been tested (Hikosaka et al. 1992; Shimai et al. 2008; Sasakura et al. 2010). To date, the upstream promoter region of the gene encoding elongation factor 1 alpha (*EF1α*) has been successfully used to ubiquitously drive various effector genes in ascidian embryos (Shimai et al. 2008; Sasakura et al. 2010; Ogura et al. 2011; Sasaki et al. 2014; Stolfi et al. 2014; Treen et al. 2014; Kawai et al. 2015; Gandhi et al. 2017). The *EF1α* promoter drives gene expression in both embryos and adults, whereas the promoter of *ATPTrl* is ubiquitously active in adults, but it does not work in embryos, as has been tested in transgenic *Ciona* lines (the Ciona Intestinalis Transgenic Line Resources [CITRES] database; http://marinebio.nbrp.jp/ciona/index.jsp).

A *cis*-regulatory region of the gene encoding heat shock protein 70 (*HSPA1/6/7-like*) has been used as a heat-inducible promoter (Kawaguchi et al. 2015). This type of *cis*-regulatory DNA allows researchers to temporarily control the expression of transgenes at any time by applying heat stimuli to embryos, larvae, and adults. The heat-inducible promoter may also be used for a spatially restricted activation of exogenous genes in vivo by combining infrared laser irradiation (Kamei et al. 2009). This system, called IR-LEGO (infrared laser-evoked gene operation), has been successfully applied to nematodes, insects, fish, and plants (Deguchi et al. 2009).

The transcription of U6 small nuclear RNA depends on RNA polymerase III (PolIII). Unlike other PolIII-dependent promoters, the U6 snRNA promoter does not require the sequence element downstream of the transcription initiation site (Das et al. 1988). Therefore, the U6 promoter is suitable for driving the expression of exogenous small RNAs, such as short hairpin RNAs (shRNAs) for RNAi and short guide RNAs (sgRNAs) for CRISPR/Cas-based genome editing, in vivo (Miyagishi and Taira 2002; Tuschl 2002). A *C. intestinalis* U6 promoter has been cloned and used for these purposes in *Ciona* (Nishiyama and Fujiwara 2008; Sasaki et al. 2014; Stolfi et al. 2014; Gandhi et al. 2017).

6.2.3 Basal or Minimal Promoters

The basal promoter contains a transcription initiation site and a DNA sequence where the general transcription factors and RNA polymerase II assembled for the initiation of transcription. The basal promoter alone is not sufficient for transcription and additional *cis*-regulatory DNAs such as enhancers and silencers are required for spatially and temporally regulated gene expression. Experimentally, a basal promoter can be identified by testing transcriptional activity of a deletion series of 5′ flanking region fused to a reporter gene; see Hikosaka et al. (1994), Corbo et al. (1997), and Di Gregorio et al. (2001) for examples of identification of basal promoters in ascidians. A short DNA sequence immediately flanking the transcription initiation site is also called "minimal promoter." In a strict sense, a minimal promoter can be defined as the minimal stretch of DNA that has an activity of the above defined basal promoter. However, the words "minimal promoter" and "basal promoter" are often used without a clear distinction. Because transcription from a basal promoter can be activated by connecting a distal *cis*-regulatory DNAs derived from different loci, basal promoters have been used as an efficient transcriptional adaptor for a variety of tissue- or cell type-specific *cis*-regulatory DNAs (Fig. 6.1c). The most widely used basal promoter is that of the *C. intestinalis FoxA.a* (*forkhead*) gene (Erives et al. 1998; Di Gregorio et al. 2001). The *FoxA.a* basal promoter originally identified and used was relatively large and contained a long coding region, and Hozumi et al. (2013) customized it by eliminating unnecessary sequences. The *FoxA.a* basal promoter can be activated by different promoters in various tissues (Erives et al. 1998; Di Gregorio and Levine 1999; Harafuji et al. 2002; Davidson and Levine 2003; Boffelli et al. 2004; Johnson et al. 2004; Christiaen et al. 2005; Hozumi et al. 2013).

Other *C. intestinalis* basal promoters that have been used for functional assays of heterologous *cis*-regulatory sequences include those of *Brachyury* (Erives et al. 1998; *Snail* (Di Gregorio et al. 2001), *Hox3* (Fanelli et al. 2003; Russo et al. 2004), *Gsx* (Bertrand et al. 2003), and *FOG* (Rothbächer et al. 2007; Roure et al. 2007; Stolfi et al. 2010; Stolfi and Levine 2011; Wagner and Levine 2012; Razy-Krajka et al. 2014; Kari et al. 2016). The basal promoter of a larval muscle-specific actin gene (*HrMA4a*) from *Halocynthia roretzi* (Kusakabe et al. 1995) is used in studies of both *Ciona* (Bertrand et al. 2003) and *Halocynthia* (Oda-Ishii et al. 2005).

Among the basal promoters mentioned above, some contain a TATA box and others are TATA-less. Interestingly, activation of the TATA-containing *Brachyury* basal promoter depends on the orientation of a *Snail* enhancer, whereas the same enhancer activates transcription in both orientations when it is placed upstream of the TATA-less *FoxA.a* basal promoter (Erives et al. 1998).

6.3 Examples of Uses for *cis*-Regulatory DNAs

Here we describe examples of typical uses for *cis*-regulatory DNAs in ascidians.

6.3.1 Visualization of Specific Cell Types and Tissues

The visualization of particular cell types and tissues has been routinely conducted with the aid of *cis*-regulatory DNAs in studies of ascidians. One of the fields in which these techniques have played a major role is neurobiology. Neurons and neural circuits were visualized in living or fixed larvae and adults by expressing fluorescent proteins under the control of neuron-specific promoters (Okada et al. 2001, 2002; Yoshida et al. 2004; Imai and Meinertzhagen 2007a, 2007b; Horie et al. 2008, 2009, 2010; Takamura et al. 2010; Stolfi and Levine 2011; Razy-Krajka et al. 2012; Kusakabe et al. 2012; Hozumi et al. 2015; Abitua et al. 2015; Stolfi et al. 2015a).

The visualization of neurites, including axons and dendrites, is important in understanding the functions of neurons in neural circuits. In ascidian larvae, the visualization of neurites can be easily achieved by simply expressing ordinary fluorescent proteins without any special modifications. This is probably due to the small size of the larva; the largest axon of *Ciona* larvae does not exceed 1 mm in length. In some experiments, however, unc-76-tagged fluorescent proteins were used to label axons, because they did this more efficiently than untagged fluorescent proteins (Stolfi and Levine 2011; Stolfi et al. 2015a). Because of the mosaicism of transgene expression in the transient transgenesis of ascidian embryos, a different subset of neurons is often labeled in individual larvae. This allows us to visualize the detailed morphology of single neu-

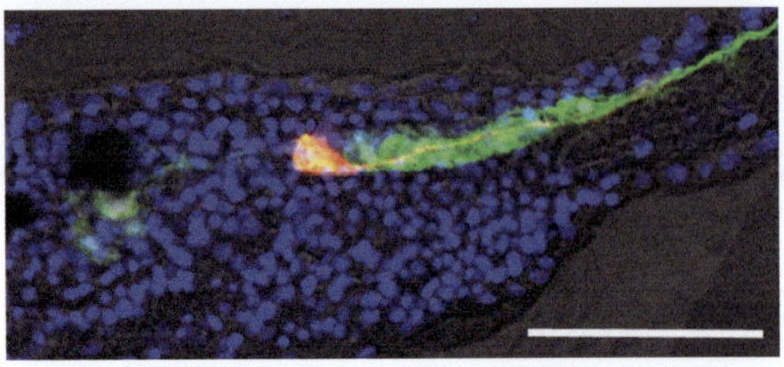

Fig. 6.2 Fluorescence labeling of neurons in a *C. intestinalis* larva. Cholinergic neurons in the motor ganglion are labeled with a green fluorescent protein driven by a cholinergic neuron-specific *VACHT* promoter (*green*) (Yoshida et al. 2004). One of these neurons is also labeled with a red fluorescent protein expressed under the control of the *cis*-regulatory region of *Dmbx* (*magenta*) (Stolfi and Levine 2011). Scale bar, 50 μm

rons (Fig. 6.2) (Imai and Meinertzhagen 2007a, b; Horie et al. 2008, 2010).

Specific antibodies against neuronal proteins have been used to determine neurotransmitter subtypes of these neurons (Horie et al. 2008, 2009, 2010). Multicolor fluorescence imaging using combinations of multiple fluorescent proteins and *cis*-regulatory DNAs has further facilitated identification of neurons in the central nervous system of ascidian larva (Stolfi and Levine 2011).

6.3.2 Visualization of Cell Membranes, Nuclei, and Other Subcellular Structures

The visualization of cell membranes, nuclei, and other organelles is useful for the identification of cells and morphometric studies of cells in embryos and larvae (Fig. 6.3). For these purposes, cloned *cis*-regulatory regions have been used to express reporter proteins connected to various localization signal sequences. For example, Corbo et al. (1997) employed the LacZ reporter fused with the nuclear localization signal (NLS) of SV40 in their first report on electroporation into *Ciona* embryos. The SV40 NLS has been used to localize fluorescence reporters in ascidian embryos (Fig. 6.3a) (Iitsuka et al. 2014; Razy-Krajka et al. 2014; Stolfi et al. 2015a; Oonuma et al. 2016). Histone H2B has also been used as a means of localizing fluorescence proteins, such as mCherry, enhanced green fluorescent protein (eGFP), cyan fluorescent protein, yellow fluorescent protein (YFP), and Kaede in the nucleus by targeting chromatin (Pasini et al. 2006; Roure et al. 2007; Sasakura et al. 2010; Wagner and Levine 2012; Razy-Krajka et al. 2014; Kawai et al. 2015; Navarrete and Levine 2016).

For the visualization of cell membranes, fluorescent proteins fused with various membrane-anchoring signals have been expressed in *Ciona* embryos by using lineage-specific *cis*-regulatory DNAs. The membrane-targeting signals used include the membrane-anchoring signal of the growth-associated protein GAP43 (Fig. 6.3b) (Roure et al. 2007; Ogura et al. 2011) and the CAAX membrane-targeting motif of Ras (Ogura et al. 2011; Wagner et al. 2014; Kawai et al. 2015; Navarrete and Levine 2016). A GFP variant containing the glycosyl phosphatidylinositol-anchoring sequence (Rhee et al. 2006) was also used to label cell membranes of cardiac progenitor cells (Cooley et al. 2011). An actin-binding GFP–moesin fusion protein was also expressed in *Ciona* embryos to visualize cell morphology (Christiaen et al. 2008). Please see Chap. 14 for a detailed description and application examples of these membrane visualization methods.

Other subcellular structures can be labeled in principle by expressing fluorescent proteins fused with an appropriate targeting sequence. For example, Roure et al. (2007) labeled basolateral cell membranes, centrosomes, and microtubules by expressing fusion proteins of *Drosophila* E-cadherin, Aurora kinase, and ensconsin (microtubule-associated protein 7) (Fig. 6.3c) respectively.

6.3.3 Discovery of Genes and Enhancers

Cis-regulatory regions have been used to experimentally detect novel genes and enhancers in the genome of ascidians. For example, the basal promoter of *FoxA.a* was used for systematic enhancer screening in *C. intestinalis* (Harafuji et al. 2002;

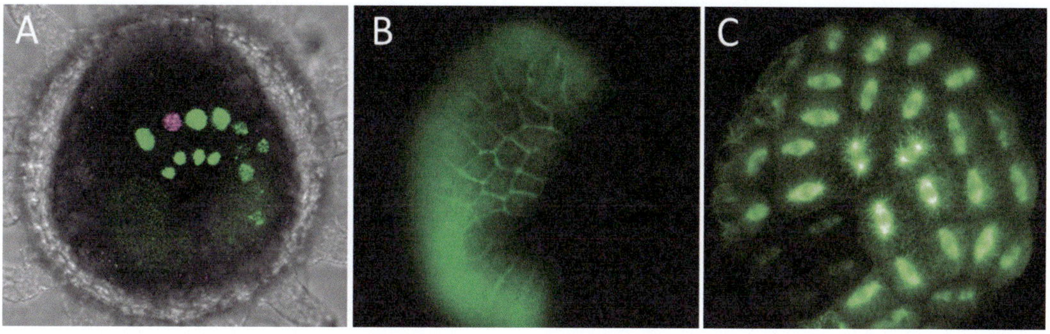

Fig. 6.3 Visualization of subcellular structures in *C. intestinalis* embryos. (**a**) Visualization of nuclei in A-lineage neural plate cells with the photoconvertible fluorescent protein Kaede expressed under the control of the *FoxB* promoter (Oonuma et al. 2016). A single nucleus was irradiated with a 405-nm laser, resulting in the conversion of fluorescence from green to red. (**b**) Fluorescent labeling of cell membranes in an embryo expressing YFP fused with GAP-43 under the control of the *FOG* promoter (Roure et al. 2007). (**c**) Visualization of the whole microtubule network in cells of an embryo expressing GFP-tagged ensconsin under the control of the *FOG* promoter (Roure et al. 2007) (Photographs: (**a**) Courtesy of Kouhei Oonuma. (**b**) and (**c**) Reproduced from Roure et al. (2007). (© Roure et al.; Creative Commons Attribution [CC BY] license)

Keys et al. 2005) and for the functional evaluation of conserved noncoding genomic sequences in *C. savignyi* (Boffelli et al. 2004). Harafuji and her colleagues (2002) placed 138 random genomic DNA fragments 5′ of the basal promoter of *FoxA.a* fused with *lacZ*, and electroporated them into *Ciona* embryos. Their study identified a tissue-specific enhancer in every 20–40 kb of random genomic DNA. Keys and his colleagues (2005) also applied this strategy to systematically identify *cis*-regulatory elements in large genomic domains containing *Hox* genes.

Another example of gene discoveries using a *cis*-regulatory DNA is enhancer trapping in the *Ciona* genomes using the *Minos* transposon (see Chap. 11 for detailed information). In germline transgenesis using a *Minos* vector containing an upstream region of an endostyle-specific gene encoding thyroid peroxidase (*TPO*), Sasakura and his colleagues obtained enhancer trap lines of *C. intestinalis* (Sasakura et al. 2003; Awazu et al. 2004, 2007) and *C. savignyi* (Matsuoka et al. 2004). The method has been improved by expressing a transposase in sperm or eggs using *cis*-regulatory DNAs that are active in germ cells (Sasakura et al. 2008; Hozumi et al. 2010).

6.3.4 Cell-Lineage and Developmental Fates

Developmental fates of embryonic cells have been examined by labeling cells with fluorescent proteins under the control of cell-type- or lineage-specific *cis*-regulatory DNAs. For example, the *cis*-regulatory region of the transcription factor gene *Phox2* was used to investigate the developmental origin of adult motor neurons in *C. intestinalis* by expressing a *YFP* transgene in the motor neurons and their progenitor cells (Dufour et al. 2006). A high-resolution visualization of the migration and division of heart precursor cells was conducted by expressing GFP using the *Mesp* enhancer (Davidson et al. 2005).

The use of the photoconvertible fluorescent protein Kaede (Ando et al. 2002; Figs. 6.3c and 6.4) allows researchers to directly chase the fate of particular cells during embryogenesis (Stolfi et al. 2015a; Oonuma et al. 2016) and metamorphosis (Horie et al. 2011; Abitua et al. 2012; Nakazawa et al. 2013; Razy-Krajka et al. 2014; Kawai et al. 2015). Live imaging of cells expressing fluorescent protein is another powerful and direct way of chasing the developmental fates of

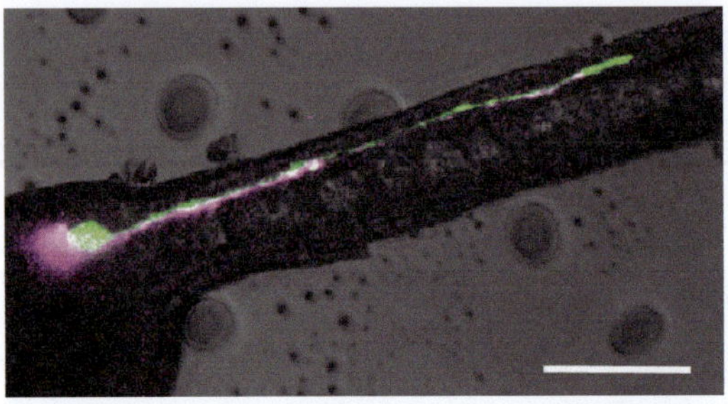

Fig. 6.4 Single-cell labeling with a photoconvertible fluorescent protein in a *C. intestinalis* larva. The photoconvertible fluorescent protein Kaede (Ando et al. 2002) was expressed in cholinergic neurons under the control of the *cis*-regulatory region of *VACHT*. A single neuron was irradiated with a 405-nm laser, resulting in the conversion of fluorescence from green to red. Scale bar, 50 μm

cells in ascidian embryos (Kawai et al. 2015; Stolfi et al. 2015a; Navarrete and Levine 2016).

6.3.5 Gene Function Analyses

Cis-regulatory DNAs have been used to express wild-type or mutant forms of proteins that regulate developmental or physiological processes (Corbo et al. 1998; Fujiwara et al. 1998; Ono et al. 1999; Takahashi et al. 1999a; Imai et al. 2000; Satoh et al. 2000; Mitani et al. 2001; Okagaki et al. 2001; Di Gregorio et al. 2002; Keys et al. 2002; Katsuyama et al. 2005; Kawai et al. 2005; Davidson et al. 2005, 2006; Shi and Levine 2008; Christiaen et al. 2009; Abitua et al. 2012, 2015; Wagner and Levine 2012; Razy-Krajka et al. 2014). For example, the *cis*-regulatory region of *Mesp* was used to drive the targeted expression of an activator form of Mesp, which induces heart formation without migration (Davidson et al. 2005). In that experiment, a constitutively active form of Mesp was made by fusing Mesp with the activation domain of the transcriptional activator protein (virion protein 16; VP16) of herpes simplex virus (Arnosti et al. 1993) at the C-terminus and selectively expressed in the B7.5 lineage. Davidson and his colleagues (2006) also examined the role of the Ets1/2 transcription factor in heart formation by selectively expressing constitutive activator and repressor forms of Ets1/2 in the B7.5 lineage using the *Mesp* enhancer. They made the constitutive activator form of Ets1/2 by adding the activator domain of VP16 at the C-terminus, whereas the repressor form was made by adding the WRPW (Trp-Arg-Pro-Trp) motif, which was known to be a transcriptional repressor domain of Hairy-related proteins (Fisher et al. 1996).

The targeted expression of dominant-negative forms of proteins by tissue or lineage-specific *cis*-regulatory DNAs is an effective way of suppressing gene function in vivo. For example, a dominant negative form of the *Ciona* fibroblast growth factor (FGF) receptor was expressed in the B7.5 lineage by the *FoxF* enhancer to determine whether FGF signaling is necessary for heart induction (Davidson et al. 2006; Beh et al. 2007). In another example, a dominant negative form of *Ciona*, Suppressor of Hairless [Su(H)], was expressed in the ectodermal cells of *Ciona* embryos by the ectodermal enhancer of *FOG* (Pasini et al. 2006).

Genome editing has now become a major strategy for addressing gene function in vivo in a wide variety of model and nonmodel organisms. In *Ciona*, genome editing techniques, both TALENs and CRISPR/Cas9, have been successfully applied to functional genomics (Treen et al. 2014; Sasaki et al. 2014; Stolfi et al. 2014; Gandhi et al. 2017). In these experiments, both ubiquitous and tissue-specific *cis*-regulatory DNAs were used to drive expression of the nucleases, and a U6 promoter was used to express sgRNAs (see Chaps. 12 and 13 for detailed information).

6.3.6 Expression of Physiological Probes (Ca^{2+} Imaging, Cell Cycle Imaging, and Optogenetic Manipulation of Neural Circuitry)

Cis-regulatory DNAs can be used to express molecular probes for monitoring the physiological activity of cells in ascidian embryos and larvae. For example, a fluorescent Ca^{2+} indicator was expressed in the neural plate of *Ciona* embryos under the control of a pan-neural *ETR1* promoter to monitor intracellular Ca^{2+} transients

(Hackley et al. 2013; Abdul-Wajid et al. 2015). In another example, cell cycle progression was visualized in embryos. Fucci is a fluorescent cell cycle indicator system that was developed to trace the cell cycle during development in vivo (Sakaue-Sawano et al. 2008). Fucci probes were expressed under the control of the *EF1α* promoter to monitor cell cycle progression in *C. intestinalis* embryos (Ogura et al. 2011).

The functions of ion channels are crucial for neuronal activities. Therefore, neuronal functions can be addressed by expressing various forms of ion channels in ascidian larvae. In *Halocynthia roretzi*, a dominant-negative form of a potassium channel was overexpressed in a group of neurons by using the *synaptotagmin* promoter (Okada et al. 2002). These larvae exhibited abnormal tail movements, suggesting that these neurons might constitute the major population of motor neurons responsible for swimming behavior. Combinations of *cis*-regulatory DNAs of neuron-specific genes and optogenetics using light-gated ion channels (Zhang et al. 2006) and a light-activated chloride pump (Zhang et al. 2007) contribute to our understanding of neuronal functions in ascidians.

6.4 Technical Tips for the Use of *cis*-Regulatory DNAs

Here, we describe the basics and technical tips for obtaining *cis*-regulatory DNAs that work in ascidian systems, for making DNA constructs, and for introducing DNAs into embryos.

6.4.1 Availability of Cloned *cis*-Regulatory DNAs

A number of *cis*-regulatory DNAs have been cloned and described, and the DNA constructs can be obtained from the authors of those publications. Some information can be seen in articles cited in this chapter and summarized in Table 6.1, and references for earlier works can be found in Kusakabe (2005). Information on published and unpublished DNA constructs, including references, nucleotide sequences, and expression patterns, are available in the DataBase of Tunicate Gene Regulation (http://dbtgr.hgc.jp/), Ascidian Network for In Situ Expression and Embryological Data (https://www.aniseed.cnrs.fr/), and CITRES (http://marinebio.nbrp.jp/ciona/) databases (Sierro et al. 2006; Sasakura et al. 2009; Tassy et al. 2010; Brozovic et al. Brozovic et al. 2016).

Guidelines have been published for the nomenclature of genetic elements in ascidians (Stolfi et al. 2015b). This nomenclature system defines rules to describe *cis*-regulatory regions and DNA constructs containing a *cis*-regulatory DNA driver and an effector coding sequence. For example, the plasmid DNA construct containing the full-length *Pitx* coding sequence fused C-terminally in frame with *EGFP* under the control of an upstream *cis*-regulatory region between −1541 and −1417 of the *Ciona intestinalis Otx* gene is described as pCiinte.REG.KH2012.C4.4313996–4315697|Otx > *Pitx::EGFP*. In the main text of an article, a simplified name *Otx*[−1541/−1417] > *Pitx::EGFP* may be used for this DNA construct, but in the materials and methods section, the formal name shown above should be mentioned. Please refer to Stolfi et al. (2015b) for details of the nomenclature.

6.4.2 Isolation of *cis*-Regulatory DNAs from the Genomes

For most ascidian genes, an upstream region of ≤5 kb from the start codon is empirically sufficient to recapitulate endogenous gene expression patterns when it is connected to a reporter gene (Yoshida et al. 2004; Kusakabe 2005), although some important *cis*-regulatory DNAs have also been found in introns (Christiaen et al. 2005; Irvine et al. 2008). For species of the order Enterogona, such as *Ciona* and *Phallusia*, genomic DNA may be readily prepared from sperm surgically obtained from the sperm duct, whereas for Pleurogona species, such as *Halocynthia* and *Styela*, genomic DNAs can be prepared from the gonad. For a species whose genome sequence has been read, the 5′ flanking sequence of a gene of interest is obtained from

Table 6.1 *Cis*-regulatory DNAs reported in the literature cited in this chapter

Tissue/organ	Cells	Species	Gene	Size (kbp)	References
Notochord		*Ci*	*Brachyury*	0.9	Corbo et al. (1997, 1998), Fujiwara et al. (1998), Yagi et al. (2004), Di Gregorio et al. (2002), Keys et al. (2002), and Kawai et al. (2005)
		Hr	*Brachyury*	0.3	Takahashi et al. (1999b)
		Ci	*Tropomyosin*	1.4	Di Gregorio and Levine (1999), Di Gregorio et al. (2002), and Keys et al. (2002)
		Ci	*Noto1*	3.1	Shi et al. (2009)
Muscle		*Hr*	*HrMA4a*	1.4	Hikosaka et al. (1992), Kusakabe et al. (1995), and Okagaki et al. (2001)
	Also TVCs	*Ci*	*TnI*	1.5 or 0.7§	Vandenberghe et al. (2001) and Davidson and Levine (2003)
		Ci	*MLC2*	0.6	Kusakabe et al. (2004)
		Ci	*MRLC1*	0.6	Kusakabe et al. (2004)
		Ci	*MA5*	0.7	Kusakabe et al. (2004)
	Also TVCs	*Ci*	*TnT*	0.4 or 0.6[a]	Davidson and Levine (2003) and Kusakabe et al. (2004)
		Ci	*TPM*	0.5	Kusakabe et al. (2004)
		Hr	*Tbx*	2.2	Mitani et al. (2001)
Intestine	Precursor cells in endodermal strand	*Ci*	*Zip*	3.6[a]	Nakazawa et al. (2013) and Kawai et al. (2015)
Endostyle		*Ci*	*TPO*	ns	Sasakura et al. (2003), Awazu et al. (2004, 2007) and Matsuoka et al. (2004)
Nervous system	Photoreceptor	*Ci*	*Arrestin*	3	Yoshida et al. (2004)
	Photoreceptor	*Ci*	*BCO*	2	Kusakabe et al. (2004)
	Brain vesicle	*Ci*	*CRALBP*	2	Kusakabe et al. (2004)
	Brain vesicle	*Ci*	*Opsin3*	2.5	Kusakabe et al. (2004) and Horie et al. (2011)
	Photoreceptor	*Ci*	*Opsin1*	5	Kusakabe et al. (2004)
	Pan-neuronal	*Hr*	*Synaptotagmin*	3.4	Okada et al. (2001), Katsuyama et al. (2002), Ono et al. (1999), Okada et al. (2002), Katsuyama et al. (2005), Imai and Meinertzhagen (2007a, b)
	Pan-neuronal	*Ci*	*Synaptotagmin*	2.6	Matsumoto et al. (2008)
	Pan-neuronal	*Ci*	*Gαi1*	4.2	Yoshida et al. (2004)
	CNS	*Ci*	*Tubulin-β2*	4.3	Kusakabe et al. (2004); Hozumi et al. (2015)
	CNS and maternal	*Ci*	*Nut*	1.0	Shimai et al. (2010), Hozumi et al. (2010), Sasakura et al. (2010), and Iitsuka et al. (2014)
	Cholinergic	*Ci*	*VACHT*	3.8	Yoshida et al. (2004), Takamura et al. (2010), Horie et al. (2010), and Hozumi et al. (2015)
	Glycinergic	*Ci*	*GlyR*	2.5	Nishino et al. (2010)

	GABAergic and glycinergic	*Ci*	*VGAT*	2.7	Yoshida et al. (2004)
	GABAergic	*Ci*	*GAD*	1.5	Takamura et al. (2010)
	Dopaminergic	*Ci*	*TY3H*	3.6	Takamura et al. (2010)
	Dopaminergic	*Ci*	*DDC*	1.5	Takamura et al. (2010)
	Dopaminergic	*Ci*	*TH*	3.5	Razy-Krajka et al. (2012)
	Dopaminergic	*Ci*	*Ptf1a*	2.4	Razy-Krajka et al. (2012)
	Glutamatergic	*Ci*	*VGLUT*	4.6	Horie et al. (2008)
	Otolith, palps	*Ci*	*βγ-crystallin*	1.0	Shimeld et al. (2005) and Abitua et al. (2012)
	Pigment cells (a9.49 lineage)	*Ci*	*Mitf*	1.0[b]	Abitua et al. (2012)
	Mesenchyme	*Ci*	*Twist*	1.5[b]	Abitua et al. (2012)
	Pharynx, sensory vesicle, neck, and motor ganglion	*Ci*	*Msx*	4.0	Russo et al. (2004) and Stolfi et al. (2015a)
	Brain, motor ganglion, palps	*Ci*	*GnRHI*	4.7	Kusakabe et al. (2012)
	Brain, motor ganglion, nerve cord	*Ci*	*GnRHII*	4.3 or 2.4[b]	Kusakabe et al. (2012) and Abitua et al. (2015)
	anterior-most neurogenic ectoderm	*Ci*	*Dll-A*	1.6[a]	Harafuji et al. (2002)
	GnRH neurons	*Ci*	*SOG/ Chemokine-like*	3.1[b]	Abitua et al. (2015)
	GnRH neurons	*Ci*	*RXFP3*	3.7[b]	Abitua et al. (2015)
	GnRH neurons	*Ci*	*CNGA*	3.8[b]	Abitua et al. (2015)

(continued)

Table 6.1 (continued)

Tissue/organ	Cells	Species	Gene	Size (kbp)	References
Neural lineage	A-lineage neural plate	*Ci*	*FoxB*	4.2	Imai et al. (2009), Navarrete and Levine (2016), and Oonuma et al. (2016)
	CNS (A8.7, A8.8, A8.15 and A8.16)	*Ci*	*Etr1*	2.8	Hackley et al. (2013), Abdul-Wajid et al. (2015), and Navarrete and Levine (2016)
	A9.13 and A9.15 derivatives	*Ci*	*Mnx*	5.1	Navarrete and Levine (2016)
	a7.10, a7.9 and a7.13 derivatives	*Ci*	*Dmrt*	1.0[b]	Wagner and Levine (2012), Abitua et al. (2015), and Oonuma et al. (2016)
	a8.18 and a8.20 derivatives	*Ci*	*FoxC*	2.1[b]	Wagner and Levine (2012) and Wagner et al. (2014)
	A12.239, A12.240	*Ci*	*Dmbx*	1.3[b]	Stolfi and Levine (2011)
	A11.117, A13.474	*Ci*	*Vsx*	0.3[b]	Stolfi and Levine (2011)
	A11.118	*Ci*	*Nk6*	2.8[b]	Stolfi and Levine (2011)
	A11.120	*Ci*	*Coe*	2.7	Stolfi and Levine (2011) and Razy-Krajka et al. (2014)
	Part of CNS	*Ci*	*Phox2*	3.0	Dufour et al. (2006)
	differentiating neurons, nerve cord	*Ci*	*Onecut*	4.3	Stolfi and Levine (2011)
	A9.30 lineage	*Ci*	*FGF8/17/18*	4.8	Imai et al. (2009) and Stolfi and Levine (2011)
	A10.64 pair, motor neurons, notochord, palps, pharyngeal mesoderm, bipolar tail neurons	*Ci*	*Islet*	3.3[b]	Stolfi et al. (2010, 2015a) Stolfi and Levine (2011), Wagner et al. (2014), and Gandhi et al. (2017)
	a-lineage neuroectoderm	*Ci*	*six1/2*	4.3[b]	Abitua et al. (2015)
	A11.117	*Ci*	*Pitx*	3.4[b]	Christiaen et al. (2005) and Stolfi and Levine (2011)
Epidermis		*Hr*	*UDP glucose-4-epimerase (EpiB)*	4	Ueki and Satoh (1995)
		Hr	*SEC61 (EpiD)*	0.2	Ueki and Satoh (1995)
		Ci	*CesA*	2.1	Sasakura et al. (2005)
		Ci	*EpiI*	2.1	Sasakura et al. (2010) and Treen et al. (2014)

Ubiquitous		*Ci*	*U6*	1.5	Nishiyama and Fujiwara (2008)
		Ci	*EF1α*	1.5	Shimai et al. (2008), Sasakura et al. (2010), Ogura et al. (2011), Sasaki et al. (2014), Stolfi et al. (2014), Treen et al. (2014), Kawai et al. (2015), and Gandhi et al. (2017)
Others	Endodermal cells	*Ci*	*Titf*	3.5	Ristoratore et al. (1999) and Fanelli et al. (2003)
	Animal blastomeres	*Ci*	*Bmp2b*	1.0[b]	Christiaen et al. (2009)
	Ventral blastomeres	*Ci*	*Chordin*	1.7[b]	Abitua et al. (2015)
	Vegetal blastomeres	*Ci*	*FoxD*	4.1[b]	Shi and Levine (2008), Christiaen et al. (2009), and Navarrete and Levine (2016)
	Endoderm and TVCs	*Ci*	*NPP*	1.0[a]	Davidson and Levine (2003)
	Endoderm and TVCs	*Ci*	*29h10*	0.5[a]	Davidson and Levine 2003
	a8.17, a8.19 and a8.25 (notochord and nerve cord)	*Ci*	*Zic-r.b (ZicL)*	2	Shi and Levine (2008)
		Ci	*FoxA.a (forkhead)*	1.7	Corbo et al. (1997), Takahashi et al. (1999a), Imai et al. (2000), Satoh et al. (2000), and Boffelli et al. (2004)
	Tail muscle, CNS, mesenchyme, TVCs,	*Ci*	*snail*	3.2	Erives et al. (1998)
	B7.5 (heart precursor cells)	*Ci*	*Mesp*	1.9	Davidson et al. (2005), Davidson et al. (2006), Christiaen et al. (2008), and Cooley et al. (2011)
	heart precursor and epidermis	*Ci*	*FoxF*	3.1[b]	Beh et al. (2007), Davidson et al. (2006), and Wagner et al. (2014)
	Animal blastomeres	*Ci*	*FOG*	2.0	Rothbächer et al. (2007) and Pasini et al. (2006)
Heat-inducible		*Ci*	*HSPA1/6/7*	2.0	Kawaguchi et al. (2015)

Ci Ciona intestinalis, Hr Halocynthia roretzi, CNS central nervous system, *TVCs* trunk ventral cells, *ns* not shown
[a]Used in combination with the *FoxA.a* basal promoter
[b]Used in combination with the *FOG* basal promoter

the genome database and used to design gene-specific primers. Genomic DNA fragments containing *cis*-regulatory DNAs can then be amplified by PCR using a pair of gene-specific primers. For a species whose genome sequence has not been determined, the 5′ flanking sequence of a gene of interest can be cloned by using a Universal GenomeWalker™ 2.0 kit (Clontech Laboratories, Mountain View, CA, USA). Use of a thermostable DNA polymerase with proofreading activity, such as PrimeSTAR® HS DNA polymerase (Takara Bio, Kusatsu, Japan) and *Pfu* DNA polymerase (Thermo Fisher Scientific, Waltham, MA, USA), is strongly recommended. Because the ascidian genomes are highly polymorphic (Dehal et al. 2002; Vinson et al. 2005), the presence of nucleotide sequence polymorphisms should be checked when the oligonucleotide primers for PCR are designed. For *C. intestinalis*, this can be achieved by looking at single nucleotide polymorphisms (SNPs) information in the Ghost genome database (http://ghost.zool.kyoto-u.ac.jp/cgi-bin/gb2/gbrowse/kh/) (Satou et al. 2005, 2008).

6.4.3 Connecting *cis*-Regulatory DNAs with a Protein Coding Sequence

A simple way of connecting a *cis* regulatory DNA to the coding sequence of a protein to be expressed is to place an upstream region, including a near entire 5′ UTR, but not including the first Met codon, upstream of the coding sequence with a short linker sequence (Fig. 6.1b). This type of connection worked well for most of the *cis*-regulatory upstream regions tested (Yoshida et al. 2004; Kusakabe et al. 2012). The 5′ end of the coding sequence preferably matches a Kozak sequence, which is a consensus sequence (RccAUGG; start codon is underlined) around translation initiation sites of vertebrate mRNAs (Kozak 1987). Alternatively, a 5′ flanking region including the start codon followed by a short coding sequence of the driver gene may be inserted in-frame into a restriction site upstream of the coding sequence of the effector gene (Fig. 6.1c). This method was also shown to work very well (Kusakabe et al. 2004; Horie et al. 2008).

Another type of connection is the use of a basal promoter as an adaptor (Fig. 6.1d). Some basal promoters, such as the basal promoter of *FoxA.a*, can work with a wide variety of heterologous distal *cis*-regulatory elements (Erives et al. 1998; Di Gregorio and Levine 1999; Harafuji et al. 2002; Boffelli et al. 2004). By connecting this type of basal promoter with the effector coding sequence in an expression vector, an enhancer of interest can then be placed upstream of the basal promoter without considering the reading frames and the presence/absence of a transcription start site.

For example, an endodermal strand enhancer of the *Zip* gene was connected to the *FoxA.a* basal promoter to generate an expression construct that drives the expression of Kaede in the endodermal strand and a part of CNS (Nakazawa et al. 2013). Roure et al. (2007) developed Gateway vector systems, in which the basal promoter of *FOG* is included in an entry clone and another entry clone contains an enhancer of interest. This system enables the simultaneous introduction of an enhancer and a basal promoter into a destination vector.

6.4.4 Introduction of DNA Constructs into Ascidian Eggs and Embryos

Microinjection was the sole method used for the introduction of exogenous DNAs into ascidian embryos until Corbo and his colleagues (1997) reported the electroporation of exogenous DNAs into *C. intestinalis* embryos. The electroporation method is faster and technically easier than microinjection and permits the simultaneous transformation of hundreds of synchronously developing embryos (Corbo et al. 2001; Zeller 2004). Electroporation has thus become the most commonly used method of introducing exogenous DNA into *Ciona* embryos. However, there are some limitations to the application of electroporation. First, the introduction of DNAs into particular blastomeres is difficult. Second, this technique has not been optimized to diverse species of ascidians. For example, electroporation has not been reported in *H. roretzi*, an important species for developmental and physiological studies. Third, electroporation is performed with dechorionated fertilized eggs, and the absence of chorion often causes morphological abnormalities, such as poor development of the larval tunic and aberrant left–right asymmetry (Sato and Morisawa 1999; Shimeld and Levin 2006). When these problems are crucial for the analysis, microinjection into intact fertilized eggs is still a good practical method to use (Oonuma et al. 2016). Please see Chaps. 2, 3, and 4 for detailed information on microinjection into *Ciona*, *Phallusia*, and *Halocynthia* eggs and Chap. 5 for electroporation methods.

6.5 Future Perspectives

With the recent advancements in molecular genetics and genomics, such as optogenetics and genome editing technologies, the application of *cis*-regulatory DNAs in biological studies of ascidians is becoming increasingly important. However, our current understanding of the functions of *cis*-regulatory DNAs is still limited. The further elucidation of *cis*-regulatory systems in ascidian genomes will boost the potential for the use of *cis*-regulatory DNAs as an important component of molecular tools for investigating the biology of ascidians.

References

Abdul-Wajid S, Morales Diaz H, Khairallah SM, Smith WC (2015) T-type Calcium Channel regulation of neural tube closure and EphrinA/EPHA expression. Cell Rep 13:829–839

Abitua PB, Wagner E, Navarrete IA, Levine M (2012) Identification of a rudimentary neural crest in a non-vertebrate chordate. Nature 492:104–107

Abitua PB, Gainous TB, Kaczmarczyk AN, Winchell CJ, Hudson C, Kamata K, Nakagawa M, Tsuda M, Kusakabe TG, Levine M (2015) The pre-vertebrate origins of neurogenic placodes. Nature 524:462–465

Ando R, Hama H, Yamamoto-Hino M, Mizuno H, Miyawaki A (2002) An optical marker based on the UV-induced green-to-red photoconversion of a fluorescent protein. Proc Natl Acad Sci U S A 99:12651–12656

Arnosti DN, Preston CM, Hagmann M, Schaffner W, Hope RG, Laughlan G, Luisi BF (1993) Specific transcriptional activation *in vitro* by the herpes simplex virus protein VP16. Nucleic Acids Res 21:5570–5576

Awazu S, Sasaki A, Matsuoka T, Satoh N, Sasakura Y (2004) An enhancer trap in the ascidian *Ciona intestinalis* identifies enhancers of its *Musashi* orthologous gene. Dev Biol 275:459–472

Awazu S, Matsuoka T, Inaba K, Satoh N, Sasakura Y (2007) High-throughput enhancer trap by remobilization of transposon *Minos* in *Ciona intestinalis*. Genesis 45:307–317

Beh J, Shi W, Levine M, Davidson B, Christiaen L (2007) FoxF is essential for FGF-induced migration of heart progenitor cells in the ascidian *Ciona intestinalis*. Development 134:3297–3305

Bertrand V, Hudson C, Caillol D, Popovici C, Lemaire P (2003) Neural tissue in ascidian embryos is induced by FGF9/16/20, acting via a combination of maternal GATA and Ets transcription factors. Cell 115:615–627

Boffelli D, Weer CV, Weng L, Lewis KD, Shoukry MI, Pachter L, Keys DN, Rubin EM (2004) Intraspecies sequence comparisons for annotating genomes. Genome Res 14:2406–2411

Brozovic M, Martin C, Dantec C, Dauga D, Mendez M, Simion P, Percher M, Laporte B, Scornavacca C, Di Gregorio A, Fujiwara S, Gineste M, Lowe EK, Piette J, Racioppi C, Ristoratore F, Sasakura Y, Takatori N, Brown TC, Delsuc F, Douzery E, Gissi C, McDougall A, Nishida H, Sawada H, Swalla BJ, Yasuo H, Lemaire P (2016) ANISEED 2015: a digital framework for the comparative developmental biology of ascidians. Nucleic Acids Res 44:D808–D818

Christiaen L, Bourrat F, Joly J-S (2005) A modular *cis*-regulatory system controls isoform-specific *pitx* expression in ascidian stomodaeum. Dev Biol 277:557–566

Christiaen L, Davidson B, Kawashima T, Powell W, Nolla H, Vranizan K, Levine M (2008) The transcription/migration interface in heart precursors of *Ciona intestinalis*. Science 320:1349–1352

Christiaen L, Stolfi A, Davidson B, Levine M (2009) Spatio-temporal intersection of Lhx3 and Tbx6 defines the cardiac field through synergistic activation of *Mesp*. Dev Biol 328:552–560

Cooley J, Whitaker S, Sweeney S, Fraser S, Davidson B (2011) Cytoskeletal polarity mediates localized induction of the heart progenitor lineage. Nat Cell Biol 13:952–957

Corbo JC, Levine M, Zeller RW (1997) Characterization of a notochord-specific enhancer from the *Brachyury* promoter region of the ascidian, *Ciona intestinalis*. Development 124:589–602

Corbo JC, Fujiwara S, Levine M, Di Gregorio A (1998) Suppressor of hairless activates *brachyury* expression in the *Ciona* embryo. Dev Biol 203:358–368
Corbo JC, Di Gregorio A, Levine M (2001) The ascidian as a model organism in developmental and evolutionary biology. Cell 106:535–538
Das G, Henning D, Wright D, Reddy R (1988) Upstream regulatory elements are necessary and sufficient for transcription of a U6 RNA gene by RNA polymerase III. EMBO J 7:503–512
Davidson B, Levine M (2003) Evolutionary origins of the vertebrate heart: specification of the cardiac lineage in *Ciona intestinalis*. Proc Natl Acad Sci U S A 100:11469–11473
Davidson B, Shi W, Levine M (2005) Uncoupling heart cell specification and migration in the simple chordate *Ciona intestinalis*. Development 132:4811–4818
Davidson B, Shi W, Beh J, Christiaen L, Levine M (2006) FGF signaling delineates the cardiac progenitor field in the simple chordate, *Ciona intestinalis*. Genes Dev 20:2728–2738
Deguchi T, Itoh M, Urawa H, Matsumoto T, Nakayama S, Kawasaki T, Kitano T, Oda S, Mitani H, Takahashi T, Todo T, Sato J, Okada K, Hatta K, Yuba S, Kamei Y (2009) Infrared laser-mediated local gene induction in medaka, zebrafish and *Arabidopsis thaliana*. Develop Growth Differ 51:769–775
Dehal P, Satou Y, Campbell RK, Chapman J, Degnan B et al (2002) The draft genome of *Ciona intestinalis*: insights into chordate and vertebrate origins. Science 298:2157–2167
Di Gregorio A, Levine M (1999) Regulation of *Ci-tropomyosin-like*, a Brachyury target gene in the ascidian, *Ciona intestinalis*. Development 126:5599–5609
Di Gregorio A, Corbo JC, Levine M (2001) The regulation of *forkhead/HNF-3beta* expression in the *Ciona* embryo. Dev Biol 229:31–43
Di Gregorio A, Harland RM, Levine M, Casey ES (2002) Tail morphogenesis in the ascidian, *Ciona intestinalis*, requires cooperation between notochord and muscle. Dev Biol 244:385–395
Di Gregorio A, Levine M (2002) Analyzing gene regulation in ascidian embryos: new tools for new perspectives. Differentiation 70:132–139
Dufour HD, Chettouh Z, Deyts C, de Rosa R, Goridis C, Joly JS, Brunet JF (2006) Precraniate origin of cranial motoneurons. Proc Natl Acad Sci U S A 103:8727–8732
Erives A, Corbo JC, Levine M (1998) Lineage-specific regulation of the *Ciona snail* gene in the embryonic mesoderm and neuroectoderm. Dev Biol 194:213–225
Fanelli A, Lania G, Spagnuolo A, Di Lauro R (2003) Interplay of negative and positive signals controls endoderm-specific expression of the ascidian *Cititf1* gene promoter. Dev Biol 263:12–23
Fisher AL, Ohsako S, Caudy M (1996) The WRPW motif of the hairy-related basic helix-loop-helix repressor proteins acts as a 4-amino-acid transcription repression and protein-protein interaction domain. Mol Cell Biol 16:2670–2677
Fujiwara S, Corbo JC, Levine M (1998) The snail repressor establishes a muscle/notochord boundary in the *Ciona* embryo. Development 125:2511–2520
Gandhi S, Haeussler M, Razy-Krajka F, Christiaen L, Stolfi A (2017) Evaluation and rational design of guide RNAs for efficient CRISPR/Cas9-mediated mutagenesis in *Ciona*. Dev Biol 425:8–20
Hackley C, Mulholland E, Kim GJ, Newman-Smith E, Smith WC (2013) A transiently expressed connexin is essential for anterior neural plate development in *Ciona intestinalis*. Development 140:147–155
Harafuji N, Keys DN, Levine M (2002) Genome-wide identification of tissue-specific enhancers in the *Ciona* tadpole. Proc Natl Acad Sci U S A 99:6802–6805
Hikosaka A, Kusakabe T, Satoh N, Makabe KW (1992) Introduction and expression of recombinant genes in ascidian embryos. Develop Growth Differ 34:627–634
Hikosaka A, Kusakabe T, Satoh N (1994) Short upstream sequences associated with the muscle-specific expression of an actin gene in ascidian embryos. Dev Biol 166:763–769
Horie T, Kusakabe T, Tsuda M (2008) Glutamatergic networks in the *Ciona intestinalis* Larva. J Comp Neurol 508:249–263
Horie T, Nakagwa M, Sasakura Y, Kusakabe TG (2009) Cell type and function of neurons in the ascidian nervous system. Develop Growth Differ 51:207–220
Horie T, Nakagawa M, Sasakura Y, Kusakabe TG, Tsuda M (2010) Simple motor system of the ascidian larva: neuronal complex comprising putative cholinergic neurons and GABAergic/glycinergic neurons. Zool Sci 27:181–190
Horie T, Shinki R, Ogura Y, Kusakabe TG, Satoh N, Sasakura Y (2011) Ependymal cells of chordate larvae are stem-like cells that form the adult nervous system. Nature 469:525–528
Hozumi A, Kawai N, Yoshida R, Ogura Y, Ohta N, Satake H, Satoh N, Sasakura Y (2010) Efficient transposition of a single *Minos* transposon copy in the genome of the ascidian *Ciona intestinalis* with a transgenic line expressing transposase in eggs. Dev Dynam 239:1076–1088
Hozumi A, Yoshida R, Horie T, Sakuma T, Yamamoto T, Sasakura Y (2013) Enhancer activity sensitive to the orientation of the gene it regulates in the chordate genome. Dev Biol 375:79–91
Hozumi A, Horie T, Sasakura Y (2015) Neuronal map reveals the highly regionalized pattern of the juvenile central nervous system of the ascidian *Ciona intestinalis*. Dev Dynam 244:1375–1393
Iitsuka T, Mita K, Hozumi A, Hamada M, Satoh N, Sasakura Y (2014) Transposon-mediated targeted and specific knockdown of maternally expressed transcripts in the ascidian *Ciona intestinalis*. Sci Rep 4:5050

Imai JH, Meinertzhagen IA (2007a) Neurons of the ascidian larval nervous system in *Ciona intestinalis*. I. Central nervous system. J Comp Neurol 501:316–334

Imai JH, Meinertzhagen IA (2007b) Neurons of the ascidian larval nervous system in *Ciona intestinalis*. II. Peripheral nervous system. J Comp Neurol 501:335–352

Imai K, Takada N, Satoh N, Satou Y (2000) β-catenin mediates the specification of endoderm cells in ascidian embryos. Development 127:3009–3020

Imai KS, Hino K, Yagi K, Satoh N, Satou Y (2004) Gene expression profiles of transcription factors and signaling molecules in the ascidian embryo: towards a comprehensive understanding of gene networks. Development 131:4047–4058

Imai KS, Levine M, Satoh N, Satou Y (2006) Regulatory blueprint for a chordate embryo. Science 312:1183–1187

Imai KS, Stolfi A, Levine M, Satou Y (2009) Gene regulatory networks underlying the compartmentalization of the *Ciona* central nervous system. Development 136:285–293

Irvine SQ, Fonseca VC, Zompa MA, Antony R (2008) *Cis*-regulatory organization of the *Pax6* gene in the ascidian *Ciona intestinalis*. Dev Biol 317:649–659

Johnson DS, Davidson B, Brown CD, Smith WC, Sidow A (2004) Noncoding regulatory sequences of *Ciona* exhibit strong correspondence between evolutionary constraint and functional importance. Genome Res 14:2448–2456

Kamei Y, Suzuki M, Watanabe K, Fujimori K, Kawasaki T, Deguchi T, Yoneda Y, Todo T, Takagi S, Funatsu T, Yuba S (2009) Infrared laser-mediated gene induction in targeted single cells *in vivo*. Nat Methods 6:79–81

Kari W, Zeng F, Zitzelsberger L, Will J, Rothbächer U (2016) Embryo microinjection and electroporation in the chordate *Ciona intestinalis*. J Vis Exp:e54313–e54313

Katsuyama Y, Matsumoto J, Okada T, Ohtsuka Y, Chen L, Okado H, Okamura Y (2002) Regulation of synaptotagmin gene expression during ascidian embryogenesis. Dev Biol 244:293–304

Katsuyama Y, Okada T, Matsumoto J, Ohtsuka Y, Terashima T, Okamura Y (2005) Early specification of ascidian larval motor neurons. Dev Biol 278:310–322

Kawaguchi A, Utsumi N, Morita M, Ohya A, Wada S (2015) Application of the *cis*-regulatory region of a heat-shock protein 70 gene to heat-inducible gene expression in the ascidian *Ciona intestinalis*. Genesis 53:170–182

Kawai N, Takahashi H, Nishida H, Yokosawa H (2005) Regulation of NF-κB/Rel by IκB is essential for ascidian notochord formation. Dev Biol 277:80–91

Kawai N, Ogura Y, Ikuta T, Saiga H, Hamada M, Sakuma T, Yamamoto T, Satoh N, Sasakura Y (2015) Hox10-regulated endodermal cell migration is essential for development of the ascidian intestine. Dev Biol 403:43–56

Keys DN, Levine M, Harland RM, Wallingford JB (2002) Control of intercalation is cell-autonomous in the notochord of *Ciona intestinalis*. Dev Biol 246:329–340

Keys DN, Lee BI, Di Gregorio A, Harafuji N, Detter JC, Wang M, Kahsai O, Ahn S, Zhang C, Doyle SA, Satoh N, Satou Y, Saiga H, Christian AT, Rokhsar DS, Hawkins TL, Levine M, Richardson PM (2005) A saturation screen for cis-acting regulatory DNA in the *Hox* genes of *Ciona intestinalis*. Proc Natl Acad Sci U S A 102:679–683

Kozak (1987) An analysis of 5′-noncoding sequences from 699 vertebrate messenger RNAs. Nucleic Acids Res 15:8125–8148

Kusakabe T (2005) Decoding *cis*-regulatory systems in ascidians. Zool Sci 22:129–146

Kusakabe TG (2017) Identifying vertebrate brain prototypes in deuterostomes. In: Shigeno S, Murakami Y, Nomura T (eds) Brain evolution by design. Springer Japan, Tokyo, pp 153–186

Kusakabe T, Makabe KW, Satoh N (1992) Tunicate muscle actin genes. Structure and organization as a gene cluster. J Mol Biol 227:955–960

Kusakabe T, Hikosaka A, Satoh N (1995) Coexpression and promoter function in two muscle actin gene complexes of different structural organization in the ascidian *Halocynthia roretzi*. Dev Biol 169:461–472

Kusakabe T, Yoshida R, Ikeda Y, Tsuda M (2004) Computational discovery of DNA motifs associated with cell type-specific gene expression in *Ciona*. Dev Biol 276:563–580

Kusakabe TG, Sakai T, Aoyama M, Kitajima Y, Miyamoto Y, Takigawa T, Daido Y, Fujiwara K, Terashima Y, Sugiuchi Y, Matassi G, Yagisawa H, Park MK, Satake H, Tsuda M (2012) A conserved non-reproductive GnRH system in chordates. PLoS One 7:e41955

Matsumoto J, Katsuyama Y, Ohtsuka Y, Lemaire P, Okamura Y (2008) Functional analysis of *synaptotagmin* gene regulatory regions in two distantly related ascidian species. Develop Growth Differ 50:543–552

Matsuoka T, Awazu S, Satoh N, Sasakura Y (2004) *Minos* transposon causes germline transgenesis of the ascidian *Ciona savignyi*. Develop Growth Differ 46:249–255

Meinertzhagen IA, Lemaire P, Okamura Y (2004) The neurobiology of the ascidian tadpole larva: recent developments in an ancient chordate. Annu Rev Neurosci 27:453–485

Mitani Y, Takahashi H, Satoh N (2001) Regulation of the muscle-specific expression and function of an ascidian T-box gene, *As-T2*. Development 128:3717–3728

Miyagishi M, Taira K (2002) U6 promoter-driven siRNAs with four uridine 3′ overhangs efficiently suppress targeted gene expression in mammalian cells. Nat Biotechnol 20:497–500

Nakazawa K, Yamazawa T, Moriyama Y, Ogura Y, Kawai N, Sasakura Y, Saiga H (2013) Formation of the digestive tract in *Ciona intestinalis* includes two distinct morphogenic processes between its anterior and posterior parts. Dev Dynam 242:1172–1183

Navarrete IA, Levine M (2016) Nodal and FGF coordinate ascidian neural tube morphogenesis. Development 143:4665–4675

Nishino A, Okamura Y, Piscopo S, Brown ER (2010) A glycine receptor is involved in the organization of swimming movements in an invertebrate chordate. BMC Neurosci 11:6

Nishiyama A, Fujiwara S (2008) RNA interference by expressing short hairpin RNA in the *Ciona intestinalis* embryo. Develop Growth Differ 50:521–529

Oda-Ishii I, Bertrand V, Matsuo I, Lemaire P, Saiga H (2005) Making very similar embryos with divergent genomes: conservation of regulatory mechanisms of *Otx* between the ascidians *Halocynthia roretzi* and *Ciona intestinalis*. Development 132:1663–1674

Ogura Y, Sakaue-Sawano A, Nakagawa M, Satoh N, Miyawaki A, Sasakura Y (2011) Coordination of mitosis and morphogenesis: role of a prolonged G2 phase during chordate neurulation. Development 138:577–587

Okada T, MacIsaac SS, Katsuyama Y, Okamura Y, Meinertzhagen IA (2001) Neuronal form in the central nervous system of the tadpole larva of the ascidian *Ciona intestinalis*. Biol Bull 200:252–256

Okada T, Katsuyama Y, Ono F, Okamura Y (2002) The development of three identified motor neurons in the larva of an ascidian, *Halocynthia roretzi*. Dev Biol 244:278–292

Okagaki R, Izumi H, Okada T, Nagahora H, Nakajo K, Okamura Y (2001) The maternal transcript for truncated voltage-dependent Ca^{2+} channels in the ascidian embryo: a potential suppressive role in Ca^{2+} channel expression. Dev Biol 230:258–277

Ono F, Katsuyama Y, Nakajo K, Okamura Y (1999) Subfamily-specific posttranscriptional mechanism underlies K^+ channel expression in a developing neuronal blastomere. J Neurosci 19:6874–6886

Oonuma K, Tanaka M, Nishitsuji K, Kato Y, Shimai K, Kusakabe TG (2016) Revised lineage of larval photoreceptor cells in *Ciona* reveals archetypal collaboration between neural tube and neural crest in sensory organ formation. Dev Biol 420:178–185

Pasini A, Amiel A, Rothbächer U, Roure A, Lemaire P, Darras S (2006) Formation of the ascidian epidermal sensory neurons: insights into the origin of the chordate peripheral nervous system. PLoS Biol 4:e225

Razy-Krajka F, Brown ER, Horie T, Callebert J, Sasakura Y, Joly J-S, Kusakabe TG, Vernier P (2012) Monoaminergic modulation of photoreception in ascidian: evidence for a proto-hypothalamo-retinal territory. BMC Biol 10:45

Razy-Krajka F, Lam K, Wang W, Stolfi A, Joly M, Bonneau R, Christiaen L (2014) Collier/OLF/EBF-dependent transcriptional dynamics control pharyngeal muscle specification from primed Cardiopharyngeal progenitors. Dev Cell 29:263–276

Rhee JM, Pirity MK, Lackan CS, Long JZ, Kondoh G, Takeda J, Hadjantonakis A-K (2006) In vivo imaging and differential localization of lipid-modified GFP-variant fusions in embryonic stem cells and mice. Genesis 44:202–218

Ristoratore F, Spagnuolo A, Aniello F, Branno M, Fabbrini F, Di Lauro R (1999) Expression and functional analysis of *Cititf1*, an ascidian NK-2 class gene, suggest its role in endoderm development. Development 126:5149–5159

Rothbächer U, Bertrand V, Lamy C, Lemaire P (2007) A combinatorial code of maternal GATA, Ets and β-catenin-TCF transcription factors specifies and patterns the early ascidian ectoderm. Development 134:4023–4032

Roure A, Rothbächer U, Robin F, Kalmar E, Ferone G, Lamy C, Missero C, Mueller F, Lemaire P (2007) A multicassette gateway vector set for high throughput and comparative analyses in *Ciona* and vertebrate embryos. PLoS One 2:e916

Russo MT, Donizetti A, Locascio A, D'Aniello S, Amoroso A, Aniello F, Fucci L, Branno M (2004) Regulatory elements controlling *Ci-msx*b tissue-specific expression during *Ciona intestinalis* embryonic development. Dev Biol 267:517–528

Sakaue-Sawano A, Kurokawa H, Morimura T, Hanyu A, Hama H, Osawa H, Kashiwagi S, Fukami K, Miyata T, Miyoshi H, Imamura T, Ogawa M, Masai H, Miyawaki A (2008) Visualizing spatiotemporal dynamics of multicellular cell-cycle progression. Cell 132:487–498

Sasaki H, Yoshida K, Hozumi A, Sasakura Y (2014) CRISPR/Cas9-mediated gene knockout in the ascidian *Ciona intestinalis*. Develop Growth Differ 56:499–510

Sasakura Y, Awazu S, Chiba S, Satoh N (2003) Germ-line transgenesis of the Tc1/*mariner* superfamily transposon *Minos* in *Ciona intestinalis*. Proc Natl Acad Sci U S A 100:7726–7730

Sasakura Y, Nakashima K, Awazu S, Matsuoka T, Nakayama A, Azuma J-I, Satoh N (2005) Transposon-mediated insertional mutagenesis revealed the functions of animal cellulose synthase in the ascidian *Ciona intestinalis*. Proc Natl Acad Sci U S A 102:15134–15139

Sasakura Y, Konno A, Mizuno K, Satoh N, Inaba K (2008) Enhancer detection in the ascidian *Ciona intestinalis* with transposase-expressing lines of *Minos*. Dev Dynam 237:39–50

Sasakura Y, Inaba K, Satoh N, Kondo M, Akasaka K (2009) *Ciona intestinalis* and *Oxycomanthus japonicus*, representatives of marine invertebrates. Exp Anim 58:459–469

Sasakura Y, Suzuki MM, Hozumi A, Inaba K, Satoh N (2010) Maternal factor-mediated epigenetic gene silencing in the ascidian *Ciona intestinalis*. Mol Gen Genomics 283:99–110

Sasakura Y, Sierro N, Nakai K, Inaba K, Kusakabe TG (2012a) Genome structure, functional genomics, and proteomics in ascidians. In: Denny P, Kole C (eds)

Genome mapping and genomics in laboratory animals: genome mapping and genomics in animals, vol 4. Springer, Berlin/Heidelberg, pp 87–132

Sasakura Y, Mita K, Ogura Y, Horie T (2012b) Ascidians as excellent chordate models for studying the development of the nervous system during embryogenesis and metamorphosis. Develop Growth Differ 54:420–437

Sato Y, Morisawa M (1999) Loss of test cells leads to the formation of new tunic surface cells and abnormal metamorphosis in larvae of *Ciona intestinalis* (Chordata, ascidiacea). Dev Genes Evol 209:592–600

Satoh G, Harada Y, Satoh N (2000) The expression of nonchordate deuterostome *Brachyury* genes in the ascidian *Ciona* embryo can promote the differentiation of extra notochord cells. Mech Dev 96:155–163

Satou Y, Kawashima T, Shoguchi E, Nakayama A, Satoh N (2005) An integrated database of the ascidian, *Ciona intestinalis*: towards functional genomics. Zool Sci 22:837–843

Satou Y, Mineta K, Ogasawara M, Sasakura Y, Shoguchi E, Ueno K, Yamada L, Matsumoto J, Wasserscheid J, Dewar K, Wiley GB, Macmil SL, Roe BA, Zeller RW, Hastings KEM, Lemaire P, Lindquist E, Endo T, Hotta K, Inaba K (2008) Improved genome assembly and evidence-based global gene model set for the chordate *Ciona intestinalis*: new insight into intron and operon populations. Genome Biol 9:R152

Shi W, Levine M (2008) Ephrin signaling establishes asymmetric cell fates in an endomesoderm lineage of the *Ciona* embryo. Development 135:931–940

Shi W, Peyrot SM, Munro E, Levine M (2009) FGF3 in the floor plate directs notochord convergent extension in the *Ciona* tadpole. Development 136:23–28

Shimai K, Hirano A, Kitaura Y, Kitano Y, Itoh A, Kiuchi A, Sasaki N, Nishikata T (2008) Novel ubiquitous promoters and expression-vector optimization in ascidian embryos. Inv Reprod Dev 51:103–110

Shimai K, Kitaura Y, Tamari Y, Nishikata T (2010) Upstream regulatory sequences required for specific gene expression in the ascidian neural tube. Zool Sci 27:76–83

Shimeld SM, Levin M (2006) Evidence for the regulation of left-right asymmetry in *Ciona intestinalis* by ion flux. Dev Dynam 235:1543–1553

Shimeld SM, Purkiss AG, Dirks RPH, Bateman OA, Slingsby C, Lubsen NH (2005) Urochordate βγ-crystallin and the evolutionary origin of the vertebrate eye lens. Curr Biol 15:1684–1689

Sierro N, Kusakabe T, Park K-J, Yamashita R, Kinoshita K, Nakai K (2006) DBTGR: a database of tunicate promoters and their regulatory elements. Nucleic Acids Res 34:D552–D555

Stolfi A, Levine M (2011) Neuronal subtype specification in the spinal cord of a protovertebrate. Development 138:995–1004

Stolfi A, Gainous TB, Young JJ, Mori A, Levine M, Christiaen L (2010) Early chordate origins of the vertebrate second heart field. Science 329:565–568

Stolfi A, Gandhi S, Salek F, Christiaen L (2014) Tissue-specific genome editing in *Ciona* embryos by CRISPR/Cas9. Development 141:4115–4120

Stolfi A, Ryan K, Meinertzhagen IA, Christiaen L (2015a) Migratory neuronal progenitors arise from the neural plate borders in tunicates. Nature 527:371–374

Stolfi A, Sasakura Y, Chalopin D, Satou Y, Christiaen L, Dantec C, Endo T, Naville M, Nishida H, Swalla BJ, Volff JN, Voskoboynik A, Dauga D, Lemaire P (2015b) Guidelines for the nomenclature of genetic elements in tunicate genomes. Genesis 53:19–14

Takahashi H, Hotta K, Erives A, Di Gregorio A, Zeller RW, Levine M, Satoh N (1999a) *Brachyury* downstream notochord differentiation in the ascidian embryo. Genes Dev 13:1519–1523

Takahashi H, Mitani Y, Satoh G, Satoh N (1999b) Evolutionary alterations of the minimal promoter for notochord-specific *Brachyury* expression in ascidian embryos. Development 126:3725–3734

Takamura K, Minamida N, Okabe S (2010) Neural map of the larval central nervous system in the ascidian *Ciona intestinalis*. Zool Sci 27:191–203

Tassy O, Dauga D, Daian F, Sobral D, Robin F, Khoueiry P, Salgado D, Fox V, Caillol D, Schiappa R, Laporte B, Rios A, Luxardi G, Kusakabe T, Joly J-S, Darras S, Christiaen L, Contensin M, Auger H, Lamy C, Hudson C, Rothbächer U, Gilchrist MJ, Makabe KW, Hotta K, Fujiwara S, Satoh N, Satou Y, Lemaire P (2010) The ANISEED database: digital representation, formalization, and elucidation of a chordate developmental program. Genome Res 20:1459 1468

Treen N, Yoshida K, Sakuma T, Sasaki H, Kawai N, Yamamoto T, Sasakura Y (2014) Tissue-specific and ubiquitous gene knockouts by TALEN electroporation provide new approaches to investigating gene function in *Ciona*. Development 141:481–487

Tuschl T (2002) Expanding small RNA interference. Nat Biotechnol 20:446–448

Ueki T, Satoh N (1995) Sequence motifs shared by the 5′ flanking regions of two epidermis-specific genes in the ascidian embryo. Develop Growth Differ 37:597–604

Vandenberghe AE, Meedel TH, Hastings KE (2001) mRNA 5′-leader *trans*-splicing in the chordates. Genes Dev 15:294–303

Vinson JP, Jaffe DB, O'Neill K, Karlsson EK, Stange-Thomann N, Anderson S, Mesirov JP, Satoh N, Satou Y, Nusbaum C, Birren B, Galagan JE, Lander ES (2005) Assembly of polymorphic genomes: algorithms and application to *Ciona savignyi*. Genome Res 15:1127–1135

Wagner E, Levine M (2012) FGF signaling establishes the anterior border of the *Ciona* neural tube. Development 139:2351–2359

Wagner E, Stolfi A, Choi YG, Levine M (2014) Islet is a key determinant of ascidian palp morphogenesis. Development 141:3084–3092

Wang W, Christiaen L (2012) Transcriptional enhancers in ascidian development. Curr Top Dev Biol 98:147–172

Yagi K, Satou Y, Satoh N (2004) A zinc finger transcription factor, ZicL, is a direct activator of *Brachyury* in the notochord specification of *Ciona intestinalis*. Development 131:1279–1288

Yoshida R, Sakurai D, Horie T, Kawakami I, Tsuda M, Kusakabe T (2004) Identification of neuron-specific promoters in *Ciona intestinalis*. Genesis 39:130–140

Zeller RW (2004) Generation and use of transgenic ascidian embryos. Methods Cell Biol 74:713–730

Zhang F, Wang L-P, Boyden ES, Deisseroth K (2006) Channelrhodopsin-2 and optical control of excitable cells. Nat Methods 3:785–792

Zhang F, Wang L-P, Brauner M, Liewald JF, Kay K, Watzke N, Wood PG, Bamberg E, Nagel G, Gottschalk A, Deisseroth K (2007) Multimodal fast optical interrogation of neural circuitry. Nature 446:633–639

7 Reporter Analyses Reveal Redundant Enhancers that Confer Robustness on *Cis*-Regulatory Mechanisms

Shigeki Fujiwara and Cristian Cañestro

Abstract

Reporter analyses of *Hox1* and *Brachyury* (*Bra*) genes have revealed examples of redundant enhancers that provide regulatory robustness. Retinoic acid (RA) activates through an RA-response element the transcription of *Hox1* in the nerve cord of the ascidian *Ciona intestinalis*. We also found a weak RA-independent neural enhancer within the second intron of *Hox1*. The *Hox1* gene in the larvacean *Oikopleura dioica* is also expressed in the nerve cord. The *O. dioica* genome, however, does not contain the RA receptor-encoding gene, and the expression of *Hox1* has become independent of RA. We have found that the upstream sequence of the *O. dioica Hox1* was able to activate reporter gene expression in the nerve cord of the *C. intestinalis* embryo, suggesting that an RA-independent regulatory system in the nerve cord might be common in larvaceans and ascidians. This RA-independent redundant regulatory system may have facilitated the *Oikopleura* ancestor losing RA signaling without an apparent impact on *Hox1* expression domains. On the other hand, vertebrate *Bra* is expressed in the ventral mesoderm and notochord, whereas its ascidian ortholog is exclusively expressed in the notochord. Fibroblast growth factor (FGF) induces *Bra* in the ventral mesoderm in vertebrates, whereas it induces *Bra* in the notochord in ascidians. Disruption of the FGF signal does not completely silence *Bra* expression in ascidians, suggesting that FGF-dependent and independent enhancers might comprise a redundant regulatory system in ascidians. The existence of redundant enhancers, therefore, provides regulatory robustness that may facilitate the acquisition of new expression domains.

Keywords

Reporter analysis · Retinoic acid · *Hox1* · Retinoic acid signaling · *Brachyury* · Fibroblast growth factor signaling · *Ciona intestinalis* · Larvacean *Oikopleura dioica* · Redundant enhancers · Shadow enhancers

S. Fujiwara (✉)
Department of Chemistry and Biotechnology, Faculty of Science and Technology, Kochi University, Kochi-shi, Kochi, Japan
e-mail: tatataa@kochi-u.ac.jp

C. Cañestro
Department de Genètica, Microbiologia i Estadística and Institut de Recerca de la Biodiversitat (IRBio), Universitat de Barcelona, Barcelona, Spain
e-mail: canestro@ub.edu

Y. Sasakura (ed.), *Transgenic Ascidians*, Advances in Experimental Medicine and Biology 1029,
https://doi.org/10.1007/978-981-10-7545-2_7

7.1 Introduction

Reporter assay is a straightforward approach to analyzing transcriptional regulation. Embryos of urochordate ascidians provide a suitable experimental system for conducting reporter analyses. The compact genome of ascidians, with relatively short intergenic regions, facilitates the identification of *cis*-regulatory elements (Dehal et al. 2002). In addition, easy electroporation and rapid embryogenesis enable the production of hundreds of transgenic embryos within a single day (Corbo et al. 1997; Zeller 2004). A large number of embryos that can be obtained from a single electroporation procedure are sufficient to subdivide them into multiple groups that can be subjected to drug treatments at different doses during different developmental stages. Thus, we can compare patterns of reporter gene expression in a large number of drug-treated and control embryos from the same batch.

Another merit of transgenic technologies is the easy access to comparative studies of *cis*-regulatory mechanisms across animal taxa. Conservation of transcriptional regulatory mechanisms can be easily visualized by the activation of the enhancer in heterologous species. The similarities and differences of reporter gene expression patterns in different species may provide circumstantial evidence suggesting evolutionary events that might have contributed to the divergence of species.

In this review, we provide two examples in which reporter assays have facilitated the identification of a robust *cis*-regulatory mechanism. First, we describe experiments in which we have identified response elements for retinoic acid (RA) within the regulatory regions of the *Hox1* gene in the ascidian *Ciona intestinalis*. Our reporter analyses identified RA-responsive and RA-independent mechanisms that activate *Hox1* expression in the nerve cord. Conservation and divergence were examined in two urochordate species, the ascidian *C. intestinalis* and the larvacean *Oikopleura dioica*. Second, we discuss the redundant regulatory pathways that activate notochord-specific expression of the *Brachyury* gene in ascidians in comparison to its vertebrate orthologs.

7.2 RA-Dependent Transcriptional Activation of *Hox1*

Retinoic acid regulates many aspects of chordate embryogenesis, mainly through transcriptional regulation of specific target genes (Mangelsdorf and Evans 1992). The nuclear RA receptor (RAR) is a transcription factor that recognizes specific DNA sequences, called the RA response element (RARE). Upon binding RA, RAR directly activates transcription of the target genes (Mangelsdorf and Evans 1992). The direct (primary) target genes include transcription factors, which in turn activate secondary target genes. The activation of the secondary targets depends on the translation of the direct targets, and is thus sensitive to translation inhibitors.

Genes orthologous to *RAR* have been identified in deuterostomes and lophotrochozoans (Cañestro et al. 2006; Albalat and Cañestro 2009). However, the contribution of RA to gene expression and organogenesis is still unclear in nonchordate animals. RA activates expression of the *Hox1* gene in the ascidian *Ciona intestinalis* (Ishibashi et al. 2003). The translation inhibitor, puromycin, does not affect the response of *Hox1* to RA (Ishibashi et al. 2005). This indicates that *Hox1* is a direct target, and that its enhancer contains RARE(s). Morpholino oligonucleotide-mediated knockdown of the RA synthase-encoding gene, *Raldh2*, suppressed the expression of *Hox1* (Imai et al. 2009; Sasakura et al. 2012). A *Hox1* null mutant did not form the atrial siphon placode, which in turn impaired the formation of atrial siphon muscle, body wall muscle and pharyngeal gill slits (Sasakura et al. 2012). Knockdown of *Raldh2* or *Hox1* resulted in similar phenotypes, suggesting that endogenous RA and *Hox1* might be required for organogenesis in ascidians (Sasakura et al. 2012).

7.2.1 Identification of RA Response Elements Within the Hox1 Enhancers

Reporter analyses carried out by ourselves and by others revealed that the 5′ flanking region and the second intron contained enhancer elements required for transcriptional activation of *Hox1* in the epidermis and nerve cord respectively (Fig. 7.1) (Kanda et al. 2009, 2013; Natale et al. 2011). Both regions contain a single RARE (Fig. 7.1a) (Kanda et al. 2009, 2013). The binding of RAR to these RAREs was confirmed by electrophoretic mobility shift assays using in vitro synthesized proteins (Kanda et al. 2009; Kanda, M. unpublished data). Disruption of the RARE sequence in the 5′ flanking region affected the enhancer activity only in the epidermis (Fig. 7.1b, c) (Kanda et al. 2009). On the other hand, transcriptional activation in the nerve cord required the RARE in the second intron (Fig. 7.1d, e) (Kanda et al. 2013). These results suggested that the tissue-specific expression of *Hox1* was not solely activated by RAR. Synergy between RAR and other tissue-specific transcription factors and/or the context of the enhancer sequences is necessary for activation in each tissue. The inhibitors of RA synthesis, citral and N,N-diethylaminobenzaldehyde, suppressed expression of the *Hox1* reporter genes in both tissues, supporting the requirement of RA for the transcriptional activation of the *Hox1* in normally developing embryos (Kanda et al. 2009).

In the above experiments, we carried out deletion analyses of the intronic enhancer. For this purpose, we relocated the second intron upstream of the 5′ flanking region (Fig. 7.1a). We did not leave the intron in its original position because deletion of the intronic sequence would affect correct splicing. The relocated intronic enhancer drove reporter gene expression normally in the anterior nerve cord (Fig. 7.1b, c). This was not surprising, as many molecular biology dictionaries state that enhancers can stimulate gene transcription from far upstream or downstream of the transcription start site, and in both forward and reversed orientations. However, activities of relocated and/or reverse-oriented enhancers should be carefully evaluated, because some enhancers are position-dependent (Mocikat et al. 1993) or orientation-sensitive (Hozumi et al. 2013; Swamynathan and Piatigorsky 2002).

7.2.2 RA-Independent Expression of Hox1 in Larvaceans

In addition to the RA-driven *Hox1* expression in ascidians described above, RA also activates *Hox1* expression in the cephalochordate amphioxus (Schubert et al. 2005, 2006). An antagonist of RAR affected hindbrain patterning and pharyngeal morphogenesis, indicating the requirement of endogenous RA for normal cephalochordate embryogenesis (Schubert et al. 2005, 2006). These findings suggested an early origin of RA/*Hox1*-dependent embryogenesis in the common chordate ancestor, as cephalochordates are basally divergent within chordates, before the vertebrate/tunicate (also known as olfactores) split (Fig. 7.2) (Delsuc et al. 2006; Putnam et al. 2008).

Larvaceans (appendicularians) are a sister group of ascidians within the tunicates (Fig. 7.2) (Tsagkogeorga et al. 2009). The pattern of *Hox1* expression in the larvacean *Oikopleura dioica* is similar to that in ascidians (Fig. 7.3a, b) (Cañestro et al. 2005). However, an excess or lack of the RA signal did not affect the expression of *O. dioica Hox1* (Cañestro and Postlethwait 2007). In fact, the *O. dioica* genome does not contain genes encoding RAR and none of the metabolic enzymes that regulate the synthesis and degradation of RA (Cañestro et al. 2006; Martí-Solans et al. 2016). This finding prompted us to test whether the enhancer of *O. dioica Hox1* was activated in *C. intestinalis* embryos, and in such a case, to test if this expression was responsive to RA.

A 5′ flanking region of *O. dioica Hox1* was fused in-frame with the translated region of *lacZ* (Fig. 7.3c). This transgene was expressed in the nerve cord of the *C. intestinalis* tailbud embryo (Fig. 7.3d) (Kanda, M., Rodriguez-Marí, A., Tagawa, M., Cañestro, C., Fujiwara, S., unpublished data). Therefore, it is clear that the *O. dioica* upstream enhancer can be activated by

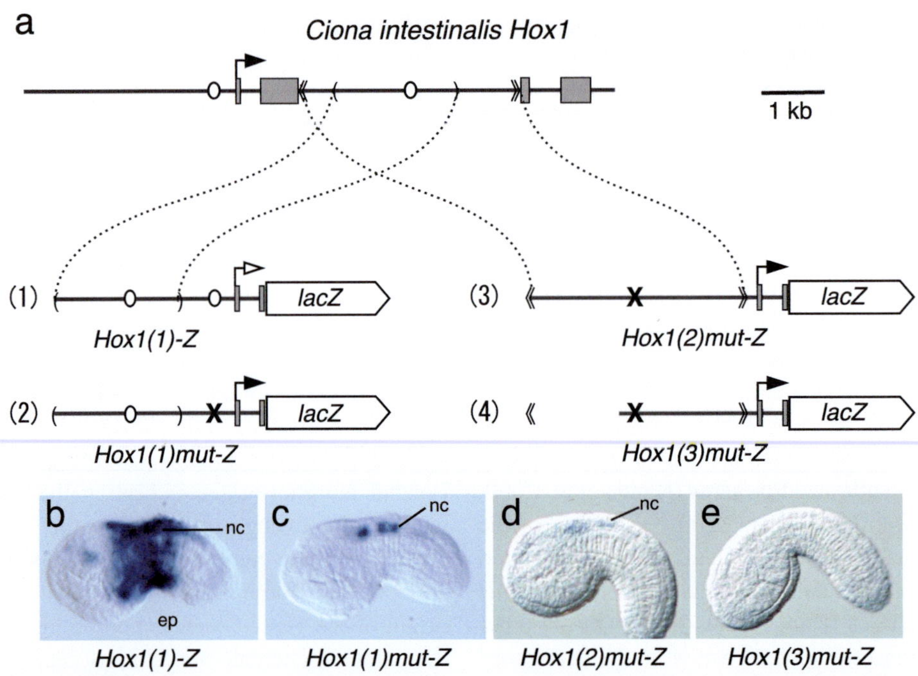

Fig. 7.1 Identification of RA-responsive and RA-independent enhancer elements of the *C. intestinalis* *Hox1*. (**a**) Structure of the *Hox1* locus in the *C. intestinalis* genome is schematized at the top. *Boxes* indicate exons. The *arrow* indicates the transcription start site. *Circles* indicate RAREs. (1) A 1.9-kb central part of the second intron was placed upstream of the 0.9-kb 5′ flanking region of *Hox1*. The construct contains the first exon, first intron, and a part of the second exon. The *lacZ* translated region was placed downstream of the initiation codon of *Hox1* in the second exon. This construct was named *Hox1(1)-Z*. (2) The transgene *Hox1(1)mut-Z* carried a point mutation in the RARE within the 5′ flanking region of *Hox1(1)-Z*. The intronic RARE remained intact. (3) The transgene *Hox1(2)mut-Z* contained the entire second intron (3.6 kb), placed upstream of a 160-bp core promoter. *Hox1(2)mut-Z* contained a point mutation in the intronic RARE. (4) A 1.3-kb fragment of the second intron was deleted from *Hox1(2)mut-Z*. The construct was named *Hox1(3)mut-Z*. These constructs were described elsewhere (Kanda et al. 2009, 2013); however, they were renamed here for this review. (**b–e**) Expression of *lacZ* was visualized by in-situ hybridization in *C. intestinalis* tailbud embryos, carrying the reporter constructs (1) ~ (4) respectively. In all panels, the anterior side of the embryo is oriented to the left, with the dorsal side up. *ep* epidermis, *nc* nerve cord. (**b**) Embryo carrying *Hox1(1)-Z* expressed *lacZ* mRNA in the epidermis and nerve cord. (**c**) *Hox1(1)mut-Z* was activated only in the nerve cord. The epidermal expression was completely silenced by a mutation of the RARE within the 5′ flanking region. (**d**) *Hox1(2)mut-Z* was weakly activated in the nerve cord. (**e**) *Hox1(3)mut-Z* was not activated in any tissue

transcription factors expressed in the nerve cord of *C. intestinalis*. Treatment with RA did not up-regulate the enhancer activity (Fig. 7.3e), suggesting that the 5′ flanking region of *O. dioica* *Hox1* might not contain any functional RARE, but it still preserved conserved enhancers able to drive the expression in the nerve cord in ascidians.

7.2.3 Redundant Enhancers Confer Robustness on the Cis-Regulatory Mechanism

A close look at the activation of the *C. intestinalis* *Hox1* neural enhancer carrying a mutation at the intronic RARE showed that the intronic neural enhancer was not completely silenced by the

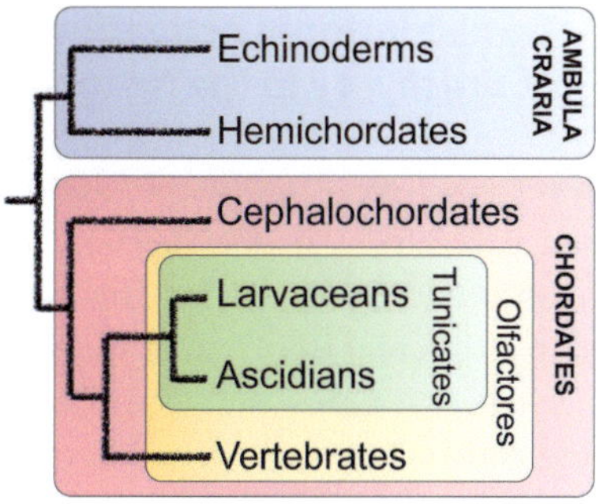

Fig. 7.2 Phylogenetic relationships of major deuterostome groups, reconstructed from Putnam et al. (2008) with a slight modification. The tree topology was supported by Bayesian and maximum likelihood methods, based on a concentrated alignment of 1,090 genes (Putnam et al. 2008). Branch length is not considered in this figure

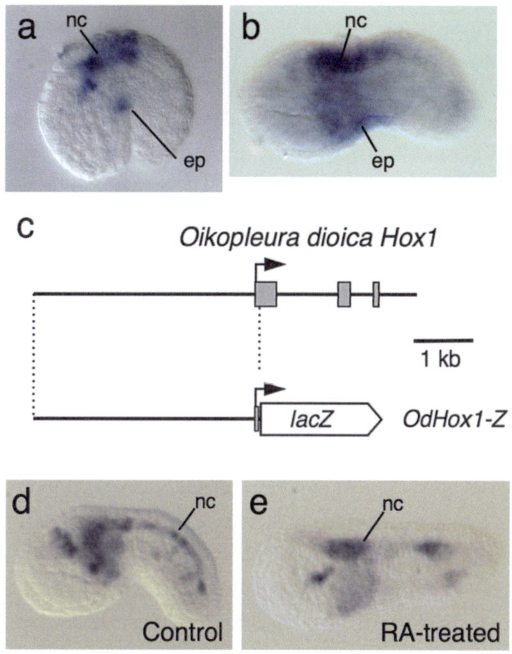

Fig. 7.3 The 5′ upstream region of the *O. dioica Hox1* shows enhancer activity in the nerve cord of the *C. intestinalis* embryo. (**a**) Expression of endogenous *Hox1* mRNA in the *O. dioica* tailbud embryo. (**b**) Expression of endogenous *Hox1* mRNA in the *C. intestinalis* tailbud embryo. (**c**) Structure of the *Hox1* locus in the *O. dioica* genome is indicated at the top. The reporter construct, *OdHox1-Z*, contained a 3.6-kb 5′ upstream region. The construct also contained the 5′ untranslated region (5′ UTR) and the initiation codon of the *O. dioica Hox1*. (**d**) Expression of *OdHox1-Z* in the nerve cord of the *C. intestinalis* embryo. This embryo was reared in artificial seawater containing 0.1% dimethylsulfoxide, which was used as a solvent for preparation of the concentrated stock solution of RA. A similar pattern of expression was also observed in untreated normal embryos. The result indicated that transcription factors in the *C. intestinalis* nerve cord can activate the upstream enhancer of the *O. dioica Hox1*. (**e**) Expression of *OdHox1-Z* in *C. intestinalis* embryo, treated with 1 μM RA. RA did not up-regulate the expression of *lacZ*, suggesting that RAR might not have been involved in transcriptional activation. In all panels, the anterior side of the embryo is oriented to the left, with the dorsal side up. *ep* epidermis, *nc* nerve cord

mutation (Fig. 7.1d) (Kanda et al. 2013). Similarly, knockdown of *RAR* using morpholino oligonucleotides did not completely suppress the neural expression of endogenous *Hox1* (Kanda et al. 2013). By contrast, mutation at the 5′ RARE almost completely silenced the epidermal enhancer (Fig. 7.1c) (Kanda et al. 2009). These observations implied that RA-independent transcription factors contributed to the transcriptional activation of *Ciona Hox1* in the nerve cord, and that plausibly, these neural transcription factors might also be responsible for the activation of the *O. dioica Hox1* enhancer in the nerve cord of both *C. intestinalis and O. dioica* embryos. The enhancer element responsible for the RA-independent activation of *Ciona Hox1* was located within the second intron (Fig. 7.1a, e) (Kanda et al. 2013). The 5′ flanking region also activated reporter gene expression in the nerve cord at later larval stages (Natale et al. 2011). This may be a maintenance element that contributes to maintaining the activated state after the initial activation. However, it is also possible that this 5′ element is another genuine redundant neural enhancer. This 5′ neural enhancer may correspond to the *O. dioica* 5′ neural enhancer.

Two or more parallel pathways contributing to the expression of *Hox1* in the same tissue may explain how the *Hox1* expression domain remained apparently unaltered in larvaceans, despite the loss of *RAR* and RA-metabolic genes during the evolution of this lineage (Albalat and Cañestro 2016; Martí-Solans et al. 2016). In the common chordate ancestor, these redundant enhancers may have activated the neural expression of *Hox1*. One is RA-dependent, and the other is RA-independent. When RA signaling was lost during the evolution of larvaceans, neural expression of *Hox1* was able to persist thanks to the robustness provided by the RA-independent enhancer. Recently, Yoshida et al. (2017) revealed that the expression of *Hox1* in the posterior endostyle of *C. intestinalis* juvenile adults also requires RA. Expression of *Hox1* was also observed in the endostyle of *O. dioica*, which was, however, not activated by RA (Cañestro et al. 2008). It is therefore plausible that the expression of *Hox1* in the endostyle has also been regulated by redundant RA-dependent and RA-independent enhancers in the common tunicate ancestor.

Redundant enhancers are similar to shadow enhancers. In *Drosophila* embryos, gene expression in the neurogenic ectoderm is controlled by a set of transcription factors, Dorsal, Twist, and Snail (Ip et al. 1992). Hong et al. (2008) found that many of the Dorsal target genes contained two clusters of binding sites for these transcription factors. Either one of the clusters was able to activate reporter gene expression in the neurogenic ectoderm (Hong et al. 2008). In such cases, the promoter proximal cluster is called "primary enhancer" and the distal one is called "shadow enhancer" (Barolo 2011). Shadow enhancers provide the *cis*-regulatory mechanisms with the robustness required to keep a normal pattern of gene expression under some environmental pressure, such as high temperatures (Perry et al. 2010). Redundant enhancers found in the *Hox1*, however, differ from shadow/primary enhancers as the latter is composed of the same set of transcription factor-binding sites (Hong et al. 2008), whereas in the case of *Hox1* they are different. Shadow enhancers can retain normal gene expression upon a sudden loss of the primary enhancer. However, a loss of function of Dorsal or Twist may severely affect both enhancers. In contrast, each of the redundant enhancers may be activated by a distinct set of transcription factors. They provide a tough regulatory framework for normal gene expression, because neither deletion of one of the enhancers nor loss of function of a transcription factor causes a critical effect on normal gene expression, which utilizes redundant enhancers. In addition, redundant enhancers may provide an opportunity to acquire new expression domains, as described below.

7.3 Transcriptional Activation of *Brachyury* in the Notochord

The *Brachyury* gene (*Bra*) in ascidians is expressed exclusively in the developing notochord (Yasuo and Satoh 1993). This pattern of expression is different from that of vertebrates, in which *Brachyury* genes are expressed in the notochord and ventral mesoderm (Herrmann and Kispert 1994). During ascidian gastrulation, fibroblast growth factor (FGF) activates *Bra* expression in the notochord (Nakatani et al. 1996; Imai et al. 2002a). Many transcription factors have been identified as responsible for the activation of *Bra* expression, including Suppressor-of-Hairless [Su(H)] (Corbo et al. 1998; Imai et al. 2002b), ZicL (Imai et al. 2002c; Yagi et al. 2004), and FoxA (Kumano et al. 2006).

7.3.1 Redundant Pathways Activate Bra Expression in Ascidians

Knockdown of *FGF9/16/20* (formerly called *FGF4/6/9*) severely affected initial activation of the *Brachyury* gene in another *Ciona* species, *C. savignyi* (Imai et al. 2002a). However, the FGF-depleted embryos finally expressed *Bra* in the notochord by the tailbud stage (Imai et al. 2002a). Similarly, suppression of the FGF signaling by a MEK inhibitor did not completely suppress the expression of *Bra* reporter genes in another ascidian species *Halocynthia roretzi* (Matsumoto et al. 2007). This was not surprising because the expression of ZicL and FoxA does not require the FGF signal (Fig. 7.4a) (Shimauchi et al. 1997; Imai et al. 2006). Nuclear localization of β-catenin activated the expression of the transcription factors FoxA and FoxD, which in turn activates the expression of ZicL (Fig. 7.4a) (Imai et al. 2002c, 2006). In contrast, the downstream effector of the FGF signaling is the transcription factor Ets (Fig. 7.4a; Miya and Nishida 2003; Matsumoto et al. 2007). These pathways independently activate the transcription of *Bra*.

The 434-bp upstream region of the *C. intestinalis Bra* gene activated reporter gene expression in the notochord cells (Corbo et al. 1997). This upstream enhancer drove detectable expression

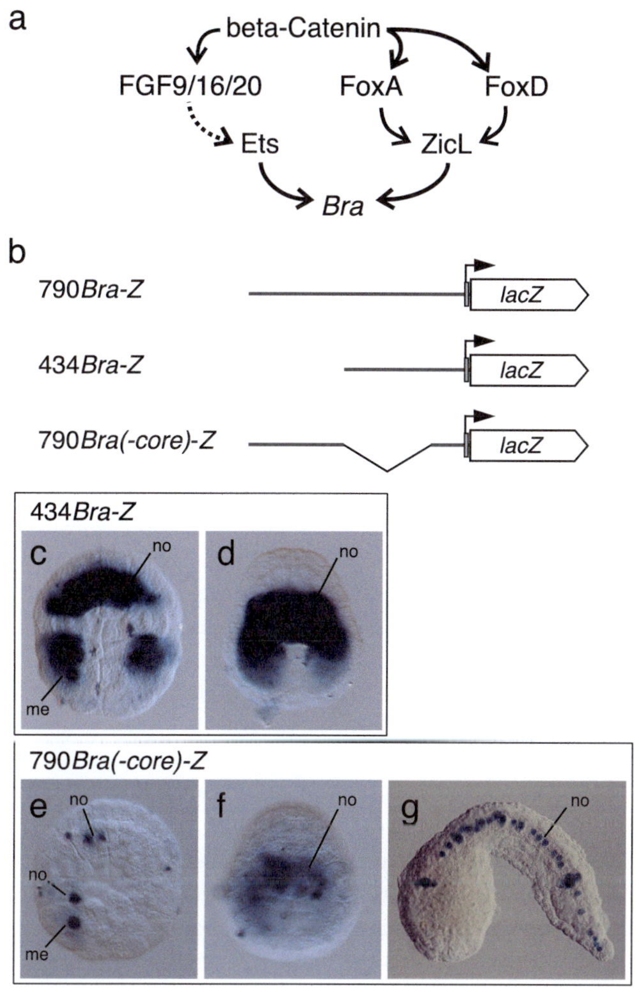

Fig. 7.4 Activation of upstream enhancers of the *C. intestinalis Bra* gene. (**a**) Redundant pathways activating the transcription of *Bra* (Imai et al. 2006). *Dashed line* indicates multiple steps of the mitogen-activated protein kinase pathway. (**b**) The 790-bp 5′ flanking region was placed upstream of the *lacZ* translated region. The construct contained the 5′ UTR and 50-bp coding region of *Bra*. This construct was named 790*Bra-Z*. The construct, 434*Bra-Z*, contained 434-bp proximal enhancer. The region between −434 ~ −61 was deleted from 790*Bra-Z*. This construct was named 790*Bra(-core)-Z*. Corbo et al. (1997) constructed 790*Bra-Z* and 434*Bra-Z*, which were re-named here for this review. (**c**) 434*Bra-Z* was activated in the notochord lineage blastomeres (*no*) at the early gastrula stage. Expression of *lacZ* mRNA was also observed in the mesenchyme lineage cells (*me*). (**d**) 434*Bra-Z* was strongly expressed at the middle gastrula stage. (**e**) 790*Bra(-core)-Z* was weakly activated in the notochord lineage (*no*) and mesenchyme lineage (*me*). (**f**) The expression of 790*Bra(-core)-Z* was also observed at the middle gastrula stage. (**g**) In this tailbud embryo, the expression of 790*Bra(-core)-Z* was visualized by enzymatic histochemical staining of the β-galactosidase activity. The expression was observed in the notochord (*no*)

of *lacZ* mRNA in gastrulating embryos (Fig. 7.4b–d). The 434-bp upstream sequence contained the binding sites for ZicL, Su(H), and Ets (Yagi et al. 2004; Farley et al. 2015). When the region −434 ~ −61 was deleted from the 790-bp upstream sequence, the resultant upstream enhancer activated *lacZ* expression in the notochord lineage cells (Fig. 7.4b, e–g) (Farley et al. 2015). The distal region (−790 ~ −435) also contained the binding sites for ZicL and Ets; therefore, the distal enhancer is regarded as a shadow enhancer (Farley et al. 2015). In the 5′ flanking region of the *C. intestinalis Bra* gene, binding sites for Ets and ZicL are intermingled (Farley et al. 2015). On the other hand, three Ets-binding sites, two Fox-binding sites, and a single ZicL-binding site appeared to be separately located in the 5′ flanking region of the *H. roretzi Bra* gene (Matsumoto et al. 2007). In such a case, these binding sites may be independent, both functionally and structurally, and may be regarded as redundant enhancers.

7.3.2 Redundant Enhancers May Provide an Opportunity to Acquire New Expression Domains

Fibroblast growth factor is an important factor that induces mesoderm in amphibians (Amaya et al. 1991). The expression of *Bra* is activated by FGF in the amphibian ventral mesoderm (Smith et al. 1991). Latinkić et al. (1997) identified an upstream enhancer of *Xenopus laevis Brachyury* that was only activated in the ventral mesoderm, but not in the notochord. The results suggested that separate enhancer elements activated *Brachyury* expression in different mesodermal tissues. In zebrafish, Nodal activates the expression of *no tail* (a zebrafish homolog of *Bra*), mainly in the dorsal mesoderm, including the notochord (Harvey et al. 2010). On the other hand, bone morphogenetic proteins (BMPs) and Wnt activate *no tail* in the ventral mesoderm (Harvey et al. 2010). Two separate enhancer elements mediate the inputs from the Nodal and BMP/Wnt signaling pathways (Harvey et al. 2010).

Pan-mesodermal expression of the *Bra* ortholog is observed in amphioxus (Holland et al. 1995). Parsimonious speculation can assert that the ancestral *Bra* gene contained two separate enhancers. One was an FGF-responsive ventral mesoderm enhancer, and the other an FGF-independent notochord enhancer. During the evolution of ascidians, the role of FGF may have changed to induce the notochord. Thus, the FGF-responsive enhancer has become an additional notochord enhancer. The ascidian *Bra* contains two (at least partially) redundant notochord enhancer elements. Conversely, one of the redundant enhancers may acquire a new expression domain without compromising the original expression pattern. According to circumstances, redundant enhancers may provide a situation similar to a transitional state of neo-functionalization after the duplication of genomic DNA segments.

7.4 Lessons from the *Bra* Reporter Analyses (Technical Tips)

The reporter analysis appears as a powerful method to analyze the mechanism of transcriptional regulation; however, some weaknesses should be kept in mind. Widely used reporter genes, such as *lacZ* and *GFP*, are easily visualized by using their enzymatic activity or fluorescence. However, there is a time lag between the initial activation of transcription and the emergence of their protein activity. This could not be overlooked when using ascidian embryos, as their embryogenesis is very rapid. For instance, the expression of the *Bra* mRNA starts at the 64-cell stage, whereas we are only able to detect the β-galactosidase activity of *Bra* > *lacZ* constructs about 2~3 h later at the late gastrula or early neurula stage (Fig. 7.5c) (Corbo et al. 1997). During this period, the notochord lineage cells divide two or three times. We can overcome the time lag problem by detecting the reporter gene expression by in situ hybridization. For instance, we can detect *lacZ* mRNA, driven by the *Bra* enhancer, at the 64-cell stage (Fig. 7.5b). With this early detection, we can be confident that the proximal *Bra* enhancer is not a maintenance element, but is responsible for initial activation.

The detection of mRNA should still be carried out carefully because the *lacZ* or *GFP* mRNAs are not subject to post-transcriptional regulations, as are endogenous genes. The *C. intestinalis Bra* mRNA is expressed in B7.3 blastomeres at the 64-cell stage (Fig. 7.5b) (Corbo et al. 1997). B7.3 undergoes asymmetric cell division to produce B8.6 (notochord-lineage) and B8.5 (mesenchyme-lineage) daughter cells. The *Bra*

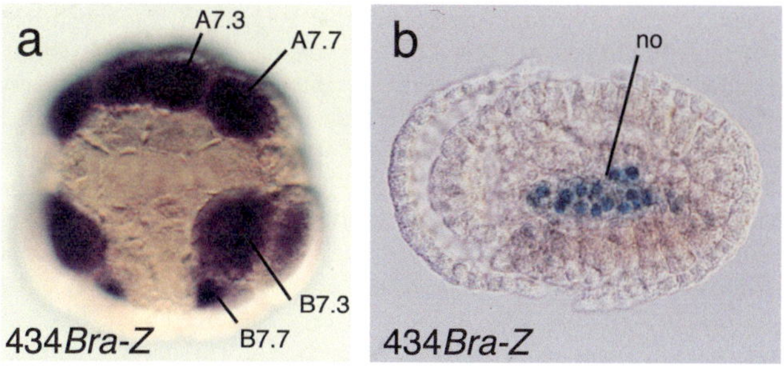

Fig. 7.5 Initial activation of 434*Bra-Z*; early detection of *lacZ* mRNA and late emergence of the β-galactosidase activity. (**a**) Expression of *lacZ* mRNA in the 64-cell embryo, carrying 434*Bra-Z*. The embryo is oriented to show the vegetal hemisphere (future dorsal side), with the anterior to the top. As the ascidian embryos are bilaterally symmetrical, names of blastomeres are only indicated on the right side. The *lacZ* mRNA was observed in A7.3, A7.7, B7.3, and B7.7 blastomeres. A7.3 and A7.7 have already been determined to give rise to the notochord. The developmental fate of B7.3 is to be determined after the next cell division. B7.3 expressed both endogenous *Bra* and 434*Bra-Z*. The expression of the *Bra* mRNA was detectable only in the notochord lineage daughter cell (B8.6). In contrast, the *lacZ* mRNA was observed in many (if not all) of the daughter cells. B7.7 has already been determined to give rise to the mesenchyme. 434*Bra-Z* was expressed in B7.7, unlike endogenous *Bra* mRNA. This was because the vector contains a sequence that mimics mesenchyme enhancers (Robert W. Zeller, personal communications). (**b**) The neurula, carrying 434*Bra-Z*. The activity of β-galactosidase was observed in the differentiating notochord cells. The enzymatic activity could not be detected until the late gastrula stage (Corbo et al. 1997)

mRNA is observed only in B8.6. There are several possibilities to explain how this occurs. (1) The *Bra* mRNA produced in B7.3 may be selectively degraded in B8.5. (2) The mRNA may be exclusively inherited by B8.6 during cell division. (3) The *Bra* mRNA has an extremely short half-life and soon disappears unless constantly produced. In this case, *Bra* mRNA is detectable only in B8.6, provided that only B8.6 continues to express *Bra*. The mRNA of *lacZ* can hardly recapitulate any of these regulations. Therefore, the *Bra* reporter genes are also expressed in the mesenchyme lineage blastomeres (Fig. 7.4c, e and Fig. 7.5a) (Corbo et al. 1997). Some aspects of the post-transcriptional regulations may be recapitulated by adding the 3′ untranslated region of the gene of interest to the translated region of reporter genes.

7.5 Conclusions

Reporter analysis is a powerful method for understanding the *cis*-regulatory network during embryogenesis. This method also allows cross-taxa experimental approaches to be made to empirically test for the conservation of *cis*-regulatory regions during the evolution of species, even from distant ones such as those between larvaceans and ascidians, which show poor sequence conservation in regulatory regions. Although developmental gene expression is regulated by diverse extracellular signals and maternally inherited factors, most of these inputs are concentrated on the enhancer elements. Reporter analyses revealed redundant enhancers that activate tissue-specific expression of developmental regulatory genes. Redundant enhancers are activated by a distinct set of transcription factors, but activate the transcription of genes in the same tissue. Redundant enhancers confer robustness on the *cis*-regulatory system, as mutation in either one of the enhancers or loss-of-function of one of the transcription factors does not severely affect normal gene expression. This provides a plausible explanation for the so-called "inverse paradox" of EvoDevo, in which apparently similar structures (phenotypic unity) show significant differences in the underlying genetic mechanisms that regulate their development (genetic diversity) (Cañestro et al. 2007). In addition, one of the redundant enhancers may also change where and/or when the regulated gene is activated, thereby contributing to the evolution of transcriptional regulatory networks.

Acknowledgments We thank Chikako Imaizumi, Reiko Yoshida, Yutaka Satou, Megumi Koutsuka, and Kazuko Hirayama (NBRP) for the animals. As this review includes our unpublished results, we thank members of our group, Miyuki Kanda, Minami Tagawa, and Adriana Rodriguez-Marí. This work was supported by a Grant-in-Aid for Scientific Research from the Japan Society for the Promotion of Science to SF, and by the grant BFU2016-80601-P from the Ministerio de Economía y Competitividad (Spain) and SGR2014-290 from Generalitat de Catalunya to CC. Our collaborative research was supported by the Heiwa Nakajima Foundation. We also thank Zenji Imoto, Kouki Tanaka, and other members of the Usa Marine Biological Institute of Kochi University for maintenance of the aquarium.

References

Albalat R, Cañestro C (2009) Identification of Aldh1a, Cyp26 and RAR orthologs in protostomes pushes back the retinoic acid genetic machinery in evolutionary time to the bilaterian ancestor. Chem Biol Interact 178:188–196

Albalat R, Cañestro C (2016) Evolution by gene loss. Nat Rev Genet 17:379–391

Amaya E, Musci TJ, Kirschner MW (1991) Expression of a dominant negative mutant of the FGF receptor disrupts mesoderm formation in *Xenopus* embryos. Cell 66:257–270

Barolo S (2011) Shadow enhancers: frequently asked questions about distributed *cis*-regulatory information and enhancer redundancy. BioEssays 34:135–141

Cañestro C, Postlethwait JH (2007) Development of a chordate anterior–posterior axis without classical retinoic acid signaling. Dev Biol 305:522–538

Cañestro C, Bassham S, Postlethwait JH (2005) Development of the central nervous system in the larvacean *Oikopleura dioica* and the evolution of the chordate brain. Dev Biol 285:298–315

Cañestro C, Postlethwait JH, Gonzàlez-Duarte R, Albalat R (2006) Is retinoic acid genetic machinery a chordate innovation? Evol Dev 8:394–406

Cañestro C, Yokoi H, Postlethwait JH (2007) Evolutionary developmental biology and genomics. Nat Rev Genet 8:932–942

Cañestro C, Bassham S, Postlethwait JH (2008) Evolution of the thyroid: anterior-posterior regionalization of the *Oikopleura* endostyle revealed by *Otx*, *Pax2/5/8*, and *Hox1* expression. Dev Dyn 237:1490–1499

Corbo JC, Levine M, Zeller RW (1997) Characterization of a notochord-specific enhancer from the *Brachyury* promoter region of the ascidian, *Ciona intestinalis*. Development 124:589–602

Corbo JC, Fujiwara S, Levine M, Di Gregorio A (1998) Suppressor of hairless activates *Brachyury* expression in the *Ciona* embryo. Dev Biol 203:358–368

Dehal P, Satou Y, Campbell RK et al (2002) The draft genome of *Ciona intestinalis*: insights into chordate and vertebrate origins. Science 298:2157–2167

Delsuc F, Brinkmann H, Chourrout D, Philippe H (2006) Tunicates and not cephalochordates are the closest living relatives of vertebrates. Nature 439:965–968

Farley EK, Olson KM, Zhang W, Rokhsar DS, Levine MS (2015) Syntax compensates for poor binding sites to encode tissue specificity of developmental enhancers. Proc Natl Acad Sci U S A 113:6508–6513

Harvey SA, Tümpel S, Dubrulle J, Schier AF, Smith JC (2010) *No tail* integrates two modes of mesoderm induction. Development 137:1127–1135

Herrmann BG, Kispert A (1994) The T genes in embryogenesis. Trends Genet 10:280–286

Holland PWH, Koschorz B, Holland LZ, Herrmann BG (1995) Conservation of *Brachyury* (*T*) genes in amphioxus and vertebrates: developmental and evolutionary implications. Development 121:4283–4291

Hong JW, Hendrix DA, Levine MS (2008) Shadow enhancers as a source of evolutionary novelty. Science 321:1314

Hozumi A, Yoshida R, Horie T, Sakuma T, Yamamoto T, Sasakura Y (2013) Enhancer activity sensitive to the orientation of the gene it regulates in the chordate genome. Dev Biol 375:79–91

Imai KS, Satoh N, Satou Y (2002a) Early embryonic expression of *FGF4/6/9* gene and its role in the induction of mesenchyme and notochord in *Ciona savignyi* embryos. Development 129:1729–1738

Imai KS, Satoh N, Satou Y (2002b) An essential role of a *FoxD* gene in notochord induction in *Ciona* embryos. Development 129:3441–3453

Imai KS, Satou Y, Satoh N (2002c) Multiple functions of a *Zic-like* gene in the differentiation of notochord, central nervous system and muscle in *Ciona savignyi* embryos. Development 129:2723–2732

Imai KS, Levine M, Satoh N, Satou Y (2006) Regulatory blueprint for a chordate embryo. Science 312:1183–1187

Imai KS, Stolfi A, Levine M, Satou Y (2009) Gene regulatory networks underlying the compartmentalization of the *Ciona* central nervous system. Development 136:285–293

Ip YT, Park RE, Kosman D, Bier E, Levine M (1992) The dorsal gradient morphogen regulates stripes of rhomboid expression in the presumptive neuroectoderm of the *Drosophila* embryo. Genes Dev 6:1728–1739

Ishibashi T, Nakazawa M, Ono H, Satoh N, Gojobori T, Fujiwara S (2003) Microarray analysis of embryonic retinoic acid target genes in the ascidian *Ciona intestinalis*. Develop Growth Differ 45:249–259

Ishibashi T, Usami T, Fujie M, Azumi K, Satoh N, Fujiwra S (2005) Oligonucleotide-based microarray analysis of retinoic acid target genes in the protochordate, *Ciona intestinalis*. Dev Dyn 233:1571–1578

Kanda M, Wada H, Fujiwara S (2009) Epidermal expression of *Hox1* is directly activated by retinoic acid in the *Ciona intestinalis* embryo. Dev Biol 335:454–463

Kanda M, Ikeda T, Fujiwara S (2013) Identification of a retinoic acid-responsive neural enhancer in the *Ciona intestinalis Hox1* gene. Develop Growth Differ 55:260–269

Kumano G, Yamaguchi S, Nishida H (2006) Overlapping expression of FoxA and Zic confers responsiveness to FGF signaling to specify notochord in ascidian embryos. Dev Biol 300:780–784

Latinkić BV, Umbhauer M, Neal KA, Lerchner W, Smith JC, Cunliffe V (1997) The *Xenopus Brachyury* promoter is activated by FGF and low concentrations of activin and suppressed by high concentrations of activin and by paired-type homeodomain proteins. Genes Dev 11:3265–3276

Mangelsdorf, D. J., Evans, R.M. (1992). Retinoid receptors as transcription factors. In "Transcriptional regulation" Ed by S. L. McKnight, K. R. Yamamoto. Cold Spring Harbor Laboratory Press, New York, pp 1137–1167

Martí-Solans J, Belyaeva OV, Torres-Aguila NP, Kedishvili NY, Albalat R, Cañestro C (2016) Co-elimination and survival in gene network evolution: dismantling the RA-signaling in a chordate. Mol Biol Evol 33:2401–2416

Matsumoto J, Kumano G, Nishida H (2007) Direct activation by Ets and Zic is required for initial expression of the *Brachyury* gene in the ascidian notochord. Dev Biol 306:870–882

Miya T, Nishida H (2003) An Ets transcription factor, HrEts, is target of FGF signaling and involved in induction of notochord, mesenchyme, and brain in ascidian embryos. Dev Biol 261:25–38

Mocikat R, Harloff C, Kütemeier G (1993) The effect of the rat immunoglobulin heavy-chain 3′ enhancer is position dependent. Gene 136:349–353

Nakatani Y, Yasuo H, Satoh N, Nishida H (1996) Basic fibroblast growth factor induces notochord formation and the expression of *As-T*, a *Brachyury* homolog, during ascidian embryogenesis. Development 122:2023–2031

Natale A, Sims C, Chiusano ML, Amoroso A, D'Aniello E, Fucci L, Krumlauf R, Branno, M, Locascio A (2011) Evolution of anterior *Hox* regulatory elements among chordates. BMC Evol Biol 11:330

Perry MW, Boettiger AN, Bothma JP, Levine M (2010) Shadow enhancers foster robustness of *Drosophila* gastrulation. Curr Biol 20:1562–1567

Putnam NH, Butts T, Ferrier DEK et al (2008) The amphioxus genome and the evolution of the chordate karyotype. Nature 453:1064–1072

Sasakura Y, Kanda M, Ikeda T, Horie T, Kawai N, Ogura Y, Yoshida R, Hozumi A, Satoh N, Fujiwara S (2012) Retinoic acid-driven *Hox1* is required in the epidermis for forming the otic/atrial placodes during ascidian metamorphosis. Development 139:2156–2160

Schubert M, Yu JK, Holland ND, Escriva H, Laudet V, Holland LZ (2005) Retinoic acid signaling acts via *Hox1* to establish the posterior limit of the pharynx in the chordate amphioxus. Development 132:61–73

Schubert M, Holland ND, Laudet V, Holland LZ (2006) A retinoic acid-*Hox* hierarchy controls both anterior/posterior patterning and neuronal specification in the developing central nervous system of the cephalochordate amphioxus. Dev Biol 296:190–202

Shimauchi Y, Yasuo H, Satoh N (1997) Autonomy of ascidian fork head/HNF-3 gene expression. Mech Dev 69:143–154

Smith JC, Price BM, Green JB, Weigel D, Herrmann BG (1991) Expression of a *Xenopus* homolog of *Brachyury* (*T*) is an immediate-early response to mesoderm induction. Cell 67:79–87

Swamynathan SK, Piatigorsky J (2002) Orientation-dependent influence of an intergenic enhancer on the promoter activity of the divergently transcribed mouse *Shsp*/αB-crystallin and *Mkbp*/*HspB2* genes. J Biol Chem 277:49700–49706

Tsagkogeorga G, Turon X, Hopcroft RR, Tilak MK, Feldstein T, Shenkar N, Loya Y, Huchon D, Douzery EJP, Delsuc F (2009) An updated 18S rRNA phylogeny of tunicates based on mixture and secondary structure models. BMC Evol Biol 9:187

Yagi K, Satou Y, Satoh N (2004) A zinc finger transcription factor, ZicL, is a direct activator of *Brachyury* in the notochord specification of *Ciona intestinalis*. Development 131:1279–1288

Yasuo H, Satoh N (1993) Function of vertebrate *T* gene. Nature 364:582–583

Yoshida, K., Nakahata, A., Treen, N., Sakuma, T., Yamamoto, T., Sasakura, Y. (2017). *Hox*-mediated endodermal identity patterns the pharyngeal muscle formation in the chordate pharynx. Development 144:1629–1634. doi:https://doi.org/10.1242/dev.144436

Zeller RW (2004) Generation and use of transgenic ascidian embryos. Methods Cell Biol 74:713–730

Investigating Evolutionarily Conserved Molecular Mechanisms Controlling Gene Expression in the Notochord

8

Julie E. Maguire, Aakarsha Pandey, Yushi Wu, and Anna Di Gregorio

Abstract

Ascidian embryos have been employed as model systems for studies of developmental biology for well over a century, owing to their desirable blend of experimental advantages, which include their rapid development, traceable cell lineage, and evolutionarily conserved morphogenetic movements. Two decades ago, the development of a streamlined electroporation method drastically reduced the time and cost of transgenic experiments, and, along with the elucidation of the complete genomic sequences of several ascidian species, propelled these simple chordates to the forefront of the model organisms available for studies of regulation of gene expression. Numerous ascidian sequences with tissue-specific enhancer activity were isolated and rapidly characterized through systematic in vivo experiments that would require several weeks in most other model systems. These *cis*-regulatory sequences include a large collection of notochord enhancers, which have been used to visualize notochord development in vivo, to generate mutant phenotypes, and to knock down genes of interest. Moreover, their detailed characterization has allowed the reconstruction of different branches of the notochord gene regulatory network. This chapter describes how the use of transgenic techniques has rendered the ascidian *Ciona* a competitive model organism for studies of notochord development, evolution, and gene regulation.

Keywords

Ascidian · Brachyury · *Ciona* · *cis*-Regulatory Module · Electroporation · Enhancer · Notochord · T-Box · Tbx2/3 · Transcription factor

Abbreviations

bp	Base pair(s)
cDNA	Complementary DNA
ChIP	Chromatin immunoprecipitation
CRM	*cis*-regulatory module
DAPI	4′,6-diamidino-2-phenylindole
FACS	fluorescence-activated cell sorting
GFP	Green fluorescent protein
GRN	Gene regulatory network
NOCE	Notochord enhancer
OBS	Orphan binding site
P3H1	prolyl 3-hydroxylase1
shRNA	Short hairpin RNA

Julie E. Maguire, Aakarsha Pandey, and Yushi Wu contributed equally to this work.

J. E. Maguire · A. Pandey · Y. Wu
A. Di Gregorio (✉)
Department of Basic Science and Craniofacial Biology, New York University College of Dentistry, New York, NY, USA
e-mail: adg13@nyu.edu

Y. Sasakura (ed.), *Transgenic Ascidians*, Advances in Experimental Medicine and Biology 1029,
https://doi.org/10.1007/978-981-10-7545-2_8

8.1 Introduction

The notochord (in Latin, *chorda dorsalis*) is the chief distinctive feature of the phylum Chordata, a large division of deuterostomes comprising two subphyla of mostly marine animals, Tunicates and Cephalochordates, in addition to the Vertebrates subphylum, which includes humans (Fig. 8.1). Owing to the scarcity of interpretable fossil records, the molecular mechanisms underlying the appearance of the notochord during the evolutionary history of multicellular animals are still under investigation (Satoh et al. 2012). Fossil remnants from the Middle Cambrian (~510–495 million years ago) have allowed the tentative identification of a structure similar to the notochord in *Pikaia gracilens*, and suggested that this extinct organism might represent the earliest stem-group chordate identified thus far (Morris and Caron 2012). However, other studies question the function of the putative notochord of *Pikaia* and the phylogenetic position of this animal within chordates (Lacalli 2012; Mallatt and Holland 2013).

In extant chordates, the notochord is an axial structure of mesodermal origin that provides support and patterning signals to the developing embryo and functions as a cornerstone for the organization of its body plan. The notochord induces the regionalization of the neural tube, patterns the paraxially located mesoderm, and influences the morphogenesis of structures ranging from endodermal derivatives to blood vessels (Fig. 8.1) (Cleaver and Krieg 2001; Reese et al. 2004; Stemple 2005). Importantly, in vertebrates, the notochord is replaced during embryonic development by the vertebral column, and its remnants form the central-most regions of the intervertebral discs, or *nuclei pulposi* (Fig. 8.1c) (Lawson and Harfe 2015). In cephalochordates and tunicates, the notochord is not replaced by the vertebral column, hence these groups are collectively known as invertebrate chordates (Fig. 8.1a, b). In the cephalochordate amphioxus (Fig. 8.1a),

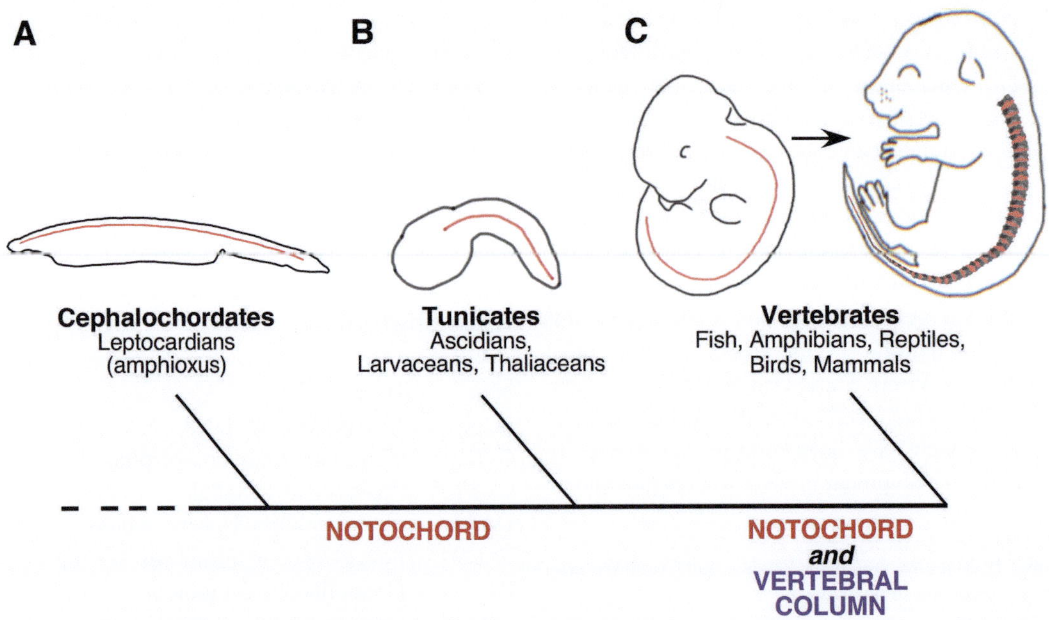

Fig. 8.1 The notochord in the three subphyla of chordates. Simplified drawings of representative animals from the three subphyla of the phylum Chordata. The notochord and notochord-derived structures are symbolized by *red lines*. (**a**) Lateral profile of the cephalochordate amphioxus (or lancelet). (**b**) Side view of an ascidian tailbud embryo; recent molecular phylogenies have indicated that tunicates are the closest living relatives of vertebrates (Delsuc et al. 2006). In (**a**) and (**b**) anterior is on the left. (**c**) Simplified drawings of developing mice at embryonic days ~E11.5 (*left*) and at ~E17 (*right*). As the backbone develops, the notochord regresses, and its remnants form the *nuclei pulposi* of the intervertebral discs (*right*)

the notochord extends to the anterior-most region of the body (e.g., Holland et al. 2004), whereas in tunicates, this structure is confined to the tail (hence the alternative name Urochordates used for this subphylum; Fig. 8.1b). Among invertebrate chordates, ascidians offer a number of experimental advantages that render them amenable to studies of notochord development. Most solitary ascidians are available and fertile year-round and can be easily fertilized in vitro. The resulting embryos develop a notochord within less than one day, and as they are relatively translucent, their notochord cells can be visualized without histological staining (Fig. 8.2).

The cell lineage of the notochord and all other larval tissues is nearly invariant and has been mapped for several cell divisions after fertilization (Conklin 1905; Ortolani 1954; Reverberi 1971; Nishida and Satoh 1983, 1985; Lemaire 2009). Moreover, tunicates feature the most compact chordate genomes. For example, the genome of the solitary ascidian *Ciona intestinalis* spans ~140 megabases (Dehal et al. 2002) and most tissue-specific enhancers, or *cis*-regulatory modules (CRMs), identified in this animal have been mapped within a few kilobases upstream of the genes that they control, or within their usually short introns (Passamaneck and Di Gregorio 2005; Stolfi and Christiaen 2012; Irvine 2013). Remarkably, orthologs of evolutionarily conserved regulators of notochord development, such as the transcription factors Brachyury and Foxa2, are present in single copy in *Ciona* (Corbo et al. 1997; Di Gregorio et al. 2001; Imai et al.

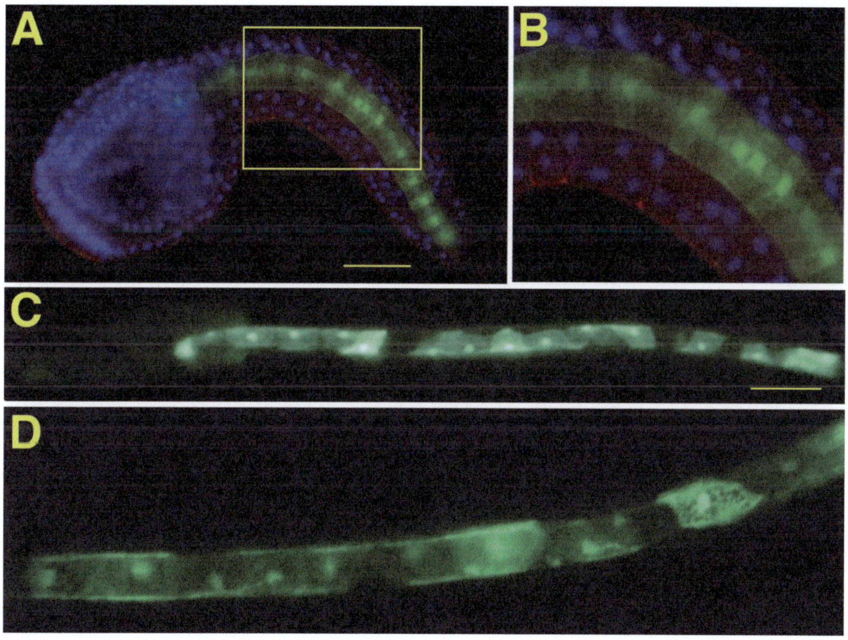

Fig. 8.2 The notochord in the ascidian *Ciona*. (**a**, **b**) *Ciona* embryo electroporated at the one-cell stage with the *Ci-Bra>green fluorescent protein* (*GFP*) plasmid, incubated until the mid-tailbud II stage (Hotta et al. 2007a), fixed, and stained with rhodamine phalloidin. The 40 definitive notochord cells express GFP; all nuclei are stained with 4′,6-diamidino-2-phenylindole (DAPI; *blue*) and the cell contours are highlighted in *red*. (**b**) High-magnification view of the region boxed in (**a**); the notochord cells have a characteristic stack-of-coins arrangement and their nuclei are centrally located. (**c**) *Ciona* embryo electroporated at the one-cell stage with the *Ci-Bra>GFP* plasmid, raised until the hatching larva stage (Hotta et al. 2007a) and fixed for imaging without counterstaining. Half of the 40 notochord cells are fluorescent, owing to mosaic incorporation of the transgene. (**d**) High-magnification view of the notochord of another *Ciona* embryo, electroporated and raised in parallel with the embryo in (**c**). The notochord cells appear considerably stretched and "wrapped" around the central lumen. Scale bars: ~100 μm

2004) and are part of relatively simplified gene regulatory networks (GRNs) (Imai et al. 2006; José-Edwards et al. 2013).

The notochord of most ascidian species is composed of 40 post-mitotic cells that stop dividing approximately around the end of neurulation, and form a definitive single-cell row in the center of the tail through convergent extension (Jiang and Smith 2007) (Fig. 8.2a). Tail elongation is achieved through extensive changes in the shape and dimensions of the 40 notochord cells (Fig. 8.2b, c). For comparison, the notochord of another tunicate, the larvacean *Oikopleura*, is initially composed of only 20 cells, but these cells continue to divide until they reach a final total of 120–160 by the third day after fertilization (Søviknes and Glover 2008). Remarkably, the ascidian species Molgula occulta and *Molgula tectiformis* are considered tailless, because they form only 20 and 10 definitive notochord cells respectively, which do not undergo convergent extension and are incapable of driving tail elongation (Takada et al. 2002).

Within the past two decades, the rapid and synchronized development of *Ciona* embryos and their amenability to transfection via a straightforward electroporation protocol have enabled the discovery of a multitude of notochord genes, in addition to live confocal imaging of notochord formation and functional studies of notochord genes. Transient transgenesis has allowed the identification of a surprising variety of *cis*-regulatory mechanisms controlling gene expression in the simple ascidian notochord. These findings, in turn, have ushered in the first comparative studies of notochord CRMs across chordates.

8.2 Identification of Novel Notochord Genes and Reconstruction of the Notochord Gene Regulatory Network

For her remarkable studies aimed at the creation of cell lineage maps, Ortolani relied upon her manual dexterity to label individual blastomeres of early ascidian embryos with minute grains of either charcoal or colored chalk, and used them to follow the localization of their daughter cells in embryonic tissues of transparent ascidians (Ortolani 1954). These studies expanded the research on cell lineage and fate determination that had been initiated by Conklin (1905) using the naturally pigmented muscle precursors of *Styela partita* to include nonpigmented ascidian embryos.

In the early 1980s, the intracellular microinjection of tracer enzymes, such as horseradish peroxidase, originally developed in leeches (Weisblat et al. 1978), was successfully used in ascidian embryos by Nishida and Satoh to determine accurate fate maps that extended to the early tailbud stage for most embryonic tissues (Nishida and Satoh 1983, 1985). The development of microinjection techniques paved the way for the first transgenic experiments in ascidians, which involved the microinjection of plasmids containing genomic fragments upstream of a muscle-specific *actin* fused to the *LacZ* reporter gene (Hikosaka et al. 1992, 1994). A few years later, the laboriousness of the microinjection techniques prompted the development of a simple and economical electroporation protocol. This method was first employed in *Ciona* for the identification and characterization of the notochord-specific *cis*-regulatory regions of *Ciona intestinalis Brachyury* (*Ci-Bra*), which encodes a transcription factor of the T-box family (Corbo et al. 1997) (Fig. 8.2) and of the *Ciona* ortholog of *forkhead/HNF-3beta/Foxa2* (*Ci-fkh/HNF-3b*, aka *Ci-FoxA.a*), which encodes a transcription factor of the forkhead/winged-helix family (Di Gregorio et al. 2001; Imai et al. 2004).

In ascidians and in other chordates, Ci-FoxA.a is expressed in notochord and in regions of the nervous system, endoderm, and additional territories, similar to its mouse counterpart *HNF-3beta* (Sasaki and Hogan 1993).

In contrast to most other chordates analyzed thus far, *Brachyury* is notochord-specific in the ascidians *Ciona* and *Halocynthia* (Corbo et al. 1997; Yasuo and Satoh 1993). This restricted notochord-specific expression of *Brachyury* provides ascidian embryos with the rare advantage of enabling studies focused on the notochord-specific function and transcriptional targets of this factor.

In an experiment aimed at the identification of *Ciona* notochord genes controlled by Ci-Bra, the *Ci-FoxA.a* promoter region, which is active early in multiple embryonic territories, including notochord, endoderm, and nervous system, was used to direct the ectopic expression of the notochord-specific Ci-Bra transcriptional activator in endodermal and neural precursors.

Transgenic embryos that ectopically express Ci-Bra in the endoderm and nervous system can be easily recognized because they display a reproducible phenotype consisting in the presence of a large mass of cells displaced to the ventral region of the tail from their normal destination. These embryos were collected alongside wild-type control embryos grown in parallel, and were subjected to RNA extraction followed by a subtractive hybridization screen (Takahashi et al. 1999; Fig. 8.3a). Whole-mount in-situ hybridizations were carried out for 501 complementary DNA (cDNA) clones, and a total of ~50 genes that are bona fide notochord transcriptional targets of Ci-Bra were identified (Fig. 8.3b–d); the function and regulation of several of these genes were analyzed in subsequent studies (Takahashi et al. 1999, 2010; Hotta et al. 1999, 2000, 2007b, 2008; Di Gregorio and Levine 1999; Dunn and Di Gregorio 2009; Katikala et al. 2013).

A similar strategy was employed to identify notochord genes controlled by Ci-Tbx2/3, the only T-box transcription factor reportedly expressed in the ascidian notochord other than Brachyury (Imai et al. 2004). In this case, a subtractive microarray screen was carried out between embryos expressing in their notochord a putative repressor form of Ci-Tbx2/3, consisting of its DNA-binding domain fused to green fluorescent protein (GFP; *Ci-Bra>Ci-Tbx2/3*^{DBD}*::GFP* (Fig. 8.3i) (José-Edwards et al. 2013). Control embryos were transfected with the *Ci-Bra>GFP* plasmid (Corbo et al. 1997). Approximately, 100–300 fluorescent transgenic embryos were selected under an epifluorescent microscope and subjected to RNA extraction, followed by hybridization to the *Ciona* Affymetrix GeneChip (Christiaen et al. 2008). Eighty-one putative Ci-Tbx2/3-downstream genes were identified, 20 of which (~29%) were expressed in the notochord, whereas others were expressed in other Ci-Tbx2/3 expression domains, including areas of the central nervous system (José-Edwards et al. 2013) (Fig. 8.3j–l). A few of these genes turned out to be shared targets of both Ci-Bra and Ci-Tbx2/3, including, in particular, *Ci-Noto4*, whose product is a protein containing a conserved phosphotyrosine binding domain required for notochord intercalation in *Ciona* embryos (Yamada et al. 2011). These results showed that in addition to regulating expression of a specific group of target genes, Ci-Tbx2/3 corroborates a crucial branch of the Ci-Bra-downstream GRN (José-Edwards et al. 2013).

Additional notochord transcription factors that had eluded previous searches were identified and provisionally positioned within the Ci-Bra-downstream gene battery through the use of a method previously developed for the isolation of heart precursors (Christiaen et al. 2008). This approach relies upon fluorescence-activated cell sorting (FACS) of the cells of interest for the identification of their specific transcriptomes, either through microarray screens, or more recently, through RNA sequencing. FACS-mediated isolation of fluorescent notochord cells was used in combination with microarray screens to identify novel notochord transcription factors, some of which are controlled either directly or indirectly by Ci-Bra (José-Edwards et al. 2011), in addition to numerous putative notochord genes (our unpublished results).

In a related approach, the cDNAs for several *Ciona* transcription factors, including Ci-Bra, were fused to the GFP coding region and cloned downstream of their respective promoters, with the goal of inducing the expression of GFP-tagged transcription factors in their native territories of activity. *En-masse* electroporations of these plasmids, followed by chromatin immunoprecipitation (ChIP) with an anti-GFP antibody and hybridization of the immunoprecipitated DNA to *Ciona* whole-genome microarrays (ChIP-chip), led to the identification of a very high number of genomic regions bound by each transcription factor at the 110-cell stage (Kubo

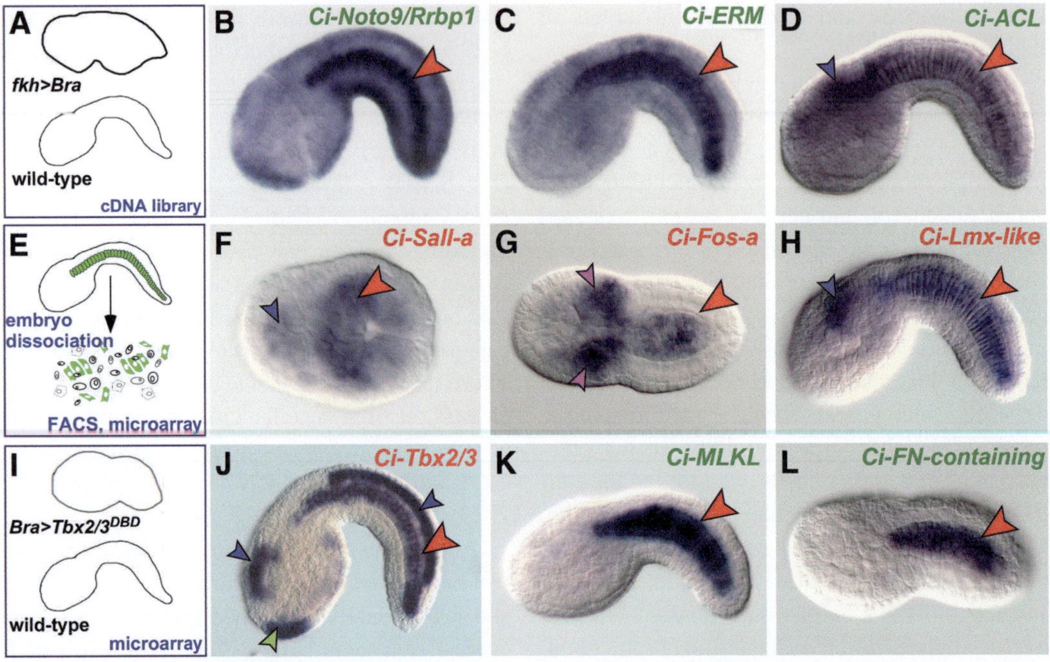

Fig. 8.3 Use of transgenesis for the identification of novel notochord genes in *Ciona*. (**a**, **e**, **i**) Schematic representations of transgenic experiments aimed at the identification of novel notochord genes in *Ciona*. (**b–d**, **h**, **k**, **l**) Whole-mount embryos at the mid-tailbud II stage hybridized in situ with digoxygenin-labeled antisense RNA probes directed against the genes annotated in the *top right* corner of each panel. (**f**, **g**, **j**) Whole-mount in-situ hybridizations of embryos at the late gastrula, initial tailbud I, and mid-tailbud/late tailbud stages respectively. (**a**) Subtractive hybridization between transgenic embryos ectopically expressing *Ci-Bra* under the control of the *Ci-FoxA.a* promoter region, which is active in notochord, neural, and endodermal cells (Takahashi et al. 1999), and wild-type control embryos. The *Ci-FoxA.a*>*Ci-Bra* plasmid is abbreviated as *fkh*>*Bra*. (**b–d**) Representative expression patterns of genes that were identified through the experiment summarized in (**a**). (**e**) Dissociation of transgenic embryos carrying the *Ci-Bra*>*GFP* plasmid, shown here at the mid-tailbud stage for simplicity, followed by fluorescence-activated sorting (FACS) of notochord cells, RNA extraction and subtractive microarray screens, in parallel with RNAs extracted from whole embryos. (**f–h**) Representative expression patterns of genes that were identified through the experiment summarized in (**e**) and encode sequence-specific notochord transcription factors (José-Edwards et al. 2011). (**i**) Subtractive microarray screen between transgenic embryos carrying the *Ci-Bra* > *Ci-Tbx2/3*^{DBD}*::GFP* plasmid, which expresses in the notochord a repressor form of Ci-Tbx2/3, and wild-type control embryos. (**j**) Whole-mount embryo hybridized in situ with a digoxygenin-labeled antisense RNA probe for *Ci-Tbx2/3*. (**k**, **l**) Two of the notochord-specific genes that were identified through the experiment are summarized in (**i**). Of note, these notochord-specific expression patterns, along with other tissue-specific patterns (not shown), suggest that Ci-Tbx2/3 might activate expression in non-notochord territories via tissue-specific co-activators (José-Edwards et al. 2013). *Red font*: genes encoding for transcription factors. *Arrowheads*: *red*, notochord; *blue*, nervous system; *purple*, mesenchyme; *green*, epidermis

et al. 2010). In particular, approximately 2,092 individual genes, including 194 transcription factors, were found to be occupied by the transgenic Ci-Bra-GFP protein in early embryos, and 3,653 were bound by Ci-FoxA.a-GFP (Kubo et al. 2010). Approximately 1,020 genes are shared between the two lists of putative targets, which suggests that these genes might be controlled synergistically by both transcription factors, as was previously suggested by the analysis of *Ci-tune* and related notochord CRMs (Passamaneck et al. 2009; José-Edwards et al. 2015) (discussed in Sect. 8.5.2).

The extensive studies of notochord genes in *Ciona* have successfully instructed related research in other chordates. A study of the expression patterns of mouse orthologs of *Ciona* notochord genes has shown that the expression of nine of these genes is conserved in the notochord cells of developing mouse embryos. These genes include the three mouse orthologs of the single-copy *Ciona leprecan/prolyl 3-hydroxylase 1* (*P3H1*) gene (Fig. 8.4), named *Leprecan/P3H1*, *Leprecan-like1/P3H2* and *Leprecan-like2/P3H3*. All three genes are expressed in notochord cells, although with different temporal onsets, and in various additional vertebrate-specific territories that are not present in ascidian embryos, such as portions of the vertebral cartilages (Capellini et al. 2008).

8.3 High-Resolution Imaging of Live Transgenic Notochord Cells

Owing to their natural translucency, most ascidian embryos are ideally suited to studies of notochord formation in vivo. In fact, notochord cells can be observed without any need for staining and/or sectioning, even though they are centrally located in the tail and are flanked by muscle cells and by the outermost epidermal cells. Moreover, the rapidity of ascidian embryogenesis allows the

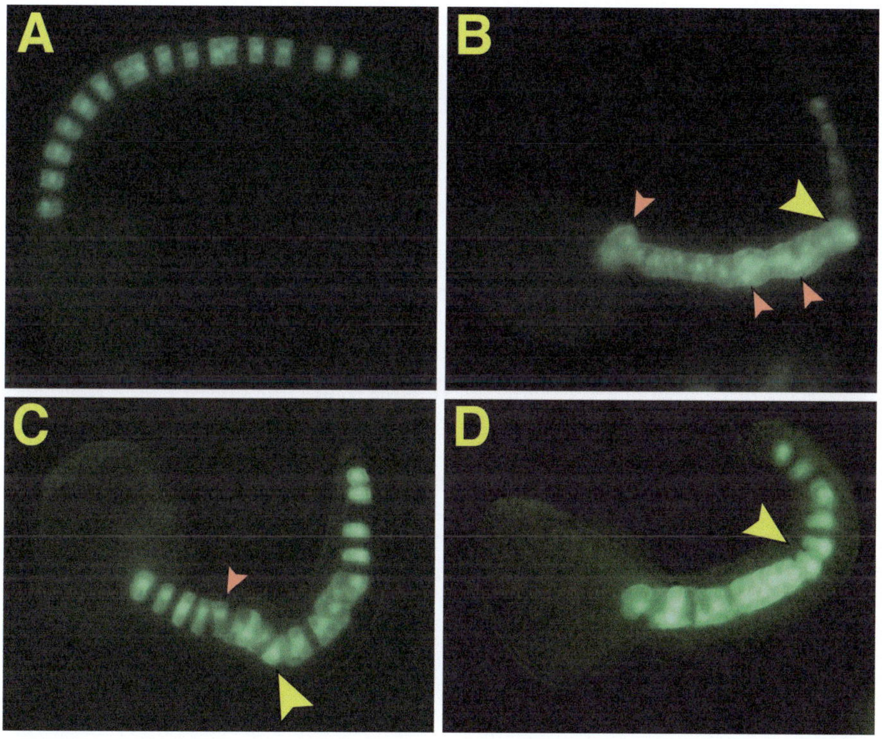

Fig. 8.4 CRISPR/Cas9-mediated knockdown of the *Ci-leprecan/P3H1* notochord gene. Specific oligonucleotide primers were designed in target sequences selected in exons one and two of the *Ci-leprecan/P3H1* coding region, through the use of the CRISPOR software (http://crispor.tefor.net), and were used for one-step overlap (OSO) PCR. The resulting PCR products were cloned into plasmids under the control of the U6 promoter (Stolfi et al. 2014). Two of these plasmids were co-electroporated with a plasmid that expresses *Cas9* in the notochord (*Ci-Bra> Cas9*), and with the *Ci-Bra>GFP* plasmid, to label notochord cells and monitor incorporation of the transgenes. For this experiment, 50 μg of each plasmid were employed. (**a**) Control embryo electroporated with 50 μg of the developmentally neutral *Ci-Bra>GFP* plasmid. Approximately half of the notochord cells display fluorescence, because of mosaic incorporation of the plasmid. (**b–d**) Embryos electroporated with the PCR products described above and the *Ci-Bra >GFP* plasmid. The shape of the notochord cells appears irregular (*small orange arrowheads*), and together with the loss of integrity of the notochordal sheath, which is predicted to be caused by the knockdown of *Ci-leprecan/P3H1*, is likely responsible for the reproducible bends in the tail (*yellow arrowheads*)

visualization of all the steps of notochord formation within less than one day, and to track morphogenetic movements as they unfold, through the use of routine differential interference contrast microscopy and time-lapse video recording. Time-lapse recording and bright-field microphotography were used in wild-type *Ciona* embryos to record morphogenesis and changes in cell shape that lead the six notochord precursors found in the 64-cell embryo to form a final rod-like structure composed of 40 post-mitotic cells (Miyamoto and Crowther 1985). In particular, these studies highlighted how the formation of cavities of increasing size between notochord cells and their progressive coalescence gradually give rise to a continuous lumen in the center of the tail (Fig. 8.2c, d; see also Figures in Chap. 15).

The lumen was described as a cavity surrounded by a basal lamina and a collagen-based notochordal sheath that is progressively built around the notochord through the extensive secretory activity of its cells (Cloney 1964). Indeed, subsequent studies proved that different *collagen* genes are expressed by notochord cells in *Ciona* (Wada et al., 2006). However, the low resolution of the imaging techniques and the lack of specific markers of cell boundaries led to the erroneous classification of the small cavities formed in between notochord cells as "intracellular vacuoles" (Miyamoto and Crowther 1985).

After the development of electroporation and the construction of the first plasmids able to induce expression of fluorescent proteins in the notochord, it was clarified through confocal imaging of fluorescent transgenic notochord cells that the "intracellular vacuoles" are rather "extracellular lumen pockets." These extracellular pockets form in between notochord cells and are sealed by tight junctions in the regions where the apical domains of adjacent notochord cells are juxtaposed (Denker et al. 2015).

The phases of lumen growth have been identified and quantified (Denker and Jiang 2012; Denker et al. 2013, 2015), and the identities of the main molecular players responsible for this process and for the elongation of notochord cells in the absence of cell division have been elucidated. In particular, using a combination of in vivo confocal studies and morphometric analyses, it has been shown that cortical actin and ezrin-radixin-moesin (ERM; Figs. 8.3c and 8.5b) are required for lumen formation, along with a microtubule network that forms at the apical cortex of the notochord cells (Dong et al. 2011), and that an equatorially positioned actomyosin ring constricts the notochord cells and promotes their elongation in the absence of cell division (Sehring et al. 2014). The actomyosin complex competes with anteriorly localized Prickle and other components of the planar cell polarity pathway for the repositioning of the cytoskeleton (Jiang et al. 2005; Newman-Smith et al. 2015; Sehring et al. 2015).

In parallel to these exquisitely detailed studies of individual notochord cells behaviors, the availability of different tissue-specific CRMs and confocal time–lapse recording allowed the visualization of notochord formation within its global embryonic context, through the labeling of muscle, notochord, and nervous system with multiple fluorophores (Rhee et al. 2005).

Last, the accurate visualization of developing notochord cells has been freed from mosaic incorporation of marker transgenes through the generation of stable transgenic *Ciona* lines. A stable transgenic line of *Ciona savignyi* was generated through the co-injection of the I-*Sce*I endonuclease and a plasmid containing a modified *Ci-Bra>GFP* sequence (Corbo et al. 1997) flanked by the I-*Sce*I recognition site (Deschet et al. 2003). This strategy induced the stable integration of multiple consecutive copies of the *Ci-Bra>GFP* transgene and the nonmosaic inheritance of the GFP labeling of notochord cells and their precursors. In turn, the pervasive and persistent fluorescence displayed by the notochord cells of stable *Ci-Bra>GFP* transgenics allowed the high-resolution visualization of the notochord defects that characterize animals heterozygous for the notochord mutation *chongmague* (*chm*) (Deschet et al. 2003). It was revealed through these analyses that embryos obtained by crossing animals from the *chm* mutant line with stable *Ci-Bra>GFP* transgenics do form the final 40 notochord cells; however, these cells fail to converge on the midline and to complete intercalation, because they continue to move and collide with each other (Deschet et al. 2003). A related transgenic line carrying the *Ci-Bra>GFP* construct was obtained in *Ciona intestinalis* through the use of the *Minos* transposon from *Drosophila hydei* and is publicly available for any future studies that might require accurate morphometric analyses of notochord development (Sasakura et al. 2010).

8.4 Functional Studies of Notochord Genes

Before the advent of transgenic techniques, functional studies of ascidian genes that required the overexpression or misexpression of genes of interest relied upon the microinjection of their corresponding in vitro synthesized messenger RNAs into eggs, zygotes or individual blastomeres. This technique has been perfected and has continued to yield several relevant results in the Japanese ascidian *Halocynthia roretzi*, where it was used, for example, to ectopically activate transcription of *Hr-FoxA*, the ortholog of *Ci-FoxA.a*, in cells of the animal region of the embryo. These studies demonstrated that Hr-FoxA is sufficient to direct ectopic expression of *Hr-Bra*, the *Brachyury* ortholog in this species (Kumano et al. 2006).

In parallel to these gain-of-function experiments, loss-of-function studies have been carried out by microinjecting morpholino oligonucleotides into fertilized eggs, in *Halocynthia* and in *Ciona*, even though the latter is characterized by smaller eggs that are more difficult to handle. A large-scale morpholino screen in *Ciona* led to the inactivation of numerous transcription factors and signaling molecules and shed light on the structure of the main tissue-specific GRNs that orchestrate development in these embryos (Imai et al. 2006). These laborious and time-consuming studies have been partly replaced by the development of plasmids able to direct notochord expression of chimeric or mutant proteins that can be simply introduced into a very large number of zygotes via electroporation.

An early example of a plasmid able to induce a gain-of-function phenotype is the fusion of the *Ci-FoxA.a* promoter region and the *Ci-Bra* coding region (Takahashi et al. 1999). As described in Sect. 8.2, electroporation of this plasmid in *Ciona* eggs elicited a dramatic rearrangement of the body plan, whereby endodermal and neural precursors that ectopically expressed Ci-Bra adopted a notochord-like phenotype (Takahashi et al. 1999; Fig. 8.3a). This suggested that Ci-Bra might play a major role in cell movements, similar to its counterparts from other chordates and from nonchordate animals (Nibu et al. 2013).

Conversely, electroporation of a plasmid that induced expression in *Ciona* notochord precursors of the *Xenopus bix* transcriptional repressor, which reportedly down-regulates *XBra* (Tada et al. 1998), caused a phenotype characterized by abnormally shaped notochord cells, a disorganized notochord and a very short tail, likely by causing the down-regulation of *Ci-Bra* (Di Gregorio et al. 2002). This phenotype is reminiscent of the loss of notochord identity observed in N-ethyl-N-nitrosourea--induced *Ci-Bra*$^{-/-}$ mutants (Chiba et al. 2009).

One experimental hindrance frequently encountered in these experiments is the mosaic incorporation of the plasmid(s) inducing expression of mutant forms of the proteins of interest. However, mosaic incorporation can be quite advantageous for functional studies, because it can induce the appearance of milder, intermediate phenotypes that are easier to interpret and can provide valuable information on the function of genes that are essential for notochord formation. For example, the expression of *Xenopus bix* in the notochord precursors of *Ciona*, caused by the *Ci-Bra>bix* plasmid, disrupts notochord formation; however, the mosaic incorporation of this plasmid allowed the formation of partial notochord fragments, and the development of a slightly longer tail (Di Gregorio et al. 2002). Similarly, the mosaic incorporation of a repressor form of Ci-Bra, obtained by fusing its DNA-binding domain with the Engrailed repression domain, enabled the detection of the down-regulation of the notochord transcription factor Ci-Fos-a in cells where Ci-Bra activity is reduced, and to use the wild-type notochord cells within the same embryos as controls. In turn, these results positioned *Ci-Fos-a* downstream of

Ci-Bra in the notochord GRN (José-Edwards et al. 2011).

Loss-of-function experiments have also been carried out through the electroporation of plasmids able to cause formation of short hairpin RNAs (shRNAs) aimed at interfering with the translation of a specific gene product (Nishiyama and Fujiwara 2008). A notochord gene that was knocked down through this method is *Ci-leprecan*, which encodes prolyl 3-hydroxylase1 (P3H1) (Hotta et al. 2000; Dunn and Di Gregorio 2009). The shRNA-mediated loss-of-function of *Ci-leprecan/P3H1* caused abnormal notochord formation and impaired tail elongation. In particular, the notochord displayed defects that ranged from the presence of one or more bends in the tail to a widened notochordal territory, whereby the notochord cells were misshapen and failed to intercalate (Dunn and Di Gregorio 2009). The function of *Ci-leprecan/P3H1* was also analyzed through the electroporation of a plasmid that induced expression of a truncated form of Ci-leprecan/P3H1 lacking the iron-binding region of its catalytic domain and therefore is presumably unable to modify collagen. The mutant Ci-leprecan/P3H1 protein likely competes with the endogenous wild-type protein by sequestering its interacting proteins, Ci-CRTAP and Ci-CYPB, which are also expressed in the *Ciona* notochord (Myllyharju and Kivirikko 1997; Dunn and Di Gregorio 2009). These experiments reproduced the phenotypes observed in the shRNA-mediated interference studies (Dunn and Di Gregorio 2009).

A related strategy has been successfully employed to interfere with the function of notochord transcription factors. In this case, a truncated form of a transcription factor of interest containing only its DNA-binding domain was expressed in the developing notochord; alternatively, the DNA-binding domain was either fused to the Engrailed repression domain, or to the *Drosophila* Hairy repression domain (Corbo et al. 1998). These mutant proteins compete with their respective endogenous counterparts by occupying their target binding sites without activating gene expression, or by repressing transcription. This approach was used, for example, to interfere with the function of Ci-Tbx2/3 and to identify the notochord genes whose expression was affected (José-Edwards et al. 2013) (Sect. 8.2, Fig. 8.3i, k, l).

The most recently developed tool for functional studies of notochord genes involves genome editing through the use of clustered regularly interspaced short palindromic repeats and the Cas9 endonuclease (CRISPR/Cas9) (Stolfi et al. 2014; Sasaki et al. 2014). The combined use of two single guide RNAs designed to target the first two exons of *Ci-leprecan/P3H1* produced different mutations, including a deletion that caused a truncation and a frame-shift in the predicted protein. Compared with control embryos (Fig. 8.4a), mutant *Ci-leprecan/P3H1* embryos display a slightly shorter tail, with one or more bends. This predominant mild phenotype indicates that most notochord cells intercalate properly, although, occasionally, misshapen cells fail to intercalate and cause the formation of kinks in the tail (Fig. 8.4b, c). These results are consistent with the effects of the shRNA-induced knockdown and with the phenotypes induced by the expression of the mutant form of Ci-leprecan/P3H1 in the notochord (Dunn and Di Gregorio 2009; Sect. 8.2). Together, these studies suggest the working model that a reduction in Ci-leprecan/P3H1 function impairs the post-translational modification of collagen molecules that compose the notochordal sheath, thus causing a reduction in its rigidity. Consequently, notochord cells that are even slightly misshapen can leave the rod-like structure and force the neighboring notochord cells to deviate (Fig. 8.4b, c) (Pandey and Di Gregorio, unpublished results).

The CRISPR/Cas9 method was also used to impair the function of another notochord gene, *Ciona fibronectin* (*Ci-FN1-containing*) (José-Edwards et al. 2013), recently renamed *Ci-Fn* in Segade et al. (2016). The results of the knockdown of this gene indicated its functional requirement for the intercalation of notochord precursors (Segade et al. 2016). However, in contrast to the *chm* mutants, in which *Ciona laminin alpha* is mutated (Veeman et al. 2008), in *Ci-Fn* mutant embryos the notochordal sheath appears unaffected, as the notochord cells that are not completely incorporated in the definitive rod-like structure are unable to cross the boundaries of the notochord territory and do not invade adjacent tissues (Segade et al. 2016).

8.5 Identification of *cis*-Regulatory Mechanisms Controlling Gene Expression in the Notochord

Cis-regulatory modules (CRMs), or enhancers, contain crucial information required for appropriate spatiotemporal gene expression and morphogenesis (e.g., Levine 2010). Compared with protein-coding regions, however, these elements are usually difficult to identify, and their elusive nature has traditionally hindered their functional analysis, especially in the case of complex genomes, such as those of vertebrate animals. In addition to its initial identification, the functional characterization of an enhancer region requires additional in vivo experiments, such as mutational analyses, that can be very challenging, time-consuming, and expensive in most multicellular model organisms.

Within the compact genome of *Ciona*, most of the enhancer regions identified thus far have been found in the proximity of the coding regions that they control, either upstream of transcription units, or within their first introns (Irvine 2013; Katikala et al. 2013). This highly desirable genomic configuration, coupled with the rapid embryonic development and the ease of transgenesis, have greatly expedited the discovery of enhancers active in all larval tissues, and the elucidation of the minimal regulatory sequences required for their function (reviewed in Irvine 2013). In particular, nearly 40 notochord CRMs have been identified and thoroughly characterized, and their minimal functional sequences have been used to uncover the transcriptional activators controlling them (Corbo et al. 1997; Di Gregorio and Levine 1999; Anno et al. 2006; Christiaen et al. 2008; Dunn and Di Gregorio 2009; Passamaneck et al. 2009; Katikala et al. 2013; José-Edwards et al. 2013, 2015; Thompson and Di Gregorio 2015; Farley et al. 2016; Segade et al. 2016).

8.5.1 Temporal Regulation of Notochord Gene Expression by *Ciona* Brachyury

A subset of crucial transcriptional activators of notochord gene expression appears to be evolutionarily conserved across chordates and to be necessary for notochord formation. Among them is *Brachyury* (Greek for "short tail"), also known as "*T*" (for "tail"). The loss of *Brachyury* function in mouse embryos homozygous for a mutation in this locus causes severe defects in the formation of posterior mesoderm, hindgut, and allantois; in particular, the notochord of these mutants is described as "nearly completely absent" (Gluecksohn-Schoenheimer 1940). Mutations in *Brachyury* orthologs have been either identified or induced in different chordates, and they all severely impaired notochord formation (Smith 1999; Chiba et al. 2009; Nibu et al. 2013).

The impact of *Brachyury* mutations on notochord development can be unequivocally evaluated in ascidian embryos, because in the ascidian species analyzed thus far, this gene is notochord-specific (Yasuo and Satoh 1993; Corbo et al. 1997; Di Gregorio 2017). The product of *Brachyury* is a sequence-specific transcription factor that was first characterized in mouse and, through electrophoretic mobility assays, was found to bind a palindromic sequence (Kispert et al. 1995). Subsequent studies in *Xenopus* and *Ciona* demonstrated that Brachyury proteins can also recognize nonpalindromic half-sites with the generic sequence TNNCAC (Casey et al. 1998; Di Gregorio and Levine 1999).

The notochord-specific expression of *Ci-Bra* permitted the subtraction screen described above (Fig. 8.3a), which yielded numerous *bona fide* Ci-Bra-downstream notochord genes (Takahashi et al. 1999). These transcriptional targets of Ci-Bra are responsible for crucial steps of notochord development, such as cell division, convergent extension, and tubulogenesis (Hotta et al. 2008), and need to be deployed in a finely regulated temporal sequence during the ~14 h that elapse between the specification of the notochord precursors and the formation of the notochordal lumen. Interestingly, although many of these genes are expressed starting at early gastrula (early-onset notochord expression), others are detected around the neural plate stage (middle-onset expression), and a third group of these genes are not detectable in the notochord before the early tailbud stage (late-onset expression) (Hotta et al. 1999; Katikala et al. 2013). The molecular mechanisms responsible for these differences in the temporal read-out of Ci-Bra-downstream gene expression were investigated by taking advantage of the compact genome of *Ciona*.

A few notochord genes were selected as representatives of the early-, middle- and late-onset Ci-Bra targets, their notochord CRMs were identified, and the minimal sequences responsible for their activity were determined. In most cases, these sequences matched published Ci-Bra binding sites (Katikala et al. 2013). However, even within the same CRM, some of these putative binding sites turned out to be dispensable, whereas others were required for activity.

Further experiments indicated that notochord CRMs that require multiple functional Ci-Bra binding sites are associated with early-onset notochord genes, such as *Ci-thrombospondin 3A* (*Ci-thbs3A*), which encodes an evolutionarily conserved extracellular matrix glycoprotein that mediates cell adhesion and migration (Katikala et al. 2013; Urry et al. 1998) (Fig. 8.5a). Additional CRMs in this category are associated with *Ci-fibrillar collagen 2A1* (*CiFColl*), which in ascidians is a component of the notochordal sheath, and in vertebrates is co-opted to cartilage (Katikala et al. 2013; Cloney 1964; Wada et al. 2006), and *Ci-ERM* (Figs. 8.3c and 8.5b), which is required in notochord cells for the acquisition of their characteristic stack-of-coins organization and for lumen formation (Katikala et al. 2013; Hotta et al. 2007b; Dong et al. 2011).

Notochord CRMs that rely upon individual functional Ci-Bra binding sites accompany middle-onset notochord genes, such as *Ci-Noto9/Rrbp1* (Figs. 8.3b and 8.5c). This gene is first detected in notochord cells by the neural plate/early neurula stage and encodes an evolutionarily conserved ribosome-binding protein that is also expressed in the notochord of *Xenopus* embryos (Katikala et al. 2013; Liu et al. 2016). In addition to the notochord CRMs directly bound by Ci-Bra, these studies identified notochord CRMs that are controlled by Ci-Bra indirectly, through a relay mechanism that involves Ci-Bra-downstream intermediary transcription factors. These CRMs are devoid of functional Ci-Bra binding sites and are associated with late-onset Ci-Bra target genes, which are first detected in notochord cells around the time of neurulation and include the ATP-citrate lyase, encoded by *Ci-ACL* (Figs. 8.3d and 8.5d), which is required for both the establishment of cell polarity along the medio-lateral axis and intercalation (Hotta et al. 2007b).

The *Ci-ACL* notochord CRM is controlled by Ci-Bra through a transcriptional activator of the homeodomain family (Fig. 8.5d), whereas another late-onset Ci-Bra target, *Ci-beta4GalT*, which encodes a beta1,4-galactosyltransferase, is controlled by Ci-Bra through a still unknown activator (Katikala et al. 2013).

Together, these results delineated a possible molecular mechanism that would ensure the appropriate timing of notochord gene expression and could be conserved in more complex chordates.

8.5.2 Synergistic Activation of Notochord CRMs by Ci-Bra and Ci-FoxA.a

In mouse and other vertebrate embryos, another major activator of notochord gene expression in addition to Brachyury is HNF-3beta/Foxa2, a member of the forkhead/winged-helix family of transcription factors (Friedman and Kaestner 2006).

Mice carrying a homozygous mutation in the *Foxa2* locus lack an organized node and fail to develop a notochord (Ang and Rossant 1994). Subsequent studies have identified numerous mouse notochord genes whose expression is influenced by Foxa2, in addition to seven Foxa2-bound mouse genomic regions that can activate reporter gene expression in the zebrafish notochord (Tamplin et al. 2011). The *Ciona* counterpart of Foxa2, *Ci-FoxA.a*, similar to its vertebrate

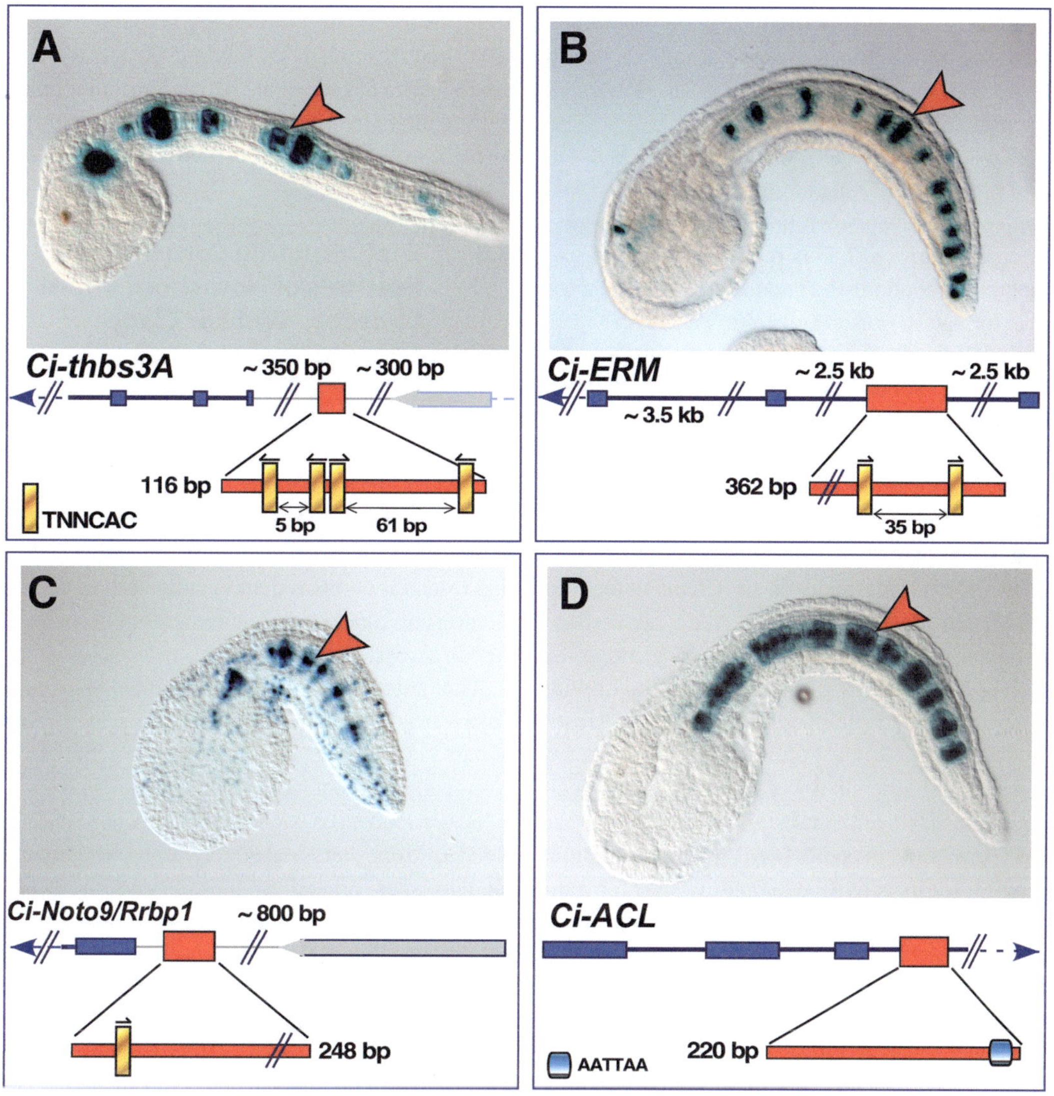

Fig. 8.5 Genomic location, organization, and activity of notochord *cis*-regulatory modules (CRMs) associated with Ci-Bra-downstream genes. *Ciona* embryos electroporated at the one-cell stage with plasmids containing the genomic regions symbolized by *red boxes* in the schematics below each microphotograph. Each of these regions was cloned upstream of the *Ci-FoxA.a* basal promoter fused to the *LacZ* reporter gene (Oda-Ishii and Di Gregorio 2007) (not depicted) and was found to act as a notochord CRM. All the respective genes are *bona fide* Ci-Bra-downstream transcriptional targets. (**a**) Among these notochord CRMs, *Ci-thrombospondin 3A* (*Ci-thbs3A*) contains multiple Ci-Bra binding sites (*yellow vertical bars* in the magnified region), with generic consensus sequence TNNCAC. (**b**) The 362-base pair (bp) *Ci-ezrin-radixin-moesin* (*Ci-ERM*) notochord CRM contains two functional Ci-Bra binding sites, both required for its activity. (**c**) The 248-bp long *Ci-Noto9* notochord CRM relies upon an individual functional Ci-Bra binding site, and its corresponding gene is activated later during notochord development (middle-onset), compared with the early-onset genes *Ci-thbs3A* and *Ci-ERM* (Katikala et al. 2013). (**d**) Last, the 220-bp *Ci-ACL* notochord CRM is devoid of canonical Ci-Bra binding sites and relies upon a sequence that resembles a binding site for a homeodomain transcription factor (*blue square*). The corresponding gene, *Ci-ACL*, is a late-onset Ci-Bra-downstream notochord gene (Katikala et al. 2013). *Red arrowheads*: representative notochord cells

counterparts, is expressed in a wide embryonic territory, which encompasses notochord, neural tube, and endoderm (Jeffery et al. 1998; Di Gregorio et al. 2001).

Knockdown experiments carried out in *Molgula oculata* using antisense oligodeoxyribonucleotides have shown that one of the *Molgula Fox* genes, *MocuFH1*, is required for the movements of notochord and endodermal cells, and for axis formation (Olsen and Jeffery 1997).

In *Ciona*, morpholino oligonucleotide-mediated knockdown of *Ci-FoxA.a* indicated that this transcription factor controls expression of numerous genes, including *Ci-Bra* (Imai et al. 2006), and that in early embryos it occupies, among numerous others, the genomic loci of 245 target genes encoding transcription factors (Kubo et al. 2010). The analysis of *Ciona* notochord CRMs has indicated that Ci-FoxA.a can activate notochord gene expression by acting alone, as well as by synergizing with either Ci-Bra or with other unrelated transcription factors (José-Edwards et al., 2015). A notochord CRM located upstream of one of the *Ci-ZicL* genes, which encode zinc-finger transcription factors that are also involved in notochord gene expression, requires for its activity two binding sites for transcription factors of the Fox family (Anno et al. 2006), whereas a single Fox binding site is required for the activity of the *Ci-quaking* notochord CRM (José-Edwards et al. 2015, and our unpublished results).

In addition to working independently, Ci-FoxA.a can synergize with Ci-Bra and activate a subset of notochord CRMs that are equally dependent on binding sites for both transcription factors. Thus far, three examples of Ci-Bra/Ci-FoxA.a-dependent CRMs have been identified; *Ci-tune*, which encodes an ascidian-specific protein of unknown function (Passamaneck et al. 2009), *Ci-CRM24*, which is located upstream of the notochord gene *discoidin domain receptor 1* (José-Edwards et al. 2015), and *Ci-CRM96*, which is associated with *Ci-pavarotti-like*, whose product is a kinesin-like protein that is still uncharacterized in ascidians, while in flies is required for proper formation of the mitotic spindle (José-Edwards et al. 2015; Adams et al. 1998). Additionally, the *Ci-FN* notochord CRM seems to rely upon a Fox binding site and a T-box binding site (Segade et al. 2016). This latter binding site might be used by Ci-Tbx2/3, as suggested by microarray results (José-Edwards et al. 2013).

8.6 Evolutionarily Conserved Features of Notochord CRMs: Chordate-Wide or Clade-Specific Mechanisms?

The availability of a large number of fully characterized *Ciona* notochord CRMs has prompted the first comparative study of their structural features, and the provisional categorization of these regulatory regions in *Ciona*. More importantly, this research compared and contrasted the architectural and functional requirements of notochord CRMs across chordates.

The notochord functions of Brachyury and Foxa2, in addition to their binding sites, are evolutionarily conserved across chordates. Similarly, despite the frequent lack of conservation in overall enhancer sequences, some of the notochord CRMs isolated from vertebrates rely upon Brachyury binding sites related to those found in *Ciona* CRMs. Thus far, notochord CRMs requiring either one or two Brachyury binding sites, similar to most *Ciona* notochord CRMs, have been found in the zebrafish *Sonic hedgehog* (*Shh*) *ar-C* intronic notochord and floor plate enhancer (Müller et al. 1999) and in the *Xenopus eFGF* promoter region (Casey et al. 1998), respectively.

Of note, in *Ciona*, the *Ci-Ephrin3* notochord CRM requires, in addition to a functional Ci-Bra binding site, an $(AC)_6$ microsatellite sequence (José-Edwards et al. 2015). Interestingly, this unusual association and interdependence of a functional Brachyury binding site with a repetitive sequence, which has been revealed through the analysis of the *Ci-Ephrin3* notochord CRM, is paralleled by related findings in mice.

Studies carried out in mouse embryonic stem cells through ChIP-chip show that mouse Brachyury can bind $(AC)_{\geq 6}$ microsatellite repeats (Evans et al. 2012). In vivo testing of the predicted enhancer activity of these genomic regions is required to determine whether these regions possess *cis*-regulatory activity and whether the functional association of Brachyury and the $(AC)_n$ microsatellite sequence uncovered in *Ciona* is conserved throughout the chordate spectrum.

Functional Foxa2 binding sites seem to be more frequent in the notochord enhancers (NOCEs) characterized from mouse embryos; however, this is likely because they were identified in screens focused on Foxa2 target genes (Tamplin et al. 2011). Notochord CRMs associated with the mouse genes *Pkd1l1–1*, *Shh*, *Biccl–1*, and *Sox9* contain five, three, two, and one Foxa2 binding sites respectively (Tamplin et al. 2011; Jeong and Epstein 2003; Bagheri-Fam et al. 2006). In zebrafish, a group of notochord CRMs have been shown to necessitate, in addition to individual Foxa2 binding sites, a sequence motif, named "motif 2," located at variable distances from it (Rastegar et al. 2008). This configuration does not have a direct counterpart in *Ciona*; however, in one of the *Ciona* notochord CRMs, a Foxa2 binding site has been found to be working cooperatively with binding sites for transcription factors of the AP1 and homeodomain families (José-Edwards et al. 2015).

In addition to the CRMs that employ a shared repertoire of various configurations of generic T-box and Fox binding sites, other notochord CRMs seem to be requiring clade-specific transcription factors and their respective binding sites for their activity. In *Ciona*, at least three notochord CRMs rely for their activity upon minimal sequences related to binding sites for transcription factors of the Myb family, whereas one CRM requires a binding site for a basic helix-loop-helix (bHLH) transcription factor (José-Edwards et al. 2015). Furthermore, the notochord CRM of *Ci-Bra* requires for its function binding sites for the early activators ZicL (Yagi et al. 2004) and Ets (Matsumoto et al. 2007; Farley et al. 2016). Combinations of these binding sites have been found to be required for the activity of a notochord CRM associated with *Ci-Mnx* (Farley et al. 2016), which encodes an early notochord transcription factor of the homeodomain family (Imai et al. 2004), and for the activity of an additional notochord CRM located upstream of the main *Ci-Bra* enhancer/promoter region (Farley et al. 2016). Binding sites for transcription factors of the Zic and Ets families are yet to be reported as functional components required for the activation of notochord CRMs in vertebrates.

On the other hand, the presence of an orphan binding site (OBS) in the node and nascent NOCE of mouse *Noto*, a gene that encodes another evolutionarily conserved notochord transcriptional activator (Alten et al. 2012), indicates the existence of additional, still uncharacterized notochord activators. Together, the presence of OBS sequences in the mouse *Foxa2* and *Sox9* notochord CRMs and in four out of seven Foxa2-downstream notochord CRMs (Tamplin et al. 2011; Alten et al. 2012), together with the absence of canonical OBS sequences in minimal *Ciona* notochord CRMs, suggest that either the OBS-binding activator might be vertebrate-specific, or, alternatively, that an additional class of notochord CRMs might exist in *Ciona* and is yet to be found. In addition, another binding site present in the minimal *Noto* NOCE matches the consensus sequence bound by transcriptional activators of the Tead family. This result is in agreement with the previously described ability of Tead proteins to activate the *Foxa2* CRM, in cooperation with a still unknown factor (Sawada et al. 2005), and suggests that binding sites for Tead activators might be additional recurring components of mouse notochord CRMs. BLAST searches suggest that the *Ciona* gene most highly related to vertebrate *Tead* genes might be *Ci-scalloped/TEF1* (gene model KH.C14.426); however, this gene is reportedly expressed in neurons of the adhesive organ and in the primordium of the oral siphon, but is not detected in notochord cells (Imai et al. 2004).

Together, these results suggest that Tead transcription factors, alone or in synergy with additional activators, might be part of a vertebrate-specific molecular mechanism for the control of notochord gene expression that evolved after the separation of ascidians from the main chordate lineage. Alternatively, this and other seemingly vertebrate-specific strategies for the control of notochord gene expression could have been present in the common chordate ancestor and would have been selectively lost in ascidians.

As novel notochord CRMs continue to be identified in both ascidians and vertebrates, we expect the number of *cis*-regulatory strategies that are found to be conserved across divergent chordates to increase. In turn, the differences between notochord CRMs in ascidians and vertebrates should point out the clade-specific regulatory mechanisms responsible for the morphological and functional differences between the notochords of these animals, and ultimately shed light on the evolutionary origins of the backbone.

Acknowledgements Thanks to all present and past laboratory members and collaborators. We are particularly indebted to Drs. Diana José-Edwards, Lavanya Katikala, Izumi Oda-Ishii, and Yale Passamaneck for their original microphotographs. Research in our laboratory is supported by the National Institute of General Medical Sciences of the National Institutes of Health under award number R01GM100466.

References

Adams RR, Tavares AA, Salzberg A, Bellen HJ, Glover DM (1998) Pavarotti encodes a kinesin-like protein required to organize the central spindle and contractile ring for cytokinesis. Genes Dev 12(10):1483–1494

Alten L, Schuster-Gossler K, Eichenlaub MP, Wittbrodt B, Wittbrodt J, Gossler A (2012) A novel mammal-specific three partite enhancer element regulates node and notochord-specific Noto expression. PLoS One 7(10):e47785

Ang SL, Rossant J (1994) HNF-3 beta is essential for node and notochord formation in mouse development. Cell 78:561–574

Anno C, Satou A, Fujiwara S (2006) Transcriptional regulation of ZicL in the *Ciona intestinalis* embryo. Dev Genes Evol 216(10):597–605

Bagheri-Fam S, Barrionuevo F, Dohrmann U, Günther T, Schüle R, Kemler R, Mallo M, Kanzler B, Scherer G (2006) Long-range upstream and downstream enhancers control distinct subsets of the complex spatiotemporal Sox9 expression pattern. Dev Biol 291(2):382–397

Capellini TD, Dunn MP, Passamaneck YJ, Selleri L, Di Gregorio A (2008) Conservation of notochord gene expression across chordates: insights from the Leprecan gene family. Genesis 46(11):683–696

Casey ES, O'Reilly MA, Conlon FL, Smith JC (1998) The T-box transcription factor Brachyury regulates expression of eFGF through binding to a non-palindromic response element. Development 125(19):3887–3894

Chiba S, Jiang D, Satoh N, Smith WC (2009) Brachyury null mutant-induced defects in juvenile ascidian endodermal organs. Development 136(1):35–39

Christiaen L, Davidson B, Kawashima T, Powell W, Nolla H, Vranizan K, Levine M (2008) The transcription/migration interface in heart precursors of *Ciona intestinalis*. Science 320(5881):1349–1352

Cleaver O, Krieg PA (2001) Notochord patterning of the endoderm. Dev Biol 234(1):1–12

Cloney RA (1964) Development of the ascidian notochord. Acta Embryol Morphol Exp 7:111–130

Conklin EG (1905) The organization and cell-lineage of the ascidian egg. J Acad Nat Sci 13:1–119

Corbo JC, Levine M, Zeller RW (1997) Characterization of a notochord-specific enhancer from the *Brachyury* promoter region of the ascidian, *Ciona intestinalis*. Development 124(3):589–602

Corbo JC, Fujiwara S, Levine M, Di Gregorio A (1998) Suppressor of hairless activates brachyury expression in the *Ciona* embryo. Dev Biol 203(2):358–368

Dehal P, Satou Y, Campbell RK, Chapman J, Degnan B, De Tomaso A, Davidson B, Di Gregorio A et al (2002) The draft genome of *Ciona intestinalis*: insights into chordate and vertebrate origins. Science 298(5601):2157–2167

Delsuc F, Brinkmann H, Chourrout D, Philippe H (2006) Tunicates and not cephalochordates are the closest living relatives of vertebrates. Nature 439(7079):965–968

Denker E, Jiang D (2012) *Ciona intestinalis* notochord as a new model to investigate the cellular and molecular mechanisms of tubulogenesis. Semin Cell Dev Biol 23(3):308–319

Denker E, Bocina I, Jiang D (2013) Tubulogenesis in a simple cell cord requires the formation of bi-apical cells through two discrete par domains. Development 140(14):2985–2996

Denker E, Sehring IM, Dong B, Audisso J, Mathiesen B, Jiang D (2015) Regulation by a TGFβ-ROCK-actomyosin axis secures a non-linear lumen expansion that is essential for tubulogenesis. Development 142(9):1639–1650

Deschet K, Nakatani Y, Smith WC (2003) Generation of ci-Brachyury-GFP stable transgenic lines in the ascidian *Ciona savignyi*. Genesis 35(4):248–259

Di Gregorio A (2017) T-box genes and developmental gene regulatory networks in ascidians. Curr Top Dev Biol 122:55–91

Di Gregorio A, Levine M (1999) Regulation of *Ci-tropomyosin-like*, a Brachyury target gene in the ascidian, *Ciona intestinalis*. Development 126(24):5599–5609

Di Gregorio A, Corbo JC, Levine M (2001) The regulation of *forkhead/HNF-3beta* expression in the *Ciona* embryo. Dev Biol 229(1):31–43

Di Gregorio A, Harland RM, Levine M, Casey ES (2002) Tail morphogenesis in the ascidian, *Ciona intestinalis*, requires cooperation between notochord and muscle. Dev Biol 244(2):385–395

Dong B, Deng W, Jiang D (2011) Distinct cytoskeleton populations and extensive crosstalk control *Ciona* notochord tubulogenesis. Development 138(8):1631–1641

Dunn MP, Di Gregorio A (2009) The evolutionarily conserved *leprecan* gene: its regulation by Brachyury and its role in the developing *Ciona* notochord. Dev Biol 328(2):561–574

Evans AL, Faial T, Gilchrist MJ, Down T, Vallier L, Pedersen RA, Wardle FC, Smith JC (2012) Genomic targets of Brachyury (T) in differentiating mouse embryonic stem cells. PLoS One 7(3):e33346

Farley EK, Olson KM, Zhang W, Rokhsar DS, Levine MS (2016) Syntax compensates for poor binding sites to encode tissue specificity of developmental enhancers. Proc Natl Acad Sci U S A 113(23):6508–6513

Friedman JR, Kaestner KH (2006) The Foxa family of transcription factors in development and metabolism. Cell Mol Life Sci 63(19–20):2317–2328

Gluecksohn-Schoenheimer S (1940) The effect of an early lethal (t) in the house mouse. Genetics 25(4):391–400

Hikosaka A, Kusakabe T, Satoh N, Makabe KW (1992) Introduction and expression of recombinant genes in ascidian embryos. Develop Growth Differ 34:631–638

Hikosaka A, Kusakabe T, Satoh N (1994) Short upstream sequences associated with the muscle-specific expression of an actin gene in ascidian embryos. Dev Biol 166:763 769

Holland LZ, Laudet V, Schubert M (2004) The chordate amphioxus: an emerging model organism for developmental biology. Cell Mol Life Sci 61(18):2290–2308

Hotta K, Takahashi H, Erives A, Levine M, Satoh N (1999) Temporal expression patterns of 39 Brachyury-downstream genes associated with notochord formation in the *Ciona intestinalis* embryo. Develop Growth Differ 41(6):657–664

Hotta K, Takahashi H, Asakura T, Saitoh B, Takatori N, Satou Y, Satoh N (2000) Characterization of Brachyury-downstream notochord genes in the *Ciona intestinalis* embryo. Dev Biol 224(1):69–80

Hotta K, Mitsuhara K, Takahashi H, Inaba K, Oka K, Gojobori T, Ikeo K (2007a) A web-based interactive developmental table for the ascidian *Ciona intestinalis*, including 3D real-image embryo reconstructions: I. From fertilized egg to hatching larva. Dev Dyn 236(7):1790–1805

Hotta K, Yamada S, Ueno N, Satoh N, Takahashi H (2007b) Brachyury-downstream notochord genes and convergent extension in *Ciona intestinalis* embryos. Develop Growth Differ 49(5):373–382

Hotta K, Takahashi H, Satoh N, Gojobori T (2008) Brachyury-downstream gene sets in a chordate, *Ciona intestinalis*: integrating notochord specification, morphogenesis and chordate evolution. Evol Dev 10(1):37–51

Imai KS, Hino K, Yagi K, Satoh N, Satou Y (2004) Gene expression profiles of transcription factors and signaling molecules in the ascidian embryo: towards a comprehensive understanding of gene networks. Development 131(16):4047–4058

Imai KS, Levine M, Satoh N, Satou Y (2006) Regulatory blueprint for a chordate embryo. Science 312(5777):1183–1187

Irvine SQ (2013) Study of cis-regulatory elements in the Ascidian *Ciona intestinalis*. Curr Genomics 14(1):56–67

Jeffery WR, Ewing N, Machula J, Olsen CL, Swalla BJ (1998) Cytoskeletal actin genes function downstream of HNF-3beta in ascidian notochord development. Int J Dev Biol 42(8):1085–1092

Jeong Y, Epstein DJ (2003) Distinct regulators of Shh transcription in the floor plate and notochord indicate separate origins for these tissues in the mouse node. Development 130(16):3891–3902

Jiang D, Smith WC (2007) Ascidian notochord morphogenesis. Dev Dyn 236(7):1748–1757

Jiang D, Munro EM, Smith WC (2005) Ascidian prickle regulates both mediolateral and anterior-posterior cell polarity of notochord cells. Curr Biol 15(1):79–85

José-Edwards DS, Kerner P, Kugler JE, Deng W, Jiang D, Di Gregorio A (2011) The identification of transcription factors expressed in the notochord of *Ciona intestinalis* adds new potential players to the *brachyury* gene regulatory network. Dev Dyn 240(7):1793–1805

José-Edwards DS, Oda-Ishii I, Nibu Y, Di Gregorio A (2013) Tbx2/3 is an essential mediator within the Brachyury gene network during *Ciona* notochord development. Development 140(11):2422–2433

José-Edwards DS, Oda-Ishii I, Kugler JE, Passamaneck YJ, Katikala L, Nibu Y, Di Gregorio A (2015) Brachyury, Foxa2 and the *cis*-Regulatory Origins of the Notochord. PLoS Genet 11(12):e1005730

Katikala L, Aihara H, Passamaneck YJ, Gazdoiu S, José-Edwards DS, Kugler JE, Oda-Ishii I, Imai JH, Nibu Y, Di Gregorio A (2013) Functional Brachyury binding sites establish a temporal read-out of gene expression in the *Ciona* notochord. PLoS Biol 11(10):e1001697

Kispert A, Koschorz B, Herrmann BG (1995) The T protein encoded by *Brachyury* is a tissue-specific transcription factor. EMBO J 14(19):4763–4772

Kubo A, Suzuki N, Yuan X, Nakai K, Satoh N, Imai KS, Satou Y (2010) Genomic cis-regulatory networks in the early *Ciona intestinalis* embryo. Development 137(10):1613–1623

Kumano G, Yamaguchi S, Nishida H (2006) Overlapping expression of FoxA and Zic confers responsiveness to FGF signaling to specify notochord in ascidian embryos. Dev Biol 300(2):770–784

Lacalli T (2012) The Middle Cambrian fossil *Pikaia* and the evolution of chordate swimming. Evodevo 3(1):12

Lawson L, Harfe BD (2015) Notochord to nucleus pulposus transition. Curr Osteoporos Rep 13(5):336–341

Lemaire P (2009) Unfolding a chordate developmental program, one cell at a time: invariant cell lineages, short-range inductions and evolutionary plasticity in ascidians. Dev Biol 332(1):48–60

Levine M (2010) Transcriptional enhancers in animal development and evolution. Curr Biol 20(17):R754–R763

Liu GH, Mao CZ, Wu HY, Zhou DC, Xia JB, Kim SK, Cai DQ, Zhao H, Qi XF (2016) Expression profile of rrbp1 genes during embryonic development and in adult tissues of *Xenopus laevis*. Gene Expr Patterns 23–24:1–6

Mallatt J, Holland N (2013) *Pikaia gracilens* Walcott: stem chordate, or already specialized in the Cambrian? J Exp Zool B Mol Dev Evol 320:247–271

Matsumoto J, Kumano G, Nishida H (2007) Direct activation by Ets and Zic is required for initial expression of the *Brachyury* gene in the ascidian notochord. Dev Biol 306(2):870–882

Miyamoto DM, Crowther RJ (1985) Formation of the notochord in living ascidian embryos. J Embryol Exp Morphol 86:1–17

Morris SC, Caron JB (2012) *Pikaia gracilens* Walcott, a stem-group chordate from the middle Cambrian of British Columbia. Biol Rev Camb Philos Soc 87:480–512

Müller F, Chang B, Albert S, Fischer N, Tora L, Strähle U (1999) Intronic enhancers control expression of zebrafish sonic hedgehog in floor plate and notochord. Development 126(10):2103–2116

Myllyharju J, Kivirikko KI (1997) Characterization of the iron- and 2-oxoglutarate-binding sites of human prolyl 4-hydroxylase. EMBO J 16(6):1173–1180

Newman-Smith E, Kourakis MJ, Reeves W, Veeman M, Smith WC (2015) Reciprocal and dynamic polarization of planar cell polarity core components and myosin. Elife 13(4):e05361

Nibu Y, José-Edwards DS, Di Gregorio A (2013) From notochord formation to hereditary chordoma: the many roles of Brachyury. Biomed Res Int 2013:826435

Nishida H, Satoh N (1983) Cell lineage analysis in ascidian embryos by intracellular injection of a tracer enzyme. I. Up to the eight-cell stage. Dev Biol 99:382–394

Nishida H, Satoh N (1985) Cell lineage analysis in ascidian embryos by intracellular injection of a tracer enzyme. II. The 16- and 32-cell stages. Dev Biol 110:440–454

Nishiyama A, Fujiwara S (2008) RNA interference by expressing short hairpin RNA in the *Ciona intestinalis* embryo. Develop Growth Differ 50(6):521–529

Oda-Ishii I, Di Gregorio A (2007) Lineage-independent mosaic expression and regulation of the *Ciona multidom* gene in the ancestral notochord. Dev Dyn 236(7):1806–1819

Olsen CL, Jeffery WR (1997) A forkhead gene related to HNF-3beta is required for gastrulation and axis formation in the ascidian embryo. Development 124(18):3609–3619

Ortolani G (1954) Risultati definitive sulla distribuzione dei territory presuntivi degli organi nel germe di Ascidie allo stadio VIII, determinati con le marche al carbone. Pubbl Staz Zool Napoli 25:161–187

Passamaneck YJ, Di Gregorio A (2005) *Ciona intestinalis*: chordate development made simple. Dev Dyn 233(1):1–19

Passamaneck YJ, Katikala L, Perrone L, Dunn MP, Oda-Ishii I, Di Gregorio A (2009) Direct activation of a notochord *cis*-regulatory module by Brachyury and FoxA in the ascidian *Ciona intestinalis*. Development 136(21):3679–3689

Rastegar S, Hess I, Dickmeis T, Nicod JC, Ertzer R, Hadzhiev Y, Thies WG, Scherer G, Strähle U (2008) The words of the regulatory code are arranged in a variable manner in highly conserved enhancers. Dev Biol 318(2):366–377

Reese DE, Hall CE, Mikawa T (2004) Negative regulation of midline vascular development by the notochord. Dev Cell 6(5):699–708

Reverberi G (1971) Ascidians. In: Reverberi G (ed) Experimental embryology of marine and freshwater invertebrates. North-Holland, Amsterdam, pp 507–550

Rhee JM, Oda-Ishii I, Passamaneck YJ, Hadjantonakis AK, Di Gregorio A (2005) Live imaging and morphometric analysis of embryonic development in the ascidian *Ciona intestinalis*. Genesis 43(3):136–147

Sasaki H, Hogan BL (1993) Differential expression of multiple fork head related genes during gastrulation and axial pattern formation in the mouse embryo. Development 118(1):47–59

Sasaki H, Yoshida K, Hozumi A, Sasakura Y (2014) CRISPR/Cas9-mediated gene knockout in the ascidian *Ciona intestinalis*. Develop Growth Differ 56(7):499–510

Sasakura Y, Suzuki MM, Hozumi A, Inaba K, Satoh N (2010) Maternal factor-mediated epigenetic gene silencing in the ascidian *Ciona intestinalis*. Mol Gen Genomics 283(1):99–110

Satoh N, Tagawa K, Takahashi H (2012) How was the notochord born? Evol Dev 14(1):56–75

Sawada A, Nishizaki Y, Sato H, Yada Y, Nakayama R et al (2005) Tead proteins activate the Foxa2 enhancer in the node in cooperation with a second factor. Development 132:4719–4729

Segade F, Cota C, Famiglietti A, Cha A, Davidson B (2016) Fibronectin contributes to notochord intercalation in the invertebrate chordate, *Ciona intestinalis*. EvoDevo 7(1):21

Sehring IM, Dong B, Denker E, Bhattachan P, Deng W, Mathiesen BT, Jiang D (2014) An equatorial contractile mechanism drives cell elongation but not cell division. PLoS Biol 12(2):e1001781

Sehring IM, Recho P, Denker E, Kourakis M, Mathiesen B, Hannezo E, Dong B, Jiang D (2015) Assembly and positioning of actomyosin rings by contractility and planar cell polarity. Elife 21(4):e09206

Smith J (1999) T-box genes: what they do and how they do it. Trends Genet 15(4):154–158

Søviknes AM, Glover JC (2008) Continued growth and cell proliferation into adulthood in the notochord of the appendicularian *Oikopleura dioica*. Biol Bull 214(1):17–28

Stemple DL (2005) Structure and function of the notochord: an essential organ for chordate development. Development 132:2503–2512

Stolfi A, Christiaen L (2012) Genetic and genomic toolbox of the chordate *Ciona intestinalis*. Genetics 192(1):55–66

Stolfi A, Gandhi S, Salek F, Christiaen L (2014) Tissue-specific genome editing in *Ciona* embryos by CRISPR/Cas9. Development 141(21):4115–4120

Tada M, Casey ES, Fairclough L, Smith JC (1998) Bix1, a direct target of *Xenopus* T-box genes, causes formation of ventral mesoderm and endoderm. Development 125(20):3997–4006

Takada N, Satoh N, Swalla BJ (2002) Expression of *Tbx6*, a muscle lineage T-box gene, in the tailless embryo of the ascidian *Molgula tectiformis*. Dev Genes Evol 212:354–356

Takahashi H, Hotta K, Erives A, Di Gregorio A, Zeller RW, Levine M, Satoh N (1999) Brachyury downstream notochord differentiation in the ascidian embryo. Genes Dev 13(12):1519–1523

Takahashi H, Hotta K, Takagi C, Ueno N, Satoh N, Shoguchi E (2010) Regulation of notochord-specific expression of Ci-Bra downstream genes in *Ciona intestinalis* embryos. Zool Sci 27(2):110–118

Tamplin OJ, Cox BJ, Rossant J (2011) Integrated microarray and ChIP analysis identifies multiple Foxa2 dependent target genes in the notochord. Dev Biol 360(2):415–425

Thompson JM, Di Gregorio A (2015) *Insulin-like* genes in ascidians: findings in *Ciona* and hypotheses on the evolutionary origins of the pancreas. Genesis 53(1):82–104

Urry LA, Whittaker CA, Duquette M, Lawler J, DeSimone DW (1998) Thrombospondins in early *Xenopus* embryos: dynamic patterns of expression suggest diverse roles in nervous system, notochord, and muscle development. Dev Dyn 211:390–407

Veeman MT, Nakatani Y, Hendrickson C, Ericson V, Lin C, Smith WC (2008) *Chongmague* reveals an essential role for laminin-mediated boundary formation in chordate convergence and extension movements. Development 135(1):33–41

Wada H, Okuyama M, Satoh N, Zhang S (2006) Molecular evolution of fibrillar collagen in chordates, with implications for the evolution of vertebrate skeletons and chordate phylogeny. Evol Dev 8:370–377

Weisblat DA, Sawyer RT, Stent GS (1978) Cell lineage analysis by intracellular injection of a tracer enzyme. Science 202:1295–1298

Yagi K, Satou Y, Satoh N (2004) A zinc finger transcription factor, ZicL, is a direct activator of Brachyury in the notochord specification of *Ciona intestinalis*. Development 131(6):1279–1288

Yamada S, Ueno N, Satoh N, Takahashi H (2011) *Ciona intestinalis* Noto4 contains a phosphotyrosine interaction domain and is involved in the midline intercalation of notochord cells. Int J Dev Biol 55(1):11–18

Yasuo H, Satoh N (1993) Function of vertebrate T gene. Nature 364(6438):582–583

Check for updates

Purification of Fluorescent Labeled Cells from Dissociated *Ciona* Embryos

9

Wei Wang, Claudia Racioppi, Basile Gravez, and Lionel Christiaen

Abstract

Genome-wide studies in *Ciona* often require highly purified cell populations. In this methods chapter, we introduce multi-channel combinatorial fluorescence activated cells sorting (FACS) and magnetic-activated cell sorting (MACS) as two sensitive and efficient tools for isolating lineage-specific cell populations from dissociated *Ciona* embryos and larvae. We present isolation of trunk ventral cell (TVC) progeny as the test case most commonly used in our laboratory. These approaches may also be applied to purify other cell populations with the proper combination of tissue-specific reporters.

Keywords

TVC · Multi-channel combinatorial FACS · MACS · RNA extraction · Cell dissociation · Cell-lineage specific population · Mosaic expression

9.1 Introduction

One of the greatest challenges for genome-wide studies in *Ciona* is to distinguish between cells and lineages of interest and the surrounding tissues, which can constitute the overwhelming majority of unsorted samples. This is crucial because most genetic elements, especially transcribed regions, show pleiotropic activity. The successful cloning of numerous enhancers for genes that are expressed exclusively in certain tissues has enabled the community to label specific lineages and tissues by using fluorescent reporters. In principle, this permits the straightforward isolation of cells of interest using fluorescence activated cell sorting (FACS).

However, in practice, most reporter constructs show "leaky" expression in other tissues, strongly interfering with cell isolation using a single tissue-specific fluorescent reporter. Besides, when using a combination of early and late reporter constructs for combinatorial labeling, depending upon the activation time and the number of divisions between the onsets of the early and late enhancers, only a certain number of cells labeled with the early reporter inherit and activate the late fluorescent reporter. This caveat is defined as the "mosaic expression" in electroporated *Ciona* embryos. Mosaic expression is inconvenient, especially when we transiently express engineered constructs with late tissue-specific enhancers. Because only a portion of the progeny

Wei Wang, Claudia Racioppi, and Basile Gravez are contributed equally to this work.

W. Wang · C. Racioppi · B. Gravez · L. Christiaen (✉)
Center for Developmental Genetics, Department of Biology, New York University, New York, NY, USA
e-mail: lc121@nyu.edu

Y. Sasakura (ed.), *Transgenic Ascidians*, Advances in Experimental Medicine and Biology 1029,
https://doi.org/10.1007/978-981-10-7545-2_9

inherits the late-active plasmid, within the isolated population, a substantial number of cells fail to inherit the perturbation construct and thus remain "wild-type." In most cases, this causes heterogeneity in experimental samples that confounds the results of whole genome assays when comparing perturbation vs control samples.

In this methods chapter, we use trunk ventral cell purification as an example of introducing a multi-channel combinatorial FACS approach (Christiaen et al. 2008), which can overcome the obstacles described above. Trunk ventral cells (TVCs) and TVC progeny of the B7.5 lineage are rare (<0.2% of the entire cellular population) in tailbud and older embryos and larvae, but they can be labeled with B7.5-lineage-specific enhancers driving fluorescent reporters, e.g., *Mesp > tagRFP* (Davidson et al. 2005). However, the leaky expression of *Mesp* enhancer into the B-line mesenchyme cells normally results in contamination of TVC progeny samples (Fig. 9.1a). This issue can be avoided by using *MyoD905 > GFP*, which is specifically expressed in mesenchyme and primary muscle cells, except for the B7.5-derived anterior tail muscles (ATMs), for counter-selection (Fig. 9.1b) (Christiaen et al. 2008). In this study, we introduced a TVC-specific reporter, *Hand-r > tagBFP* (Fig. 9.1b), and used a third FACS channel to co-select the TVC progeny for two reasons. First, the extra layer of co-selection of FACS can further improve the purity of sorted TVC progeny. Second, we used *Hand-r* (Woznica et al. 2012; and Christiaen laboratory, unpublished observations) or *FoxF* (Beh et al. 2007) enhancer-driven constructs to express reporters and perturbation constructs (e.g., Wang et al. 2013; Razy-Krajka et al. 2014) selectively in the TVC progeny, thus avoiding precocious effects in the B7.5 lineage cells. Because of the mosaic inheritance of *Hand-r* and *FoxF* enhancer constructs, only a portion of TVC progeny labeled with *Mesp > tagRFP* also expresses these constructs. Previous estimates suggest that 20–30% of the *Mesp > tagRFP*+ embryos can show mosaic expression of *FoxF*-driven reporters (Gline et al. 2015). Using the same enhancer for the perturbation plasmid to perform the co-selection avoids sorting cells that did not inherit the perturbation constructs because of mosaicism.

Although FACS is a sensitive and accurate method for purifying specific cells from the rest of the embryonic cell population (Christiaen et al. 2008; Christiaen et al. 2009; Racioppi et al. 2014; Razy-Krajka et al. 2014), it requires specialized equipment and skilled operators that are usually not available to individual laboratories, and as cells are processed through high-pressure fluids, they can easily be damaged. To circumvent

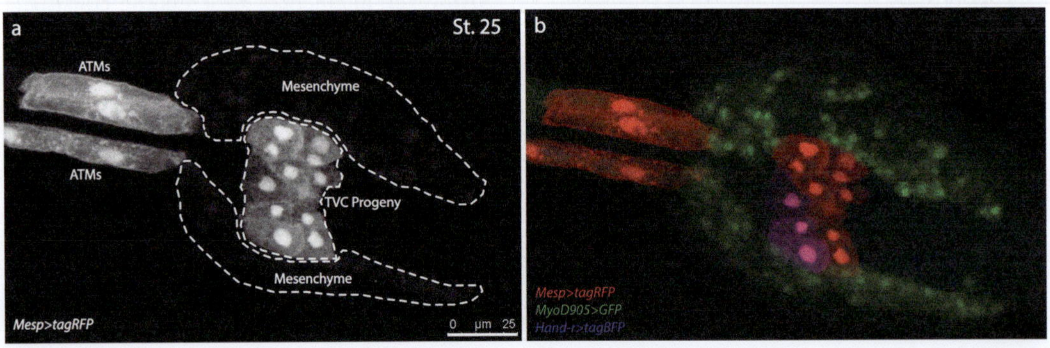

Fig. 9.1 *Ciona* late tailbud for multi-channel combinatorial FACS. *Ciona* late tailbud expresses *Mesp > tagRFP*, *MyoD905 > GFP*, and *Hand-r > tagBFP* fluorescent reporters. Embryos are fixed at St. 25 and imaged using Leica TCS SP8 X confocal microscopy. (**a**) Isolated single channel showing the *Mesp > tagRFP* labelled B7.5 lineage (TVC progeny and ATMs are in *gray*) and the leaky expression in mesenchyme cells (faint gray signal). (**b**) The overlapped channels of *Mesp > tagRFP*, *MyoD905 > GFP*, and *Hand-r > tagBFP* reporters. *Mesp > tagRFP* labeled B7.5 lineage cells are in *red*. *MyoD905 > GFP* labelled mesenchyme cells are in *green*. *Hand-r > tagBFP* reporter labels only a portion of TVC progeny because of the mosaicism. The tagRFP-tagBFP+ cells, which appear in *magenta* (*red* overlaid with *blue*), are sorted for further analysis

these problems, affinity-based cell separation technologies are also adapted to isolating *Ciona* cells using membrane markers.

Magnetic-activated cell sorting (MACS; Miltenyi Biotec) separates specific cells from the rest of the embryonic cell population using antibody-coated superparamagnetic nanoparticles (50-nm microbeads). The antibodies are specific to particular cell surface markers, either expressed in the population of interest for positive selection, or expressed on undesired cell types for negative selection. After incubating dissociated cells expressing a trans-membrane marker with antibody-coated microbeads, the suspension is added to a separation column, which contains a matrix composed of ferromagnetic spheres. This induces the formation of a magnetic gradient, leading to microbead-laden cells to stick to the column, whereas unlabeled cells flow through.

If a negative selection is performed, the flow-through is the population of interest and the bound cell–microbead complexes are discarded. If a positive selection is performed, the cells of interest are bound to the microbeads. The column is washed several times to ensure that unbound or weakly bound cells are washed away. Then the column is removed from the magnetic separator and the cell–microbead complexes are eluted.

The MACS technology also permits indirect magnetic labeling, where the cells are labeled with a primary antibody directed against a cell surface marker in a first step; in a second step, the cells are magnetically labeled with microbeads, which bind either to the primary antibody or to a molecule conjugated to the primary antibody.

Separation using MACS presents several advantages over FACS. It does not require an expensive, dedicated instrument (although automatic platforms are available), it is generally easier to set up than FACS, large numbers of cells can be processed simultaneously (although it is not possible to count the number of cells separated during the MACS process), and it allows multiplex experiments with parallel separations from up to 24 samples in a single run.

However, one disadvantage of MACS concerns the microbeads. Most of them are only available from Miltenyi Biotec; thus, if there are no microbeads with the particular marker of interest available, the conjugation must be done unsupported. Moreover, although combinatorial logics can, in principle, be implemented by using different membrane markers, such as human Cluster of Differentiation 4 (hCD4) and mouse Cluster of Differentiation 8 (mouse CD8), in practice, because all purifications rely on the same microbeads, this requires successive incubation and column purification, which render these combinatorial experiments particularly impractical.

To isolate transfected cells, for example to assay CRISPR/Cas9-mediated mutagenesis rates (Stolfi et al. 2014), we have been using either glycoprotein hCD4, or mouse CD8 as cell surface markers. In our hands, hCD4 appeared to be better trafficked to the plasma membrane, usually generating more specific membrane labeling when fusing to fluorescent proteins such as GFP or mCherry (Gline et al. 2015).

9.2 Methods

9.2.1 Cell Dissociation

Mechanical and enzymatic digestion of *Ciona* embryos electroporated with the combination of tissue specific enhancers driving different fluorescent reporter releases fluorescent cells into suspension. This allows cells of interest that uniformly express a specific fluorescent reporter to be successfully enriched.

To proceed to cell dissociation:

1. Collect *Ciona* stage-selected embryos in borosilicate glass tubes 12x75 mm (rinse them overnight with distilled water and dry before use, to avoid embryos sticking to the glass).
2. Put the tubes on ice and let the embryos settle.
3. Rinse the embryos three times in artificial calcium and magnesium-free sea water (CMF-ASW: 449 mM NaCl, 33 mM Na2SO4, 9 mM KCl, 2.15 mM NaHCO3, 10 mM Tris-Cl pH 8.2, 2.5 mM EGTA).

4. Remove the supernatant and keep the tubes on ice.
5. Dilute the stock solution of trypsin (w/v, Sigma, T-4799, usually prepared at 2.5% in CMF-ASW) to a final concentration of 0.2% in CMF-ASW.
6. Add 2 mL of CMF-ASW supplemented with 0.2% trypsin and dissociate embryos by thorough pipetting (2–3 min for the tailbud stage, 5–8 min for the larval stage).
7. Block the trypsin activity by adding 2 mL of ice-cold CMF-ASW supplemented with 0.05% bovine serum albumin (BSA, Sigma, A-3311).
8. Keep the cell suspensions on ice and filter into 5-mL polystyrene round bottomed tubes equipped with a 35-μm cell strainer cap (Falcon).
9. Centrifuge the cell suspensions for 3 min at 800 relative centrifugal force (rcf) at 4 °C.
10. Remove the supernatant and rinse in CMF-ASW supplemented with 0.05% BSA.
11. Repeat the centrifugation and rinse steps 9 and 10 at least once
12. Filter the cell suspensions into 5-mL polystyrene round bottomed tubes equipped with a 35-μm cell strainer cap before cell sorting.

9.2.2 Fluorescence Activated Cell Sorting

The successful multi-channel FACS relies on the clean separation of the spectrum of fluorescent reporters that are co-expressed in the same cell population. Thus, this requires careful selection of reporters and multiple single or dual color control samples to calibrate the gate and compensation conditions. The parameters given below were determined for a BD FACSAria™ flow cytometer equipped with a 488-nm blue laser, a 561-nm yellow/green laser, a 633-nm far-red laser, and a 407-nm ultra violet laser.

Four single color samples are prepared for calibration of gating and compensation on the flow cytometry sorter:

Non-electroporated sample (no fluorescent reporter).
Single-color sample 1:electroporated with *MyoD905 > GFP*.
Single-color sample 2: electroporated with *Mesp > tagRFP*.
Single-color sample 3: electroporated with *Hand-r > tagBFP*.
Three dual-color samples to test the separation of channels as pairs.
Dual-color sample 1: electroporated with *MyoD905 > GFP* and *Mesp > tagRFP*.
Dual-color sample 2: electroporated with *MyoD905 > GFP* and *Hand-r > tagBFP*.
Dual-color sample 3: electroporated with *Mesp > tagRFP* and *Hand-r > tagBFP*.
One triple-color sample electroporated with *MyoD905 > GFP*, *Mesp > tagRFP*, and *Hand-r > tagBFP* to test the separation of all three channels.

1. Before loading the samples into the sorter, filter the cell suspension again on the cell-strainer cap of a 5-mL round-bottomed tube and gently pipette the cell suspension 3–5 times.
2. Load the non-electroporated sample into the sorter and count ~10,000 cells. Distinguish the well-dissociated single cells from undissociated tissues and debris based on size and internal complexity, which can be defined by forward scatter (FS) and side scatter (SS). Gate the single cell population as P1 (Fig. 9.2a).
3. Load the single-color sample 1 into the sorter and count ~10,000 cells. Gate the GFP+ population using the 488-nm blue laser, PMT (Photomultiplier Tubes) Detector E, LP (Long Pass) Filter 505, and BP (Band Pass) Filter 530/30 for FITC (Fig. 9.2b).
4. Load the single-color sample 2 into the sorter and count ~10,000 cells. Gate the tagRFP+ population using 561-nm yellow/green laser, PMT Detector B and BP Filter 586/15 for DsRed (Fig. 9.2c).
5. Load the single-color sample 3 into the sorter and count ~10,000 cells. Gate the tagBFP+ population using 407-nm (violet) laser, PMT Detector B, and BP Filter 450/50 for Pacific Blue (Fig. 9.2d).

6. Mix the single-color samples 1 and 2 by 1:1, load the sample mixture into the sorter, and count ~10,000 cells.
7. Adjust the compensation, ensure clean separation of the GFP+ and tagRFP⁺ populations.
8. Repeat the same steps using mixtures of single-color samples 1 and 3, and 2 and 3, adjust the compensation to ensure the clean separation of GFP⁺ vs tagBFP⁺ populations and tagRFP⁺ vs tagBFP⁺ populations.
9. Mix the single-color samples 1, 2, and 3 in 1:1:1 proportions. Load the sample mixture into the sorter and count ~10,000 cells, making sure the GFP⁺, tagRFP⁺, and tagBFP⁺ cell populations are still well separated.

   Load the dual-color sample 1 and count ~10,000 cells. Use the GFP⁺ and tagRFP⁺ gate, which was defined by a single-color sample to identify the cell population with only GFP expression, only tagRFP expression, and with both GFP and tagRFP expression. Gate the GFP-tagRFP⁺ population.
10. Repeat the same steps for two-color samples 2 and 3, define the gates for GFP-tagBFP⁺ and tagRFP-tagBFP⁺ populations.

Note: calibration of gating (steps 1–10) is recommended every time before sorting. However, the BD FACSAria™ flow cytometer has proved to be consistent and repeatable across daily operations. Therefore, the calibration of gating can be skipped when

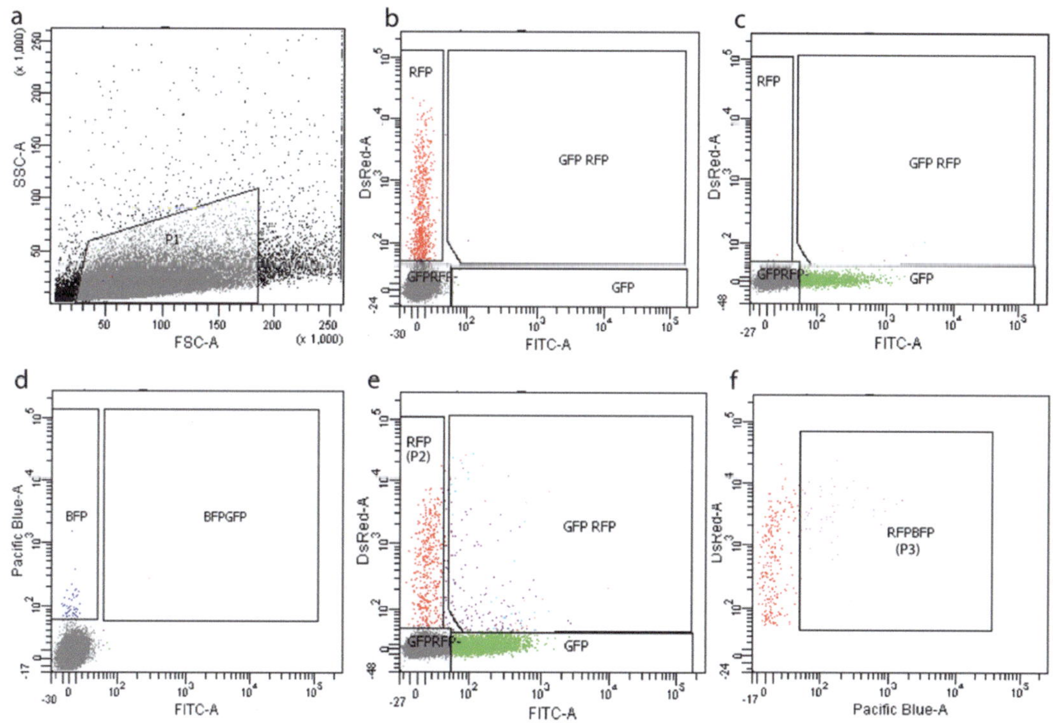

Fig. 9.2 FACS gating strategy for TVC progeny isolation. Samples electroporated with blank buffer, *Mesp > tagRFP*, *MyoD905 > GFP*, and *Hand-r > tagBFP* fluorescent reporters are used to gate the TVC progeny populations. (**a**) Gate P1 for single-cell population based on FS and SS from the blank sample. (**b**) Gate the tagRFP⁺ cell population using the sample electroporated with *Mesp > tagRFP*. (**c**) Gate the GFP⁺ cell population using the sample electroporated with MyoD905 > GFP. (**d**) Gate the tagBFP⁺ cell population using the sample electroporated with Hand-r > tagBFP. (**e**) Define the cells that only have tagRFP as P2 in the sample co-electroporated with *Mesp > tagRFP*, *MyoD905 > GFP*, and *Hand-r > tagBFP*. (**f**) Analyze P2 with gates set for tagRFP and tagBFP, define the tagRFP-tagBFP⁺ cell population as P3. P3 is TVC progeny that we sort for downstream genome-wide analysis

sorting the same type of samples frequently within a short period, such as 1 month.

Load the triple-color sample and count ~10,000 cells.

11. Analyze the sample with the lasers and filters for FITC and DsRed, select the population with only tagRFP signal, mark it as P2 (Fig.9.2e).
12. Analyze the P2 population with the lasers and filters for DsRed and Pacific Blue. Select the cell population that falls into the tagRFP- $tagBFP^{+}$ gate and give it the P3 identification (Fig.9.2f).

In our experiments, this P3 population consists of TVC progeny that co-express tagRFP and tagBFP (co-selection), without the expression of GFP (counter selection of mesenchyme cells with leaky *Mesp > tagRFP* and *Hand-r > tagBFP* expression). These cells only represent 0.1%–0.2% of the P1 population from the whole embryo.

9.2.3 Magnet Activated Cell Sorting

After the last step of cell dissociation from the *Ciona* embryos protocol:

1. Transfer each cell suspension into 2-mL non adhesive tubes.
2. Centrifuge for 3 min at 800 rcf and at 4 °C.
3. Discard supernatant and resuspend cells in 90 μL of CMF-ASW supplemented with 0.05% BSA.
4. Add 10 μL of superparamagnetic microbeads coupled with the anti-human CD4 antibody (Miltenyi Biotec) to select for positive hCD4 cells.
5. Combine each sample into one 2 mL non-adhesive tube to have a total of 20 μL of microbeads in a final volume of 200 μL.
6. Incubate samples for 1 h on agitation at 4 °C.
7. Place the MS columns inside the permanent magnet OctoMACs separator.
8. Equilibrate the MS columns with 500 μL of CMF-ASW. *Note that it is important to degas all the buffers, for example, by filtration using a vacuum system, just before using them, otherwise microbubbles can interfere with the process and cause hCD4-negative cells to be retained in the column.*
9. Assemble pre-separation filters (70 μm; Miltenyi Biotec) onto the MS columns (to remove larger debris for a smooth flow-through of the columns and an efficient magnetic hold of the $hCD4^{+}$ cells labeled with the microbeads).
10. Add 1 mL of CMF-ASW supplemented with 0.05% BSA to the samples (after incubation to rinse the cell suspension).
11. Centrifuge for 3 min at 1,000 rcf and at 4 °C.
12. Suspend each sample in 500 μL of CMF-ASW supplemented with 0.05% BSA.
13. Load samples onto the pre-separation filters over the MS columns.
14. Collect flow-through in 2-mL non-adhesive tubes as hCD4-negative cells.
15. Wash with 500 μL CMF-ASW supplemented with 0.05% BSA and collect in the same tube.
16. Repeat this step twice more.
17. Remove each MS column from the OctoMACS separator and place above another 2-mL non-adhesive tube.
18. Add 1 mL of CMF-ASW supplemented with 0.05% BSA to the top of the column.
19. Gently flush out with a plunger supplied by the kit, on ice.
20. Collect flow-through containing positive hCD4 cells.
21. Centrifuge for 3 min at 1,000 rcf and at 4 °C.

9.2.4 RNA Extraction and Cell Purity Check

From typical FACS experiments, 200 to 2,000 cells are collected from each sample, for RNA extraction if transcriptome profiling by RNA-seq is the desired outcome.

Both RNAqueous®-Micro Total RNA Isolation Kit (Thermo Fisher Scientific, Catalog number: AM1931) and AGENCOURT® RNACLEAN® XP (Beckman Coulter, Part Number: A63987) have shown good efficiency for extraction and purifying total RNA from the small amounts of sorted *Ciona* cells.

The quality and quantity of total RNA are checked by 2200 TapeStation system (Agilent), using High Sensitivity RNA ScreenTape (5067–5579) with High Sensitivity RNA ScreenTape Sample Buffer (5067–5580), and High Sensitivity RNA ScreenTape Ladder (5067–5581). The range of total RNA input is 500–10,000 pg/μL, and the total RNA integrity is estimated using the software RNA Integrity Number or RIN. RNA samples pass the quality control if RIN > 8 and are then used for downstream cDNA synthesis and high-throughput sequencing. As an alternative, especially when the input total RNA concentration is low (extracted from less than 400 cells, for example), Agilent 2100 Bioanalyzer also serves well for quality and quantity checking with Agilent RNA 6000 Pico Kit (5067–1511), which has a quantitative range of 50 to 5,000 pg/μl.

Checking sample purity is indispensable following either FACS- and MACS-based purification. When sorting B7.5-lineage-derived trunk ventral cells, the main contaminations are normally contributed by the mesenchyme and tail muscle. Thus, we performed semi-quantitative PCR (real-time PCR is even better) using primers designed for TVC-specific genes (e.g., *Hand-r*), mesenchymal cell-specific genes (e.g., *Twist1/2*), and tail muscle-specific genes (e.g., *Mox*) to check the purity of TVC progeny after sorting. Our recent single-cell RNA-seq experiments also provided evidence that multi-channel FACS can enrich the purity of TVC progeny up to 85% out of the sorted cells (Wang et al., in preparation).

Acknowledgements We are grateful to Pui-Leng Ip, Ken Birnbaum, and the personnel of GenCore at NYU for their expert assistance with the FACS, ScreenTape, and high-throughput sequencing. We thank Nicole Kaplan for sharing the image of *Mesp > tagRFP*, *MyoD905 > GFP*, and *Hand-r > tagBFP* labelled *Ciona* late tailbud. Research in the laboratory of L.C. is supported by R01 awards HL108643 and GM096032 from the NIH/NHLBI and NIH/NIGMS respectively; and by grant 15CVD01 from the Leducq Foundation. C.R. has been supported by a long-term fellowship ALTF 1608-2014 from EMBO.

References

Beh J, Shi W, Levine M, Davidson B, Christiaen L (2007) FoxF is essential for FGF-induced migration of heart progenitor cells in the Ascidian Ciona intestinalis. Development 134(18):3297–3305

Christiaen L, Davidson B, Kawashima T, Powell W, Nolla H, Vranizan K, Levine M (2008) The transcription/migration interface in heart precursors of Ciona intestinalis. Science 320(5881):1349–1352

Christiaen L, Wagner E, Shi W, Levine M (2009) Isolation of individual cells and tissues from electroporated sea squirt (Ciona) embryos by Fluorescence-Activated Cell Sorting (FACS). Cold Spring Harb Protoc 2009(12):db.prot5349

Davidson B, Shi W, Levine M (2005) Uncoupling heart cell specification and migration in the simple chordate Ciona intestinalis. Development 132(21):4811–4818

Gline S, Kaplan N, Bernadskaya Y, Abdu Y, Christiaen L (2015) Surrounding tissues canalize motile Cardiopharyngeal progenitors towards collective polarity and directed migration. Development 142(3):544–554

Racioppi C, Kamal AK, Razy-Krajka F, Gambardella G, Zanetti L, di Bernardo D, Sanges R, Christiaen LA, Ristoratore F (2014) Fibroblast growth factor signalling controls nervous system patterning and pigment cell formation in Ciona intestinalis. Nat Commun 5(September):4830

Razy-Krajka F, Lam K, Wang W, Stolfi A, Joly M, Bonneau R, Christiaen L (2014) Collier/OLF/EBF-dependent transcriptional dynamics control pharyngeal muscle specification from primed cardiopharyngeal progenitors. Dev Cell 29(3):263–276

Stolfi A, Gandhi S, Salek F, Christiaen L (2014) Tissue-specific genome editing in Ciona embryos by CRISPR/Cas9. Development 141(21):4115–4120

Wang W, Razy-Krajka F, Siu E, Ketcham A, Christiaen L (2013) NK4 antagonizes Tbx1/10 to promote cardiac versus pharyngeal muscle fate in the ascidian second heart field. PLoS Biol 11(12):e1001725

Woznica A, Haeussler M, Starobinska E, Jemmett J, Li Y, Mount D, Davidson B (2012) Initial deployment of the cardiogenic gene regulatory network in the basal chordate, Ciona intestinalis. Dev Biol 368(1):127–139

Germline Transgenesis in *Ciona* 10

Yasunori Sasakura

Abstract

Transgenesis is an indispensable method for elucidating the cellular and molecular mechanisms underlying biological phenomena. In *Ciona*, transgenic lines that have a transgene insertion in their genomes have been created. The transgenic lines are valuable because they express reporter genes in a nonmosaic manner. This nonmosaic manner allows us to accurately observe tissues and organs. The insertions of transgenes can destroy genes to create mutants. The insertional mutagenesis is a splendid method for investigating functions of genes. In *Ciona intestinalis*, expression of the *gfp* reporter gene is subjected to epigenetic silencing in the female germline. This epigenetic silencing has been used to establish a novel method for knocking down maternal expression of genes. The genetic procedures based on germline transgenesis facilitate studies for addressing gene functions in *Ciona*.

Keywords

Germline transgenesis · Transposon · *Minos* · *Sleeping beauty* · I-*Sce*I · Cellulose · Maternal

Y. Sasakura (✉)
Shimoda Marine Research Center, University of Tsukuba, Shimoda, Shizuoka, Japan
e-mail: sasakura@shimoda.tsukuba.ac.jp

10.1 Introduction: Stable and Transient Transgenesis

Transgenesis is an indispensable method for elucidating the cellular and molecular mechanisms underlying biological phenomena. The previous chapters have described techniques for creating transgenic ascidians through microinjection and electroporation. By means of these methods, researchers can introduce artificial DNA constructs into ascidian embryos for various purposes such as labeling cells, monitoring gene expression, and modifying gene functions (Corbo et al. 1997; Takahashi et al. 1999; Zeller et al. 2006). Over the past decade, transgenesis has often been used for the labeling of live cells with a fluorescent reporter gene (Rhee et al. 2005). In some ascidians, such as *Ciona*, *Phallusia*, and *Halocynthia*, the transgenic techniques are relatively easy, and the techniques are therefore routinely used in these model tunicate species (Lemaire 2011).

The introduction of exogenous DNAs into embryos is classified as transient transgenesis, because the DNAs introduced are usually unstable and are usually lost during development. The instability is due to the epichromosomal nature of the exogenously introduced DNA: because in most cases the DNA is not integrated into the chromosomes, the epichromosomal DNA is not delivered accurately to the sister cells during mitoses. Therefore, to make the exogenous DNA

Y. Sasakura (ed.), *Transgenic Ascidians*, Advances in Experimental Medicine and Biology 1029,
https://doi.org/10.1007/978-981-10-7545-2_10

stable, it is necessary to integrate the DNA into chromosomes. If such an integration event occurs in a cell of germ cell lineage, the eggs and/or sperm derived from the cell have the exogenous DNA in their chromosomes. Indeed, the progeny animals from the eggs or sperm possess the exogenous DNA in all the cells constituting their bodies. In such cases, the exogenous DNA can behave like endogenous DNA, and can be transmitted stably to subsequent generations. This manner of transgenesis is called germline transgenesis, and the animal groups inheriting the exogenous DNA to subsequent generations are often called transgenic lines.

Methods of stable transgenesis have been reported in *Ciona intestinalis* and *Ciona savignyi* (Deschet et al. 2003; Sasakura et al. 2003a, b; Matsuoka et al. 2004), and these are discussed later. There are advantages and disadvantages to both transient and stable transgenesis, and thus the appropriate form of transgenesis must be selected when planning an experiment. Stable transgenesis requires longer than transient transgenesis, because the process includes steps for establishing and culturing the transgenic lines. Once they are created, however, the transgenic lines can be used repeatedly. In addition, the use of the same line for different experiments helps to assure their reproducibility. Another advantage of stable transgenesis is that it ensures that all cells constituting the targeted tissues or organs are labeled, because all cells of stable transgenic animals can be assumed to possess the transgene. Therefore, the judgment as to the presence/absence of cells is much simpler than in transient transgenesis; if a number of cells lose their fluorescence in a stable transgenic animal, it means that the cells are either absent or have lost their reporter expression. In transient transgenesis, there is a third possibility—namely, that the fluorescence-negative cells do not harbor reporter constructs because of the loss of the transgene. Therefore, stable transgenesis is suitable for experiments requiring the labeling of the whole tissue or organ: for example, for counting how many cells constitute a tissue/organ, or observing how tissues/organs behave during development (Chiba et al. 2009; Horie et al. 2011; Hozumi et al. 2015). Stable transgenesis is also more effective for the labeling of cells/tissues/organs at later developmental stages compared with transient transgenesis, as transiently introduced transgenes can be successively lost during development, suggesting that longer development could increase the chances of the cells losing a transgene (Sasakura 2007).

On the other hand, stable transgenesis is inferior for observing the detailed shape of individual cells (Hatta et al. 2006; Hozumi et al. 2015), because the expression of a reporter gene in the whole tissue/organ could mask the cellular boundary. To observe the cell shape, labeling a few cells in a specimen by transient transgenesis may be better. Moreover, transient transgenesis is a faster method than stable transgenesis. We can carry out an experiment at the G0 generation by transient transgenesis, which is one generation earlier than stable transgenesis. Finally, no special effort is required to maintain the transgenic lines for transient transgenesis. Therefore, transient transgenesis can be regarded as a time-saving method.

10.2 Integration of Exogenous DNA into Chromosomes

There are several ways of creating stable transgenic animals; in this section the major methods currently in use are discussed. As mentioned above, the integration of exogenous DNAs into chromosomes is a required step for stable transgenesis. In general, the integration events are initiated simply by the introduction of linearized DNAs into the cells. The integration of exogenous DNAs occurs via the system for repairing the double-strand breaks (DSBs) of genomic DNAs. DSBs occur when the double helix of genomic DNAs is broken by factors that damage DNAs. Because DSBs are harmful to cells, the cells attempt to repair them using two major mechanisms. One of them, nonhomologous end-joining, simply ligates the two ends of the DNAs, without reference to the sequence of sister chromatins (Bruma et al. 2006). When multiple copies of linearized exogenous DNAs are available in the cells, the ends of the exogenous DNAs could be ligated to the broken ends of chromosomal DNA during the repair, causing the exogenous DNAs to be incorporated into chromosomes. In many model organisms, transgenic lines have thus been created by the simple introduction of linearized exogenous DNAs into embryos (e.g., Liu 2013).

It is known that transgenes are often inserted into chromosomes as tandem repeats of multiple copies, probably because their ends are repeatedly ligated together between each other before or during the integration into chromosomes. Such tandem repeats are usually less stable than single-copy integrations, because repeat sequences of DNAs could be deleted by the recombination between homologous DNA stretches. The repeated nature of DNAs makes it difficult to identify the insertion sites of the exogenous DNAs, because PCR is difficult to apply for identification, as the primers designed in the exogenous DNAs could amplify the inside of the repeats. Moreover, the introduction of linearized DNA is not efficient for transgenesis because the method uses double strand breaks of chromosomes that occur by chance.

In *Ciona intestinalis*, electroporation of expression vectors into the one-cell embryos can yield germ cell transgenesis. Approximately 20–30% of the electroporated animals become founders that transmit transgene insertion(s) to progeny. A curious point with regard to the germ cell transgenesis in *Ciona* is that the electroporated DNAs are not linearized before electroporation: the circular plasmid vectors extracted from *Escherichia coli* can be used without treatment. Although the mechanism has not been studied in depth, it seems likely that some plasmid DNA copies could be linearized during electroporation or after introduction into *Ciona* embryos.

To achieve a higher efficiency of transgenesis, I-*Sce*I endonuclease has been adopted by several organisms (Thermes et al. 2002; Pan et al. 2006). I-*Sce*I is a type of restriction enzyme recognizing a long DNA sequence (5′-tagggataacagggtaat-3′). The long recognition site indicates that this endonuclease rarely digests chromosomal DNAs. When performing transgenesis with I-*Sce*I, the DNAs intended for delivery into the chromosome are flanked by two I-*Sce*I recognition sites. A mixture of the vector containing this construct and the I-*Sce*I enzyme are introduced into embryos, and the germ cells from the animals are collected to test the transmission of exogenous DNAs to the germ cell genome.

It is not well known how I-*Sce*I increases the chance of DNA integration into chromosomes. The exogenous DNA molecules are digested by I-*Sce*I to form linear DNAs in embryos. The linear DNAs, when ligated by an endogenous system, usually regain the I-*Sce*I recognition site, and thereby become the target of I-*Sce*I. The repetitive digestion of I-*Sce*I may be sufficient to maintain a high number of DNA ends, and may increase the chance of integration into chromosomes.

In several animals, including *Ciona savignyi* and *Ciona intestinalis*, the successful generation of stable transgenic lines has been reported (Deschet et al. 2003; Awazu et al. 2004). Like the method of linearized DNA introduction, the transgenic lines created by the I-*Sce*I-mediated method tend to have multiple copies of exogenous DNA at a single integration site, making it difficult to characterize the integration site.

The DNA transposon vectors provide us with solutions to overcome the weak points of the methods of stable transgenesis mentioned above. DNA transposons are mobile DNA elements that have inverted repeats at their ends and encode a protein named transposase between the repeats. Transposase is an enzyme that catalyzes the excision of transposons from one DNA followed by their insertion into the target sequence of another DNA. By this mechanism, DNA transposons can be mobilized from one location in DNA to another in a cut-and-paste manner.

Researchers can modify transposon vectors by removing the genes flanked by inverted repeats to replace them with reporter constructs, because the DNA sequence of the transposase gene are not necessary for the integration events. These artificial transposon vectors, when given transposase from a different source, can be mobilized between DNAs.

Usually, the transposon vectors are constructed on a plasmid vector backbone, and are introduced into embryos with mRNAs encoding transposase or an expression vector of transposase (Ivics et al. 1997; Kawakami and Shima 1999). The remobilization of transposon vectors can occur between plasmid DNA and chromosomal DNAs, and thus exogenous DNAs can be inserted into the genome of organisms. If this event occurs in the germ line, we can generate transgenic lines.

Because transposase catalyzes the remobilization event, the efficiency of chromosomal integration may be higher than that by plasmid insertions. Moreover, a lower copy number of exogenous DNAs, often a single copy, could be integrated into a target site. Therefore, the transposon insertions are usually stable among generations, and characterization of the insertion sites can be conducted by several PCR methods (Liu et al. 1995). In the next section, the transposon-based methods in ascidians are discussed.

10.3 Transposon-Mediated Transgenesis in *Ciona*

In *Ciona intestinalis*, germ cell transgenesis has been realized using the Tc1/*mariner* superfamily transposon *Minos* (Sasakura et al. 2003a, 2003b). *Minos* is derived from the fly *Drosophila hydei* (Franz and Savakis 1991); however, the transposon and transposase can catalyze remobilization of DNA in multiple organisms (Loukeris et al. 1995; Arca et al. 1997; Klinakis et al. 2000a, b; Shimizu et al. 2000; Pavlopoulos and Averof 2005). Another Tc1/*mariner* superfamily transposon, *Sleeping beauty* (SB), has also been used in *Ciona intestinalis* (Hozumi et al. 2013). SB is an artificially resurrected transposon originally derived from salmonid fish (Ivics et al. 1997). In both transposons, the initial experiments used in vitro synthesized mRNAs of transposase as the source of the proteins. When a transposon vector and transposase mRNA are simultaneously microinjected or electroporated, the transposon copies can be integrated into the chromosomes of germ lines, thereby creating transgenic lines. The basic strategy is shown in Fig. 10.1. Approximately 30% of animals into which transposon/transposase has been introduced are estimated to become founders and transmit the transposon insertions to the next generation.

Generally speaking, the fluorescent signals in transgenic lines tend to be weaker than those seen in transient transgenic animals, even though the same vector is used. The reason for the weaker signals is thought to be that the copy number of reporter genes in a cell is lower in stable transgenic lines than in transient transgenic animals. The strength of fluorescence also differs among transgenic lines with the same vector (Sasakura 2007). This is because of the local effects of genomic regions around the insertion sites; the expression of genes from transposon vectors could be influenced by the surrounding chromosomal DNAs. For example, vectors inserted into a heterochromatin region may have lower transcription efficiency than vectors inserted into relaxed euchromatic regions. For this reason, it is recommended that several transgenic lines be established, followed by selection of lines with the brightest reporter proteins.

10.4 Remobilization of Transposon

The *Minos* or SB transposon copies inserted into a *Ciona* chromosome cannot jump again because transposons and transposases have a strict compatibility, and *Ciona* does not possess a transposase that can catalyze mobilization of *Minos* or SB (Sasakura et al. 2007). However, the transposons inserted into the *Ciona* genome can be remobilized by providing their own transposases. By this technique, it becomes possible to generate new transgenic lines with brighter reporter gene-expression signals than the original lines (Hozumi et al. 2010).

The different expression efficiency of reporter genes between the original and new transgenic lines is caused by the difference in insertion sites between them. To induce the remobilization of transposons, transposase mRNA can be introduced into unfertilized eggs of the wild-type

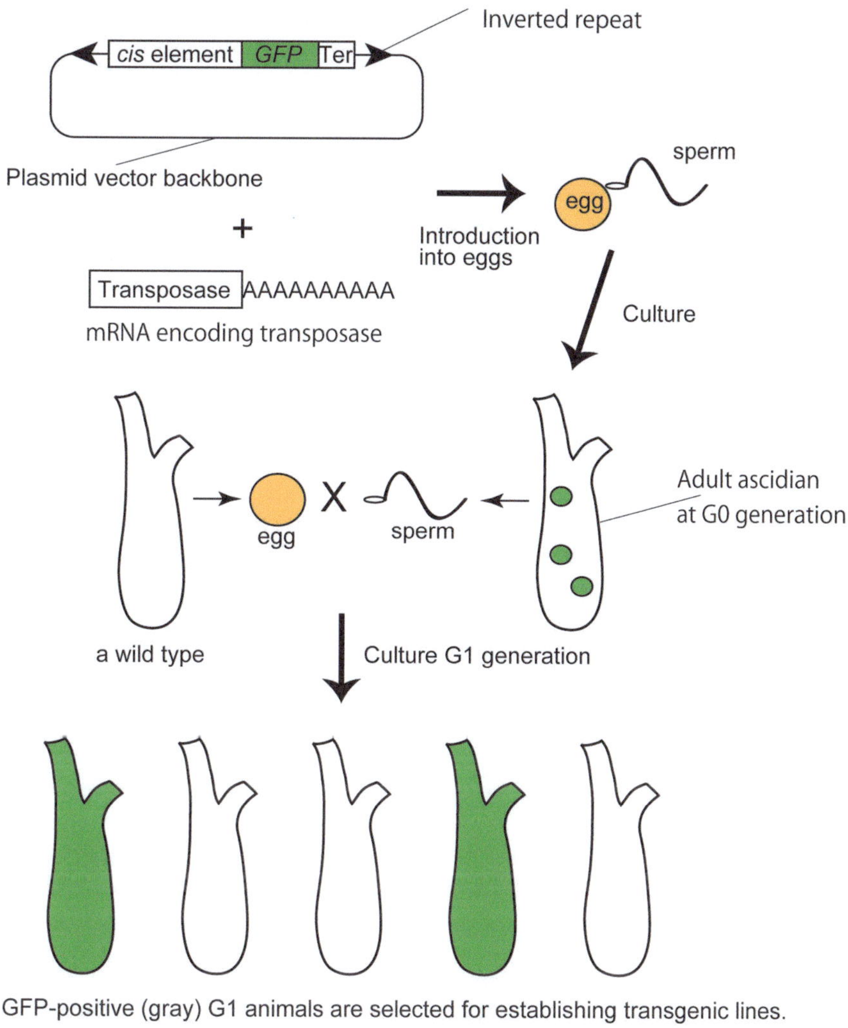

Fig. 10.1 Procedures for the establishment of transgenic lines. This figure illustrates transposon-based transgenesis as an example. The A stretch in the transposase mRNA represents the poly A tail. The *green spots* in the G0 ascidian indicate the mosaic state of transgenes. *Ter* transcription termination signal sequence

strain by means of microinjection, followed by insemination of the microinjected eggs with sperm of a transgenic line to generate hemizygous transgenic animals that express transposase (Awazu et al. 2007).

Alternatively, transgenic lines that express transposase in germ cells can be used for the remobilization technique (Sasakura et al. 2008; Hozumi et al. 2010, 2013). These transposase lines are called "jump-starter" lines. The advantage of jump-starter lines is that remobilization events can be induced without laborious procedures such as microinjection of transposase mRNA. Indeed, the only steps necessary to create new transgenic lines are crossing a transposase line and a transposon donor line to create double transgenic animals, crossing the double transgenic animals with wild types, and the subsequent screening of progeny with new transposon insertions. The double transgenic lines can then be maintained for future screenings of new transgenic lines.

In *Ciona*, jump-starter lines expressing transposase in sperm or eggs have been created (Sasakura et al. 2008; Hozumi et al. 2010, 2013). A hermaphroditic *Ciona* usually matures sperm more quickly than eggs, and the number of sperm cells in an animal is incomparably greater than the number of eggs. For these reasons, the advantage of sperm transposase lines over egg transposase lines is that it is easier to generate a high number of progeny, which increases the chance of obtaining a progeny with the new transposon insertions. By contrast, sperm transposase lines have weaker activity for remobilizing transposon insertions than egg transposase lines (Hozumi et al. 2010). This inferiority is generally attributed to the sperm transposase lines using the *cis* element of a gene encoding protamine: this gene is expressed at a later stage of spermatogenesis just before the condensation of chromatins, and therefore the transposon insertions may have a shorter time to remobilize (the packaging of chromatins by protamine may inhibit remobilization).

Egg transposase lines express transposase during oogenesis, and the chromatin condensation that occurs in sperm does not occur in oocytes, allowing the transposon copies a greater chance to jump.

The generation of mature eggs under our inland culturing condition is less efficient than sperm maturation; the number of progeny from the egg transposase lines (strictly speaking, from the double transgenic animals between an egg transposase line and a transposon donor line) tends to be small. It is estimated that less than 8% of progeny from the double transgenic animals possess remobilized transposon insertions when egg transposase lines and a transposon donor line harboring a single transposon insertion in the genome are used (Hozumi et al. 2010). By contrast, sperm transposase lines do not exhibit the detectable activity to remobilize a single transposon insertion in the *Ciona* genome (Hozumi et al. 2010); therefore, sperm transposase lines have been used to remobilize transposons from tandem arrays of transposons created in *Ciona* genome. The tandem arrays harbor multiple transposon copies; thus, there is a high chance of a few transposon copies being remobilized (Awazu et al. 2007; Sasakura et al. 2008).

10.5 Useful Transgenic Lines

One advantage of stable transgenesis over transient transgenesis is the nonmosaic state of transgenes. The nonmosaic state allows us to observe the structure of tissues more accurately than in experiments with transient transgenic animals. For this reason, stable transgenic lines have been used to trace tissue formation during normal development and compare it with that in mutant animals to address gene functions (Sasakura et al. 2012a). Currently, transgenic lines that specifically label several different tissues, such as the notochord, muscle, epidermis, mesenchyme, nervous system, and endodermal tissues, are available in *Ciona* (Fig. 10.2; Joly et al. 2007). These marker lines are valuable resources for the tunicate community.

Nonmosaicism theoretically ensures that all cells belonging to the target tissues and organs are labeled. This advantage allows us to accurately count the number of cells constituting these tissues and organs.

Ascidians are the best models for studying neuronal circuits in the chordate body, because the number of neurons in their nervous systems is exceptionally small, particularly at the larval stage (Nicol and Meinertzhagen 1991). For the central nervous system (CNS) of *Ciona* larvae, estimations of the cell number have been carried out using microscopic techniques in addition to transient transgenesis (Imai and Meinertzhagen 2007a, b). However, no such attempt has been made for the juvenile stage (the stage soon after metamorphosis).

Our group recently counted the number of cells constituting the juvenile CNS of *Ciona intestinalis* by using transgenic lines expressing fluorescent proteins in the CNS (Hozumi et al. 2015). We found that the juvenile CNS contains 50–60 cells 4 days after metamorphosis, and then this number gradually increases to about 70–80 cells 7 days after metamorphosis.

In *Ciona*, transgenic lines that specifically label a neuron subtype secreting a neurotransmitter have been created (Horie et al. 2011). These transgenic lines were used to characterize the position of cholinergic, glutamatergic, and

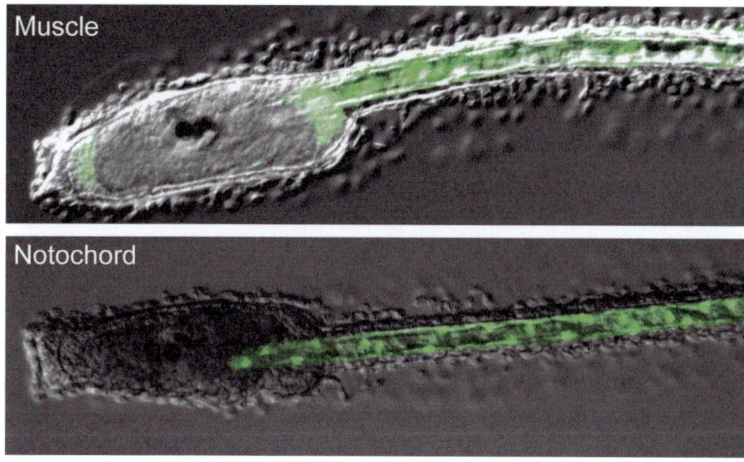

Fig. 10.2 Transgenic lines. *Top*: Larva of the transgenic line expressing green fluorescent protein (GFP) in the muscle. *Bottom*: Larva of a transgenic line expressing GFP in the notochord

GABA/glycinergic neurons in the juvenile CNS. The CNS of ascidian juveniles and adults is oval-shaped with no characteristic structure along the anterior–posterior (A-P) axis. This feature is in contrast with the CNS of ascidian larvae and vertebrates, both of which possess region-specific structural characteristics (Wada et al. 1998; Dufour et al. 2006; Ikuta and Saiga 2007; Sasakura et al. 2012b).

Despite these structural differences, the juvenile CNS of *Ciona* has regional identity along the A-P axis according to the positions of specific neuronal subtypes. For example, glutamatergic neurons are preferentially located at the anterior side, cholinergic neurons tend to be in the middle, and most GABA/glycinergic neurons are in the anterior and posterior regions, but barely present in the middle region. The positioning of the neuron subtypes is similar to that in the larval CNS (Horie et al. 2009; Sasakura et al. 2012b), suggesting the presence of a shared mechanism in the formation of the larval and juvenile CNS.

If transient transgenesis had been used for this study, it would have been more difficult to accurately count the cells making up the juvenile CNS, as introduced reporter vectors are frequently lost during metamorphosis, probably because of the extensive cell proliferation during this period. Thus, the use of stable transgenesis is a recommended approach for studies of this type.

Currently, fluorescent proteins are applied as markers of subcellular organelles, cellular events, and physiological reactions (Pologruto et al. 2004; Tsutsui et al. 2008; Shimozono et al. 2013). For example, green fluorescent proteins (GFPs) and red fluorescent proteins that are fused with a subset of proteins expressed during particular cell-cycle phases are used as sensors of the cell-cycle phases of cells. These sensors, named Fucci, short for fluorescent ubiquitination-based cell cycle indicators, have been widely used because they allow researchers to observe cell-cycle progressions in live cells and embryos (Sakaue-Sawano et al. 2008; Ogura et al. 2011).

In *Ciona*, G1-Fucci and S/G2/M-Fucci expressing transgenic lines are used to monitor cell-cycle progression in juveniles treated with gonadotropin-releasing hormones (GnRHs; Kamiya et al. 2014). GnRHs play a role in promoting metamorphosis in *Ciona*. Moreover, tGnRH-3 and tGnRH-5 inhibit growth during metamorphosis. The cell-cycle progression in tGnRH-3- and tGnRH-5-treated juveniles was observed with Fucci transgenic lines, revealing that tGnRH-3 and tGnRH-5 inhibit cell proliferation by causing G1 arrest.

10.6 Insertional Mutagenesis

When an insertion site of a transposon copy corresponds to an exon, a critical element of an intron or a *cis* element for transcription of a gene, the insertion has the potential to disrupt gene function and generate mutants. The insertional mutagenesis has an advantage over mutagenesis with chemical mutagens (Nakatani et al. 1999) in that the disrupted genes can be easily identified with the aid of inserted transposons. Therefore, transposon-based mutagenesis is called transposon tagging.

The estimated efficiency of the insertional mutagenesis of *Ciona intestinalis* with *Minos* transposons is about 1% per insertion site (Sasakura et al. 2007). In *Ciona intestinalis*, the studies using two insertional mutants resulted in the identification of several gene functions.

One mutant, named *swimming juvenile* (Fig. 10.3) (Sasakura et al. 2005), has a transposon insertion at the proximal region of the *cis* element of the gene encoding cellulose synthase (Nakashima et al. 2004).

In *Ciona*, transposons are frequently inserted as a long concatemer (the ends of the concatemer are the inverted repeats of transposons, and the region sandwiched by two inverted repeats includes tens of copies of the reporter gene, *cis* element, transposon element, and vector backbone). The transposon insertion of *swimming juvenile* fits into the case, and this long insertion may efficiently disrupt the function of the *cis* element that is usually so flexible that the insertion of a single, short transposon copy could not sufficiently disrupt the *cis* element function.

Swimming juvenile mutants exhibit malformed tunic morphology because cellulose is a major component in the tunic. Moreover, *swimming juvenile* mutant larvae exhibit an abnormal order of metamorphic events; the mutants start trunk metamorphosis while retaining their tails (tail regression is one of the earliest events in ascidian metamorphosis) (Cloney 1982). Therefore, the mutant larvae look like young juveniles with tails, suggesting that cellulose might be necessary for proper metamorphosis. The other mutant is discussed in the next chapter.

10.7 Maternal-Specific Knockdowns

A surprising finding of stable transgenesis in *Ciona* is the epigenetic gene silencing in female germ lines (Sasakura et al. 2010; Iitsuka et al. 2014). When transgenic lines that express GFP in eggs and oocytes are created with the help of a *cis* element of maternally expressed genes, GFP expression frequently becomes mosaic in the ovary; the ovary contains GFP-positive eggs and -negative eggs (Fig. 10.4). The GFP-negative eggs possess the GFP gene in their genome, suggesting that epigenetic gene silencing might occur in the germ cells.

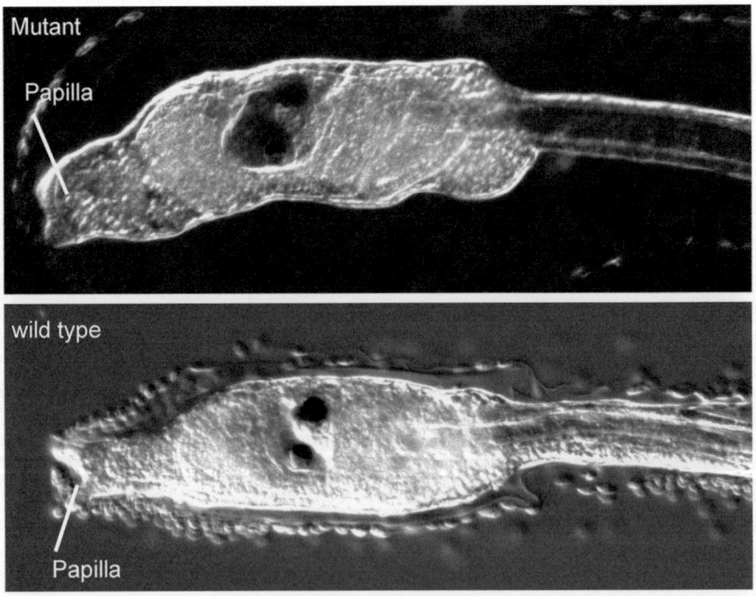

Fig. 10.3 A mutant of *Ciona intestinalis* generated by the transposon-based technique. The *upper image* shows the trunk of a *swimming juvenile* mutant larva, and the *lower image* the trunk of a wild-type larva. The anterior region, including the papillae, is thickened and rounded in the mutant larva

Fig. 10.4 Mosaic GFP expression in the ovary of a GFP transgenic line

In addition to GFP, the maternal gene whose *cis* element is used in the vector for the transgenic line is specifically silenced in the GFP-negative eggs at the transcriptional or post-transcriptional level. This phenomenon is called MASK, an acronym for maternal-specific knockdown (Iitsuka et al. 2014). The mechanism of MASK has not been well characterized, but a part of the 5′ untranslated region of the target gene needs to be joined with *GFP* for the knockdowns, suggesting that the transcription of the gene part with *GFP* might somehow inhibit expression of the gene in the manner of RNA interference (Fire et al. 1998).

In MASK, the zygotic expression of target genes at the zygotic stage is not silenced, suggesting that the phenotype seen in the MASK embryos might precisely reflect the maternal function of the gene. This feature is useful to distinguish the maternal and zygotic functions of genes if the genes have both expressions. Moreover, the maternal function of the genes could be addressed even if the zygotic expression of the genes plays an essential role. This is difficult in the case of conventional mutagenesis, as homozygous mutant females created by conventional mutagenesis could exhibit abnormal phenotypes at developmental stages and would not reach the reproductive stage because of the abnormalities. The MASK method facilitates the study of maternal gene function in ascidians, a historical subject in this animal group (Conklin 1905a, b, c).

10.8 Concluding Remarks

There are several approaches to generating transgenic lines in *Ciona*, including electroporation of a plasmid DNA, I-*Sce*I-mediated transgenesis, and transposon-based transgenesis. The resulting transgenic lines are valuable resources as tissue-specific markers and molecular tools for observing cellular events, physiology, and mutants. With the aid of the transgenic lines, sophisticated genetic procedures can be conducted in *Ciona* that facilitate studies on addressing gene functions. In the next chapter, the other famous application of transposon-based techniques, the enhancer trap, is discussed.

References

Arca B, Zabalou S, Loukeris TG, Savakis C (1997) Mobilization of a *Minos* transposon in *Drosophila melanogaster* chromosomes and chromatid repair by heteroduplex formation. Genetics 145:267–279

Awazu S, Sasaki A, Matsuoka T, Satoh N, Sasakura Y (2004) An enhancer trap in the ascidian *Ciona intestinalis* identifies enhancers of its *Musashi* orthologous gene. Dev Biol 275:459–472

Awazu S, Matsuoka T, Inaba K, Satoh N, Sasakura Y (2007) High-throughput enhancer trap by remobilization of transposon *Minos* in *Ciona intestinalis*. Genesis 45:307–317

Bruma S, Chen BPC, Chen DJ (2006) Role of non-homologous end joining (NHEJ) in maintaining genomic integrity. DNA Repair 5:1042–1048

Chiba S, Jiang D, Satoh N, Smith WC (2009) Brachyury null mutant-induced defects in juvenile ascidian endodermal organs. Development 136:35–39

Cloney RA (1982) Ascidian larvae and the events of metamorphosis. Am Zool 22:817–826

Conklin EG (1905a) The organization and cell lineage of the ascidian egg. J Acad. Nat Sci 13:1–119

Conklin EG (1905b) Organ forming substances in the eggs of ascidians. Biol Bull 8:205–230

Conklin EG (1905c) Mosaic development in ascidian eggs. J Exp Zool 2:146–223

Corbo JC, Levine M, Zeller RW (1997) Characterization of a notochord-specific enhancer from the *Brachyury* promoter region of the ascidian, *Ciona intestinalis*. Development 124:589–602

Deschet K, Nakatani Y, Smith WC (2003) Generation of *Ci-Brachyury-GFP* stable transgenic lines in the ascidian *Ciona savignyi*. Genesis 35:248–259

Dufour HD, Chettouh Z, Deyts C, de Rosa R, Goridis C, Joly JS, Brunet JF (2006) Precraniate origin of cranial motoneurons. Proc Natl Acad Sci U S A 103:8727–8732

Fire A, Xu S, Montgomery MK, Kostas SA, Driver SE, Mello CC (1998) Potent and specific genetic interference by double-stranded RNA in *Caenorhabditis elegans*. Nature 391:806–811

Franz G, Savakis C (1991) *Minos*, a new transposable element from *Drosophila hydei*, is a member of the Tc1-like family of transposons. Nucleic Acids Res 19:6646

Hatta K, Ankri N, Faber DS, Korn H (2006) Visualizing neurons one-by-one in vivo: optical dissection and reconstruction of neural networks with reversible fluorescent proteins. Dev Dyn 235:2192–2199

Horie T, Nakagawa M, Sasakura Y, Kusakabe TG (2009) Cell type and function of neurons in the ascidian nervous system. Develop Growth Differ 51:207–220

Horie T, Shinki R, Ogura Y, Kusakabe TG, Satoh N, Sasakura Y (2011) Ependymal cells of chordate larvae are stem-like cells that form the adult nervous system. Nature 469:525–528

Hozumi A, Kawai N, Yoshida R, Ogura Y, Ohta N, Satake H, Satoh N, Sasakura Y (2010) Efficient transposition of a single *Minos* transposon copy in the genome of the ascidian *Ciona intestinalis* with a transgenic line expressing transposase in eggs. Dev Dyn 239:1076–1088

Hozumi A, Mita K, Miskey C, Mates L, Izsvak Z, Ivics Z, Satake H, Sasakura Y (2013) Germline transgenesis of the chordate Ciona intestinalis with hyperactive variants of sleeping beauty transposable element. Dev Dyn 242:30–43

Hozumi A, Horie T, Sasakura Y (2015) Neuronal map reveals the highly regionalized pattern of the juvenile central nervous system of the ascidian *Ciona intestinalis*. Dev Dyn 244:1375–1393

Iitsuka T, Mita K, Hozumi A, Hamada M, Satoh N, Sasakura Y (2014) Transposon-mediated targeted and specific knockdown of maternally expressed transcripts in the ascidian *Ciona intestinalis*. Sci Rep 4:5050

Ikuta T, Saiga H (2007) Dynamic change in the expression of developmental genes in the ascidian central nervous system: revisit to the tripartite model and the origin of the midbrain-hindbrain boundary region. Dev Biol 312:631–643

Imai JH, Meinertzhagen IA (2007a) Neurons of the ascidian larval nervous system in *Ciona intestinalis*: I. Central nervous system. J Comp Neurol 501:316–334

Imai JH, Meinertzhagen IA (2007b) Neurons of the ascidian larval nervous system in *Ciona intestinalis*: II. Peripheral nervous system. J Comp Neurol 501:335–352

Ivics Z, Hackett PB, Plasterk RH, Izsvak Z (1997) Molecular reconstruction of sleeping beauty, a Tc1-like transposon from fish, and its transposition in human cells. Cell 91:501–510

Joly JS, Kano S, Matsuoka T, Auger H, Hirayama K, Satoh N, Awazu S, Legendre L, Sasakura Y (2007) Culture of *Ciona intestinalis* in closed systems. Dev Dyn 236:1832–1840

Kamiya C, Ohta N, Ogura Y, Yoshida K, Horie T, Kusakabe TG, Satake H, Sasakura Y (2014) Nonreproductive role of gonadotropin-releasing hormone in the control of ascidian metamorphosis. Dev Dyn 243:1524–1535

Kawakami K, Shima A (1999) Identification of the *Tol2* transposase of the medaka fish Oryzias latipes that catalyzes excision of a nonautonomous *Tol2* element in zebrafish *Danio rerio*. Gene 240:239–244

Klinakis AG, Loukeris TG, Pavlopoulos A, Savakis C (2000a) Mobility assays confirm the broad host-range activity of the *Minos* transposable element and validate new transformation tools. Insect Mol Biol 9:269–275

Klinakis AG, Zagoraiou L, Vassilatis DK, Savakis C (2000b) Genome-wide insertional mutagenesis in human cells by the *Drosophila* Mobile element *Minos*. EMBO Rep 1:416–421

Lemaire P (2011) Evolutionary crossroads in developmental biology: the tunicates. Development 138:2143–2152

Liu C (2013) Strategies for designing transgenic DNA constructs. Methods Mol Biol 1027

Liu YG, Mitsukawa N, Oosumi T, Whittier RF (1995) Efficient isolation and mapping of *Arabidopsis thaliana* T-DNA insert junctions by thermal asymmetric interlaced PCR. Plant J 8:457–463

Loukeris TG, Livadaras I, Arca B, Zabulou S, Savakis C (1995) Gene transfer into the Medfly, *Ceratitis capitata*, with a *Drosophila hydei* transposable element. Science 270:2002–2005

Matsuoka T, Awazu S, Satoh N, Sasakura Y (2004) *Minos* transposon causes germline transgenesis of the ascidian *Ciona savignyi*. Develop Growth Differ 46:249–255

Nakashima K, Yamada L, Satou Y, Azuma J, Satoh N (2004) The evolutionary origin of animal cellulose synthase. Dev Genes Evol 214:81–88

Nakatani Y, Moody R, Smith WC (1999) Mutations affecting tail and notochord development in the ascidian *Ciona savignyi*. Development 126:3293–3301

Nicol D, Meinertzhagen IA (1991) Cell counts and maps in the larval central nervous system of the ascidian *Ciona intestinalis* (L.) J Comp Neurol 309:415–429

Ogura Y, Sakaue-Sawano A, Nakagawa M, Satoh N, Miyawaki A, Sasakura Y (2011) Coordination of mitosis and morphogenesis: role of a prolonged G2 phase during chordate neurulation. Development 138:577–587

Pan F, Chen Y, Loeber J, Henningfeld K, Pieler T (2006) I-*Sce*I meganuclease-mediated transgenesis in *Xenopus*. Dev Dyn 235:247–252

Pavlopoulos A, Averof M (2005) Establishing genetic transformation for comparative development studies in the crustacean *Parhyale hawaiensis*. Proc Natl Acad Sci U S A 102:7888–7893

Pologruto TA, Yasuda R, Svoboda K (2004) Monitoring neural activity and [Ca2+] with genetically encoded Ca2+ indicators. J Neurosci 24:9572–9579

Rhee JM, Oda-Ishii I, Passamaneck YJ, Hadjantonakis AK, Di Gregorio A (2005) Live imaging and morphometric analysis of embryonic development in the ascidian *Ciona intestinalis*. Genesis 43:136–147

Sakaue-Sawano A, Kurokawa H, Morimura T, Hanyu A, Hama H, Osawa H, Kashiwagi S, Fukami K, Miyata T, Miyoshi H, Imamura T, Ogawa M, Masai H, Miyawaki A (2008) Visualizing spatiotemporal dynamics of multicellular cell-cycle progression. Cell 132:487–498

Sasakura Y (2007) Germline transgenesis and insertional mutagenesis in the ascidian *Ciona intestinalis*. Dev Dyn 236:1758–1767

Sasakura Y, Awazu S, Chiba S, Kano S, Satoh N (2003a) Application of *Minos*, one of the Tc1/*mariner* superfamily transposable elements, to ascidian embryos as a tool for insertional mutagenesis. Gene 308:11–20

Sasakura Y, Awazu S, Chiba S, Satoh N (2003b) Germline transgenesis of the Tc1/*mariner* superfamily transposon *Minos* in *Ciona intestinalis*. Proc Natl Acad Sci U S A 100:7726–7730

Sasakura Y, Nakashima K, Awazu S, Matsuoka T, Nakayama A, Azuma J, Satoh N (2005) Transposon-mediated insertional mutagenesis revealed the functions of animal cellulose synthase in the ascidian *Ciona intestinalis*. Proc Natl Acad Sci U S A 102:15134–15139

Sasakura Y, Oogai Y, Matsuoka T, Satoh N, Awazu S (2007) Transposon-mediated transgenesis in a marine invertebrate chordate, *Ciona intestinalis*. Genome Biol 8:S3

Sasakura Y, Konno A, Mizuno K, Satoh N, Inaba K (2008) Enhancer detection in the ascidian *Ciona intestinalis* with transposase-expressing lines of *Minos*. Dev Dyn 237:39–50

Sasakura Y, Suzuki MM, Hozumi A, Inaba K, Satoh N (2010) Maternal factor-mediated epigenetic gene silencing in the ascidian *Ciona intestinalis*. Mol Gen Genomics 283:99–110

Sasakura Y, Kanda M, Ikeda T, Horie T, Kawai N, Ogura Y, Yoshida R, Hozumi A, Satoh N, Fujiwara S (2012a) Retinoic acid-driven *Hox1* is required in the epidermis for forming the otic/atrial placodes during ascidian metamorphosis. Development 139:2156–2160

Sasakura Y, Mita K, Ogura Y, Horie T (2012b) Ascidians as excellent chordate models for studying the development of the nervous system during embryogenesis and metamorphosis. Develop Growth Differ 54:420–437

Shimizu K, Kamba M, Sonobe H, Kanda T, Klinakis AG, Savakis C, Tamura T (2000) Extrachromosomal transposition of the transposable element *Minos* in embryos of the silkworm *Bombyx mori*. Insect Mol Biol 9:277–281

Shimozono S, Iimura T, Kitaguchi T, Higashijima S, Miyawaki A (2013) Visualization of an endogenous retinoic acid gradient across embryonic development. Nature 496:363–366

Takahashi H, Hotta K, Erives A, Di Gregorio A, Zeller RW, Levine M, Satoh N (1999) Brachyury downstream notochord differentiation in the ascidian embryo. Genes Dev 13:1519–1523

Thermes V, Grabher C, Ristratore F, Bourrat F, Choulika A, Wittbrodt J, Joly JS (2002) I-*Sce*I meganuclease mediates highly efficient transgenesis in fish. Mech Dev 118:91–98

Tsutsui H, Karasawa S, Okamura Y, Miyawaki A (2008) Improving membrane voltage measurements using FRET with new fluorescent proteins. Nat Methods 5:683–685

Wada H, Saiga H, Satoh N, Holland PWH (1998) Tripartite organization of the ancestral chordate brain and the antiquity of placodes: insights from ascidian *Pax-2/5/8*, *Hox* and *Otx* genes. Development 125:1113–1122

Zeller RW, Weldon DS, Pellatiro MA, Cone AC (2006) Optimized green fluorescent protein variants provide improved single cell resolution of transgenic expression in ascidian embryos. Dev Dyn 235:456–467

The Enhancer Trap in *Ciona*

11

Yasunori Sasakura

Abstract

Enhancer trap is a famous application of transposons. This method is useful for the creation of marker transgenic lines that express a reporter gene in tissue- or organ-specific manner, characterization of enhancers in the genome, finding novel patterns of gene expression, and mutagenesis. In *Ciona intestinalis*, efficient enhancer traps with *Minos* and *Sleeping beauty* transposons have been reported. With the enhancer trap lines, the intronic enhancers regulating the expression of the *Musashi* gene, the compartment in the digestive tube, the presence of enhancers sensitive to the orientation of the gene that they regulate, and the functions of the *Hox1* gene have been revealed. The enhancer trap lines generated with the transposon vectors are valuable resources for use as visual markers.

Keywords

Transposon · Enhancer · Enhancer trap · *Hox* · Mutagenesis · *Minos*

11.1 Introduction: Mechanisms of the Enhancer Trap

The enhancer trap is a well-known application of transposon-mediated transgenesis (O'Kane and Gehring 1987). As stated in the previous section, the expression of a gene in a transposon vector can be influenced by the status of the surrounding genomic DNA where the transposon is inserted. The enhancers are a type of genomic DNA element that regulates the efficiency, timing, and spatial patterns of gene expression (Alberts et al. 2008). The position and orientation of enhancers with regard to their regulating genes are not necessarily fixed: enhancers can be present at the 3′ downstream region, at the distal and proximal regions of the 5′ upstream region, and in introns. Because of this flexibility of position, it can be difficult to characterize enhancers in the genome.

When a transposon vector is inserted at a genomic region that is under the control of an enhancer, the expression pattern of the (reporter) gene in the transposon vector can be altered according to the activity of the enhancer. This alteration of expression pattern is the signature of an enhancer trap. Through an enhancer trap, we can confirm the presence of an enhancer that regulates gene expression near the insertion site of a transposon, although sometimes enhancers are located at a region distal to the insertion sites. Moreover, we can see the nature of the enhancer by monitoring the expression pattern of the

Y. Sasakura (✉)
Shimoda Marine Research Center, University of Tsukuba, Shimoda, Shizuoka, Japan
e-mail: sasakura@shimoda.tsukuba.ac.jp

Y. Sasakura (ed.), *Transgenic Ascidians*, Advances in Experimental Medicine and Biology 1029,
https://doi.org/10.1007/978-981-10-7545-2_11

reporter gene. Therefore, an enhancer trap is useful for identifying enhancer elements.

The enhancer trap in animals was first reported in the fruit fly *Drosophila melanogaster* (O'Kane and Gehring, 1987), and similar phenomena have since been revealed in several organisms in which transgenic techniques are available, including zebrafish (Kawakami 2005). In *Ciona intestinalis*, our group reported efficient enhancer trap events with *Minos* and *Sleeping beauty* transposon vectors (Awazu et al. 2004; Hozumi et al. 2013). This technique is discussed in this chapter.

11.2 The Enhancer Trap in *Ciona intestinalis*

The genome project of *Ciona intestinalis* (Dehal et al. 2002) revealed the compact genome organization of this ascidian. The *Ciona* genome encodes about 16,000 genes in 160 megabase pairs of nucleotides per haploid, indicating that a gene is present in every 10 kilobase pair of the genome. The condensed location of genes in the *Ciona* genome means that enhancers are also densely located in the genome, suggesting a high probability of transposon insertions entrapping an enhancer. Our estimation suggested that about 15% of transgenic lines are enhancer trap lines (Sasakura et al. 2008).

An ideal transposon vector for the enhancer trap uses a promoter with a weak, nondetectable level of the activity for driving gene expression in nonselective tissues. This is because the presence of an enhancer trap can be easily monitored with the vector: enhanced reporter gene expression throughout the body means the entrapment of a ubiquitous and strong enhancer, whereas strong reporter gene expression in a tissue-specific manner suggests the presence of a tissue-specific enhancer. The tissue-independent nature of the promoter activity ensures the entrapment of enhancers for all tissues. For this reason, a promoter from a ubiquitously expressed gene, such as heat shock protein, has been applied to enhancer traps in several animal models such as *Drosophila* (Duffy 2002). In *Ciona*, however, such an enhancer trap vector has not been developed. Instead, a promoter that can express genes in a very restricted area is used for this experimental purpose (Sasakura et al. 2003). The promoter of the gene encoding thyroid peroxidase (TPO; Ogasawara et al. 1999) expresses genes at the anterior and posterior ends of the endostyle (Fig. 11.1), and this promoter has been used for an enhancer trap.

Currently, about 50 enhancer trap lines have been described in *Ciona*, with the enhancer trap vector using *TPO* promoter (Fig. 11.2) (Awazu et al. 2007; Sasakura et al. 2008; Hozumi et al. 2010; Ohta et al. 2010; Yoshida and Sasakura 2012). Almost all of these enhancer trap lines exhibit reporter gene expression after metamorphosis, but only a few enhancer trap lines have been isolated that has reporter gene expression at the larval stage (Awazu et al. 2007; EJ[MiTSAdTPOG]76 available at http://marinebio.nbrp.jp/ciona/index.jsp). The mechanism underlying this bias of expression timing has not been understood. Because the activity of transcription from the *TPO* promoter can be upregulated by enhancers responsible for the expression at the larval stage (Sasakura et al. 2016), it may be that the *TPO* promoter is not incompatible with early enhancers. A possible explanation is that enhancers responsible for the expression after metamorphosis are much more present in the *Ciona* genome than those for the larval stage. Because the complexity of the ascidian body is dramatically increased by metamorphosis, ascidians may require many enhancers for the proper regulation of gene expression in the complex adult body.

11.3 Using the Enhancer Trap to Create Marker Lines

One advantage of the enhancer trap is that it does not require the isolation of tissue-specific enhancers to create marker lines for the tissue. In *Ciona*, many *cis* elements for specifying the tissue-specific expression in larvae have been isolated

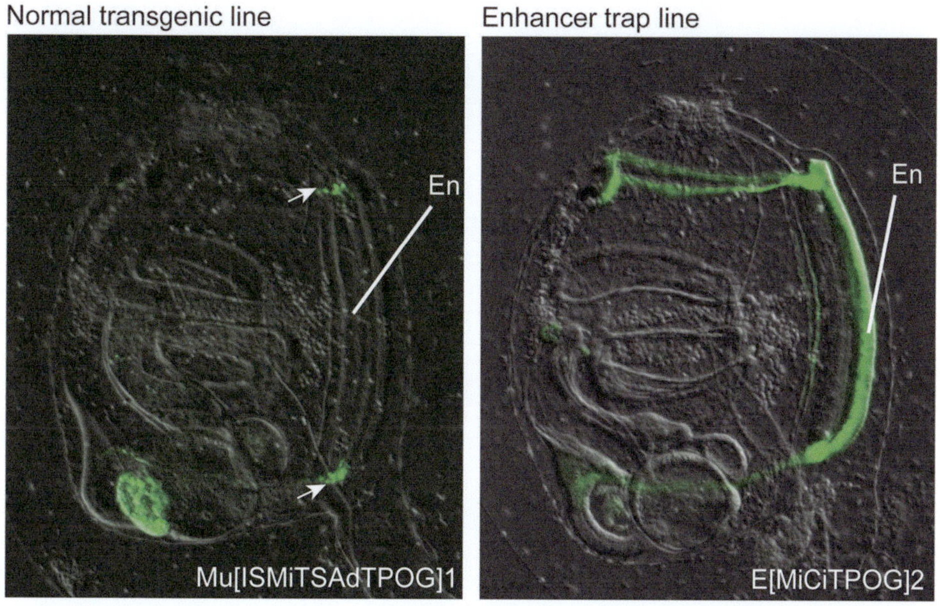

Fig. 11.1 An enhancer trap in *Ciona*. *Left panel*: Basal GFP expression from an enhancer trap vector. *Arrows* illustrate the GFP expression at the anterior and posterior poles of the endostyle (*En*). *Right*: GFP expression of an enhancer trap line. Both animals harbor the same enhancer trap vector. In the animal on the right, GFP expression is enhanced in the endostyle. The names of the enhancer trap lines, which can be used to search for the transgenic lines in our database, are shown at the bottom side of the panels

(Corbo et al. 1997). However, very little information has been accumulated about enhancers for adult tissues. The advantage of an enhancer trap is that it enables us to establish marker lines for adult tissues without the information about enhancers for adult tissues. As stated above, transposon insertion sites of enhancer trap lines provide us with clues toward the isolation of enhancers for adult tissues.

Yoshida and Sasakura (2012) attempted to exhaustively isolate enhancer trap lines for the digestive tube (Fig. 11.2). The digestive tube, or gut, is a long tubular organ for the digestion, absorption, and excretion of food. Like vertebrate digestive tubes, the digestive tube of *Ciona* can be divided along the anterior–posterior axis into several regions with distinct structures and functions (Willey 1983; Ikuta et al. 2004); however, the mechanism for specifying the identity of the region is not known in *Ciona* or other ascidians. A plausible reason for this is the unavailability of markers that can reveal the differentiation of a specific region in the digestive tube. Large-scale screening isolated ten enhancer trap lines that exhibit GFP expression in the digestive tube and its accessory organ, named the pyloric gland (Yoshida and Sasakura 2012). The enhancer trap lines showed that the digestive tube of *Ciona* can be divided into ten compartments (Fig. 11.2e), which exceeds the number of digestive tube boundaries based on morphological characteristics, suggesting that the regional division of the digestive tube based on gene expression might be more complicated than that deduced by the structure. The enhancer trap lines for the digestive tube are valuable resources for the analysis of gene functions in the organ. Indeed, some enhancer trap lines for the digestive tube were used in the functional study of *Hox10*, which is necessary for forming the posterior part of the digestive tube (Kawai et al. 2015).

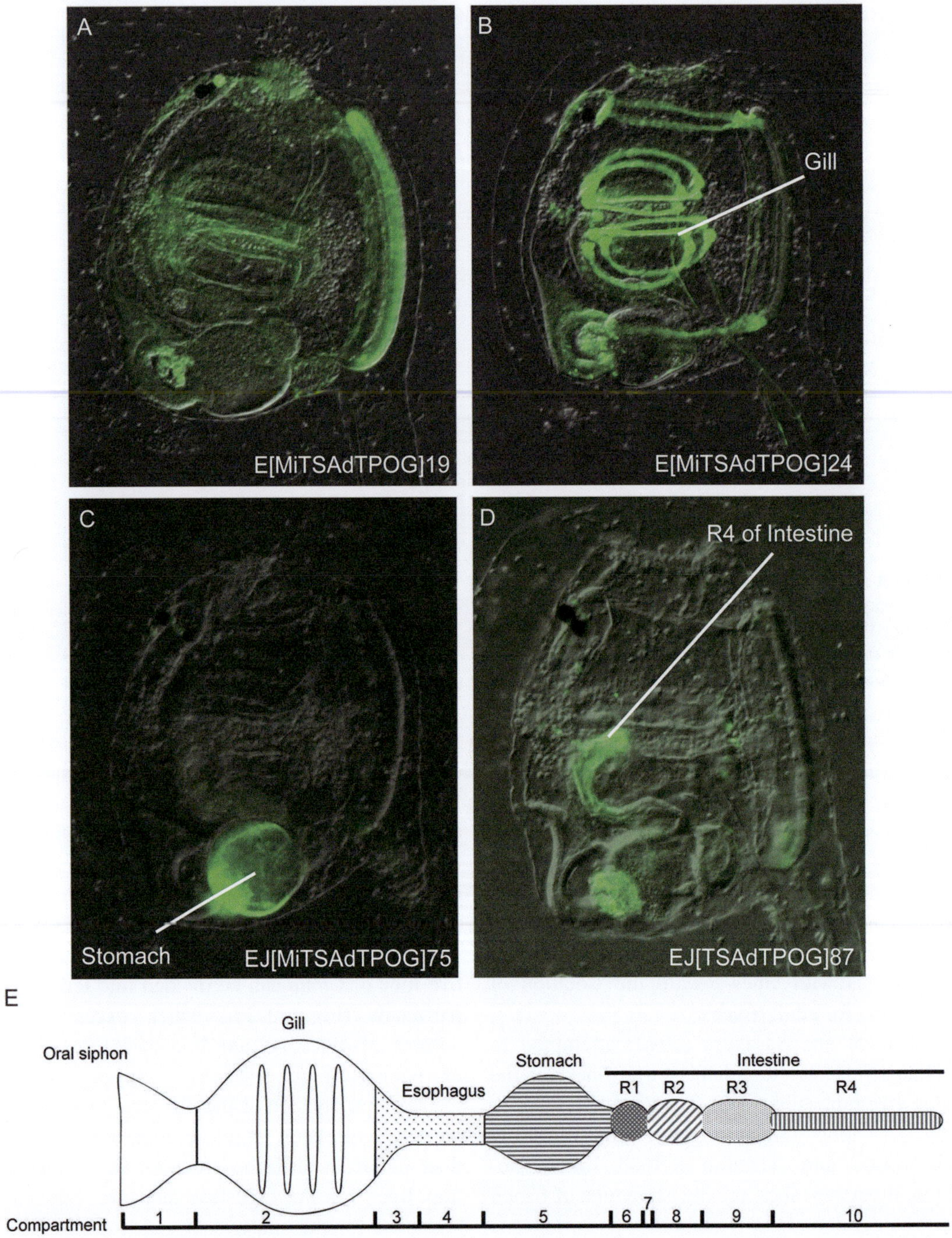

Fig. 11.2 Various enhancer trap lines. The four enhancer trap lines harbor the same enhancer trap vector. The names of the enhancer trap lines are shown in the panels. (**a**) GFP is expressed in a ubiquitous fashion. (**b**) GFP is expressed in the gills. (**c**) GFP is expressed in the stomach. (**d**) GFP is expressed in the intestine. (**e**) A schematic diagram showing the morphology and the ten compartments of the digestive tube of *Ciona intestinalis* (Yoshida and Sasakura 2012)

11.4 Using the Enhancer Trap to Characterize Enhancers

Enhancer trap lines provide us with clues about the location of enhancers in the genome with the aid of transposon insertion sites, thereby allowing us to characterize enhancers. In *Ciona*, detailed characterization of enhancers has been reported for a gene encoding a Musashi homolog (Awazu et al. 2007). Musashi is a kind of RNA-binding protein widely conserved among animals (Nakamura et al. 1994). The enhancer trap line of *Musashi* exhibits GFP expression in the pharyngeal endodermal tissues such as the endostyle and gill (Fig. 11.3). The enhancer trap line has a transposon insertion at the 8th intron of *Musashi*. Several introns of this gene have the enhancer activity responsible for the expression in the endostyle and gill, suggesting that plural intronic enhancers are responsible for the expression of this gene.

Enhancers are usually not influenced by the orientation of target genes. This is a major difference compared with promoters, which regulate genes in a unidirectional manner. However, an enhancer that is sensitive to the orientation of the target gene has been identified in the *Ciona* genome through the analysis of an enhancer trap line (Fig. 11.4a, b; Hozumi et al. 2013). This particular enhancer trap line expresses GFP specifically in the central nervous system (CNS). Indeed, there is a CNS enhancer near the insertion site. In addition, there are endoderm enhancers near the insertion site (Fig. 11.4d), and a gene near the insertion site is expressed in the endoderm in addition to the CNS. Therefore, the endodermal enhancers are not entrapped by the enhancer trap line.

We found that the endodermal enhancers are sensitive to the orientation of the genes. The *gfp* gene in the enhancer trap line is in the opposite direction to the endogenous gene regulated by the endodermal and CNS enhancers. When we constructed DNA vectors whose genes were organized in a similar manner to those in the enhancer trap line, the endodermal enhancers could not induce endodermal expression when the reporter genes were placed in the opposite orientation from the endogenous gene (Fig. 11.4a). By contrast, the endodermal enhancer could induce endodermal expression of

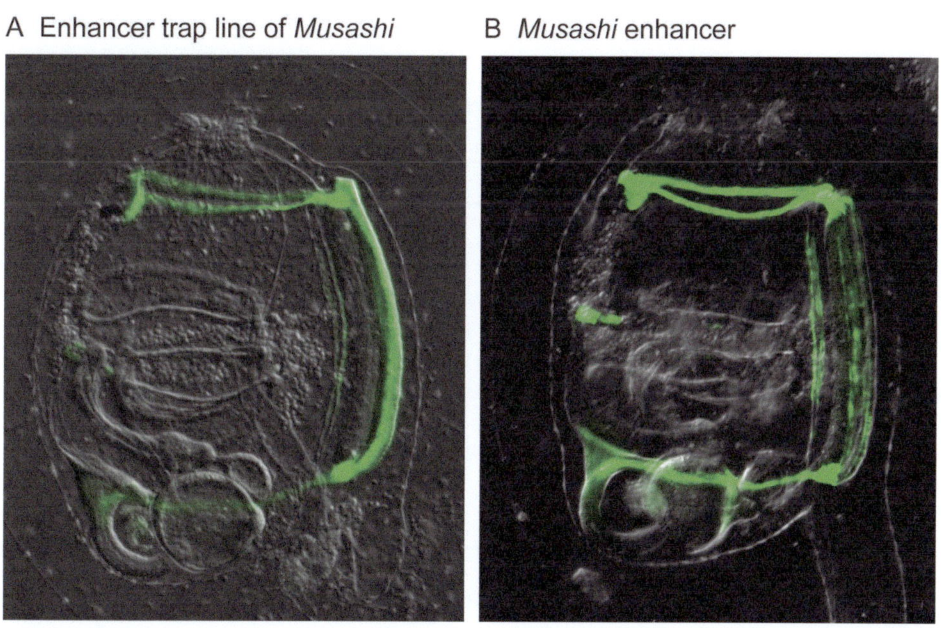

Fig. 11.3 An enhancer trap for isolating enhancers. (**a**) The enhancer trap line for the *Musashi* gene. The panel is the same as in Fig. 11.1a. (**b**) Expression of GFP from a vector having an enhancer element isolated from *Musashi*. The animals in (**a**) and (**b**) exhibit very similar GFP expression patterns

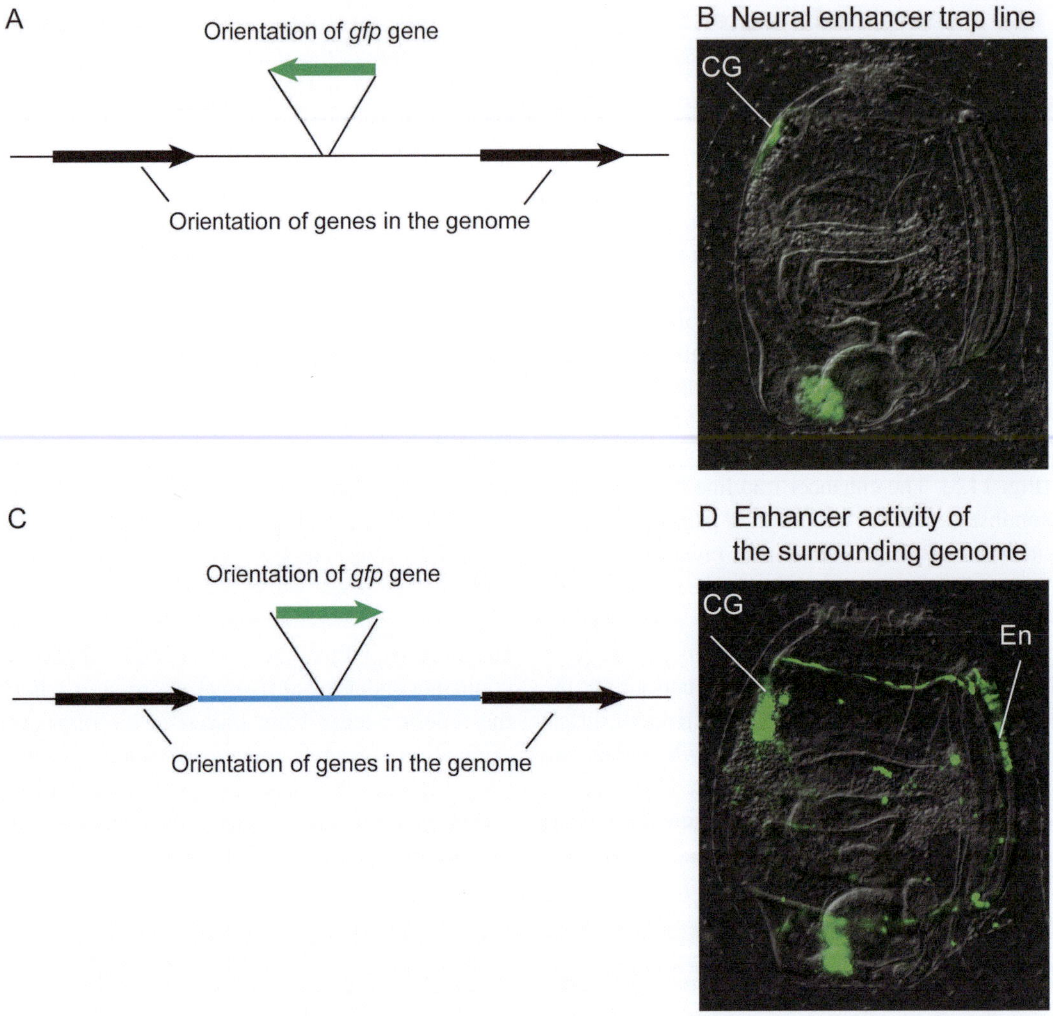

Fig. 11.4 The enhancer sensitive to the orientation of gene. (**a**) The schematic illustration of the enhancer trap event in the enhancer trap line shown in (**b**). The *arrows* indicate the orientation of genes. (**b**) The enhancer trap line EJ[MiTSAdTPOG]15 showing GFP expression in the cerebral ganglion (CG, the brain of the juvenile). (**c**) The schematic illustration of the DNA construct electroporated into the animal shown in **d**. The *arrows* indicate the orientation of the genes. The DNA region shown in *blue* was subcloned into a plasmid vector, and a *gfp* gene was inserted at the region corresponding to that of *gfp* in (**a**). Note that the orientation of the *gfp* gene is opposite to that in (**a**). (**d**) Expression of GFP from a vector having an enhancer element isolated from the genomic region near the transposon insertion site of the enhancer trap line in (**a**). GFP is expressed in the endostyle (*En*) in addition to the CG

the reporter genes when the orientation of the reporter genes was the same as that of the endogenous gene (Fig. 11.4c, d). The CNS enhancer does not have such sensitivity to the orientation of target genes, and thus the enhancer trap line shows CNS-specific GFP expression (Fig. 11.4b). Although the mechanism of orientation sensitivity is not known, a plausible explanation is that the orientation of genes influences the formation of the secondary structure of DNA, and the enhancer loses the accessibility to promoters in an orientation-dependent manner.

11.5 Using the Enhancer Trap to Elucidate Gene Expression Patterns

Enhancer trap lines can be used to predict the expression patterns of genes that are located near the causative insertion sites of enhancer traps, because the enhancers entrapped by the transposons are thought to regulate the genes near the insertion site in the same manner as the reporter gene in the transposon. In a study on the isolation of digestive tube trap lines, the expression patterns of genes near the insertion sites were investigated by means of in-situ hybridization so as to compare the expression patterns with those of GFP in the enhancer trap lines (Yoshida and Sasakura 2012). Among 10 enhancer trap lines, five had a gene with an expression pattern similar or identical to the expression pattern of GFP in the same lines. The coincidence of reporter expression and gene expression indicates that enhancer trap lines would be valuable for isolating genes expressed in a tissue or organ of interest.

In some enhancer trap lines, however, the genes that are in close proximity to the insertion sites do not exhibit an expression pattern similar to or identical to those of reporter genes (Ohta et al. 2010). In this case, the enhancers responsible for the enhancer trap events may not be the ones regulating the expression of genes near the insertion sites. This suggests that the target genes of these enhancers might not be close to the insertion sites, or the enhancers do not regulate the expression of an endogenous gene. Transposon insertions could add promoters at genomic points where the normal genome does not have a promoter. The artificially added promoters may recruit enhancers in an unexpected manner, and this kind of artificial modification of the genome could result in the entrapment of enhancers in an unusual fashion.

11.6 Using the Enhancer Trap for Mutagenesis

The compact genome of *Ciona* enables transposons to be inserted into genomic regions encoding genes with high probability. This suggests that transposon insertions causative for an enhancer trap have a high chance of being inserted into a critical element of genes (such as exons and transcription factor-binding sites) to disrupt gene functions. Therefore, enhancer trap lines can be a good source of mutants.

In *Ciona*, an enhancer trap line has been isolated that entraps the enhancer(s) regulating the expression of *Hox1* (Sasakura et al. 2012). The GFP expression of this line recapitulates the expression of *Hox1*. This enhancer trap line has the transposon insertion at the proximal region of the *Hox1* promoter. Animals homozygous for this transposon insertion line exhibit a dramatic reduction in the expression of *Hox1*, suggesting that the transposon insertion disrupts a core element of the *Hox1 cis* element. The homozygous *Hox1* enhancer trap animals show abnormality in the formation of atrial siphon primordia at the larval stage, and after metamorphosis they exhibit abnormal pharynges (Fig. 11.5), indicating that *Hox1* plays a critical role in the formation of these structures (Sasakura et al. 2012).

Compared with other splendid model organisms for genetics, such as *Drosophila* and *C. elegans*, *Ciona intestinalis* does not readily lend itself to large-scale culturing with an inland culturing system (the inland culturing system is needed to isolate transgenic lines without accidentally spreading them in the wild). For this reason, an efficient method of isolating mutant lines with higher probability is needed, and the enhancer trap could provide a solution to this problem.

11.7 The Enhancer Trap as a Resource for the Tunicate Community

Enhancer trap lines and other transgenic lines are valuable resources for use as visual markers. To facilitate their sharing among researchers, a database of transgenic lines has been constructed and can be accessed at the CITRES website (http://marinebio.nbrp.jp/ciona/). Using this database, we can see the

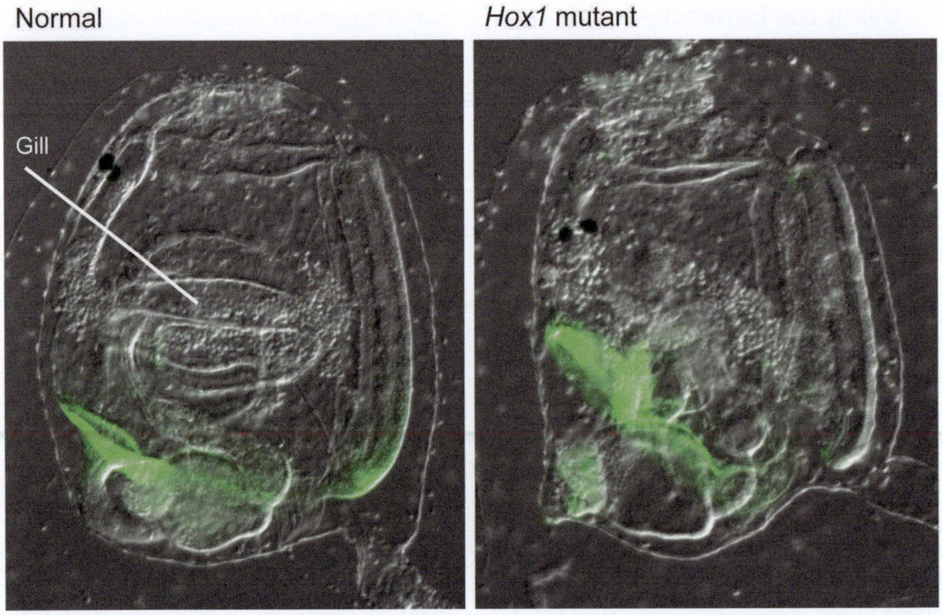

Fig. 11.5 A *Hox1* mutant line generated using an enhancer trap. *Left panel* is a heterozygous mutant juvenile of *Hox1*. The name of the line is EJ[MiTSAdTPOG]124. *Green* represents GFP expression regulated by the *Hox1* enhancer. Its shape is normal, and gills are present at the center of the body. The image in the *right panel* shows a homozygous mutant juvenile of *Hox1*. Gills are absent

expression patterns of fluorescent proteins, DNA sequences of reporter constructs, and insertion sites of transposons, in addition to the various ways in which the transgenic lines were used (Sasakura et al. 2009). The database is equipped with a keyword search system, making it possible to identify the transgenic lines most appropriate for different experimental purposes. Moreover, the website allows its users to "shop" for transgenic lines, which can then be mailed directly.

11.8 Conclusions

An efficient enhancer trap is the representative characteristic of transposon techniques in *Ciona*. This technique is useful for creating marker lines, and for the characterization of enhancers in the genome and mutagenesis. Therefore, enhancer trap lines are a valuable resource for uncovering novel biological phenomena in *Ciona*.

References

Alberts B, Johnson A, Lewis J, Raff M, Roberts K, Walter P (2008) Molecular biology of the cell, 5th ed. Garland Science, New York

Awazu S, Sasaki A, Matsuoka T, Satoh N, Sasakura Y (2004) An enhancer trap in the ascidian *Ciona intestinalis* identifies enhancers of its *Musashi* orthologous gene. Dev Biol 275:459–472

Awazu S, Matsuoka T, Inaba K, Satoh N, Sasakura Y (2007) High-throughput enhancer trap by remobilization of transposon *Minos* in *Ciona intestinalis*. Genesis 45:307–317

Corbo JC, Levine M, Zeller RW (1997) Characterization of a notochord-specific enhancer from the *Brachyury* promoter region of the ascidian, *Ciona intestinalis*. Development 124:589–602

Dehal P, Satou Y, Campbell RK, Chapman J, Degnan B, De Tomaso A, Davidson B, Di Gregorio A, Gelpke M, Goodstein DM, Harafuji N, Hastings KE, Ho I, Hotta K, Huang W, Kawashima T, Lemaire P, Martinez D, Meinertzhagen IA, Necula S, Nonaka M, Putnam N, Rash S, Saiga H, Satake M, Terry A, Yamada L, Wang HG, Awazu S, Azumi K, Boore J, Branno M, Chin-Bow S, DeSantis R, Doyle S, Francino P, Keys DN, Haga S, Hayashi H, Hino K, Imai KS, Inaba K, Kano S, Kobayashi K, Kobayashi M, Lee BI, Makabe

KW, Manohar C, Matassi G, Medina M, Mochizuki Y, Mount S, Morishita T, Miura S, Nakayama A, Nishizaka S, Nomoto H, Ohta F, Oishi K, Rigoutsos I, Sano M, Sasaki A, Sasakura Y, Shoguchi E, Shin-I T, Spagnuolo A, Stainier D, Suzuki MM, Tassy O, Takatori N, Tokuoka M, Yagi K, Yoshizaki F, Wada S, Zhang C, Hyatt PD, Larimer F, Detter C, Doggett N, Glavina T, Hawkins T, Richardson P, Lucas S, Kohara Y, Levine M, Satoh N, Rokhsar DS (2002) The draft genome of *Ciona intestinalis*: insight into chordate and vertebrate origins. Science 298:2157–2167

Duffy JB (2002) GAL4 system in *Drosophila*: a fly geneticist's Swiss army knife. Genesis 34:1–15

Hozumi A, Kawai N, Yoshida R, Ogura Y, Ohta N, Satake H, Satoh N, Sasakura Y (2010) Efficient transposition of a single *Minos* transposon copy in the genome of the ascidian *Ciona intestinalis* with a transgenic line expressing transposase in eggs. Dev Dyn 239:1076–1088

Hozumi A, Yoshida R, Horie T, Sakuma T, Yamamoto T, Sasakura Y (2013) Enhancer activity sensitive to the orientation of the gene it regulates in the chordate genome. Dev Biol 375:79–91

Ikuta T, Yoshida N, Satoh N, Saiga H (2004) *Ciona intestinalis Hox* gene cluster: its dispersed structure and residual colinear expression in development. Proc Natl Acad Sci U S A 101:15118–15123

Kawai N, Ogura Y, Ikuta T, Saiga H, Hamada M, Sakuma T, Yamamoto T, Satoh N, Sasakura Y (2015) *Hox10*-regulated endodermal strand cell migration is essential for development of the ascidian intestine. Dev Biol 403:43–56

Kawakami K (2005) Transposon tools and methods in zebrafish. Dev Dyn 234:244–254

Nakamura M, Okano H, Blendy JA, Montell C (1994) Musashi, a neural RNA-binding protein required for *Drosophila* adult external sensory organ development. Neuron 13:67–81

Ogasawara M, Di Lauro R, Satoh N (1999) Ascidian homologs of mammalian thyroid peroxidase genes are expressed in the thyroid-equivalent region of the endostyle. J Exp Zool 285:158–169

Ohta N, Horie T, Satoh N, Sasakura Y (2010) Transposon-mediated enhancer detection reveals the location, morphology and development of the cupular organs, which are putative hydrodynamic sensors, in the ascidian *Ciona intestinalis*. Zool Sci 27:842–850

O'Kane C, Gehring WJ (1987) Detection in situ of genomic regulatory elements in *Drosophila*. Proc Natl Acad Sci U S A 84:9123–9127

Sasakura Y, Awazu S, Chiba S, Satoh N (2003) Germ-line transgenesis of the Tc1/*mariner* superfamily transposon *Minos* in *Ciona intestinalis*. Proc Natl Acad Sci U S A 100:7726–7730

Sasakura Y, Konno A, Mizuno K, Satoh N, Inaba K (2008) Enhancer detection in the ascidian *Ciona intestinalis* with transposase-expressing lines of *Minos*. Dev Dyn 237:39–50

Sasakura Y, Inaba K, Satoh N, Kondo M, Akasaka K (2009) *Ciona intestinalis* and *Oxycomanthus japonicus*, representatives of marine invertebrates. Exp Anim 58:459–469

Sasakura Y, Kanda M, Ikeda T, Horie T, Kawai N, Ogura Y, Yoshida R, Hozumi A, Satoh N, Fujiwara S (2012) Retinoic acid-driven *Hox1* is required in the epidermis for forming the otic/atrial placodes during ascidian metamorphosis. Development 139:2156–2160

Sasakura Y, Ogura Y, Treen N, Yokomori R, Park SJ, Nakai K, Saiga H, Sakuma T, Yamamoto T, Fujiwara S, Yoshida K (2016) Transcriptional regulation of a horizontally transferred gene from bacterium to chordate. Proc Biol Sci 283:20161712

Willey A (1983) Studies on the Protochordata. I. On the origin of the branchial stigmata, preoral lobe, endostyle, atrial cavities, etc. in *Ciona intestinalis*, Linn., with remarks on *Clavelina lepadiformis*. Q J Micr Sci 34:317–360

Yoshida R, Sasakura Y (2012) Establishment of enhancer detection lines expressing GFP in the gut of the ascidian Ciona Intestinalis. Zool Sci 29:11–20

Check for updates

12 TALEN-Based Knockout System

Keita Yoshida and Nicholas Treen

Abstract

Targeted mutagenesis of genes-of-interest is a powerful method of addressing the functions of genes. Genome editing techniques, such as transcriptional activator-like effector nucleases (TALENs) and clustered regularly interspaced short palindromic repeats (CRISPR)/Cas9 systems, have enabled this approach in various organisms because of their ease of use. In the ascidian, *Ciona intestinalis*, recent studies show that TALEN-based knockout can be applied to establishing both mutant lines and tissue-specific knockout for addressing gene functions. Here, we introduce recent updates to the TALEN toolkit that facilitate detailed functional analysis of genes in ascidians.

Keywords

Ascidian · *Ciona intestinalis* · Genome editing · TALEN · Conditional knockout

K. Yoshida (✉)
Shimoda Marine Research Center, University of Tsukuba, Shimoda, Shizuoka, Japan
e-mail: yoshida.keita@twmu.ac.jp

N. Treen
Lewis-Sigler Institute for Integrative Genomics, Princeton University, Princeton, NJ, USA

12.1 Introduction

Transcriptional activator-like effector nucleases (TALENs) are artificially constructed nucleases that combine an array of DNA-binding TALE domains with a DNA-cutting FokI domain. *Xanthamonas* bacteria are known to manipulate gene expression of their host plants during infection by introducing TALE transcription factors into host cells (Kay and Bonas 2009). Predicting the exact DNA sequence that a DNA binding protein will bind to can be challenging, but it was shown that TALEs have a 1:1 TALE domain:nucleotide binding code (Boch et al. 2009; Moscou and Bogdanove 2009). This property showed a high potential for application to genome editing tools. Zinc-finger nucleases have successfully been used for genome editing, including in *Ciona* (Kawai et al. 2012). However, the extensive screening that was needed to design a functional Zinc finger nuclease pair limited the application of this technology. The construction of TALEN pairs is much simpler; although it does require experience in molecular biology, it is perfectly feasible for a small laboratory to produce a limited number of TALEN pairs that typically have enough mutational efficiency to observe the effects of gene knockout of a zygotically expressed gene in the G0 generation.

There are several published TALEN construction protocols. We use the Platinum Gate TALEN kit (Sakuma et al. 2013a), which is

Y. Sasakura (ed.), *Transgenic Ascidians*, Advances in Experimental Medicine and Biology 1029,
https://doi.org/10.1007/978-981-10-7545-2_12

freely distributed by Addgene (kit #1000000043). TALEN mutagenesis works on a vast range of organisms (Carroll 2014). The only limitation appears to be the ability to get the TALEN pair into the nucleus where it binds to the DNA and produces mutations because of the imperfect nature of the cell's DNA repair machinery.

Well-established methods of introducing DNA or RNA into *Ciona* embryos by microinjection or electroporation are described in this book. We have previously taken advantage of these tools to introduce TALENs into *Ciona* embryos to perform either ubiquitous or tissue-specific knockouts.

The first example of using TALENs in *Ciona* used plasmids that target *Fgf3* and *Hox12* (Treen et al. 2014). Using this approach, we were capable of reproducing previously described morpholino knockdowns (Ikuta et al. 2010; Shi et al. 2009). By specifically knocking out *Fgf3* in mature neurons, we showed that this signaling ligand is required for tail regression during metamorphosis.

In another study, *Hox10* knockout in the endodermal strand was shown to disrupt its migration (Kawai et al. 2015). Electroporation-mediated knockouts are a powerful tool, but the microinjection of TALEN mRNAs has advantage that it can act earlier in development. We found that injection of TALEN mRNA from the one-cell stage was comparably as effective at introducing mutations as electroporations, and was much more efficient at mutating the germline, which can be difficult to mutate using TALEN expression vectors, because it is transcriptionally quiescent during embryogenesis. Therefore, TALEN mRNA microinjection is highly effective at constructing knockout lines (Yoshida et al. 2014) and we have published the successful production of knockout lines of *Hox4* and *Hox5* genes, and several others have been produced that will be described soon.

The scientific community in general has overwhelmingly accepted clustered regularly interspaced short palindromic repeats (CRISPR) as the genome editing technology of choice in the past few years. CRISPR expression constructs are undoubtedly easier to produce than TALENs and can be reliably used in *Ciona*, as described in the next chapter, but there are still some advantages to using TALENs and we describe three recent updates to the *Ciona* TALEN toolkit that we are currently employing.

12.2 2A-TALEN Systems

In our original description of TALENs, we employed two separate gene expression cassettes on a single plasmid. One cassette to express TALEN, and the other to express a fluorescent protein to confirm the success of electroporation. Although the above construct does not require an artificial peptide sequence to be added to TALENs that might risk decreasing their activities, the vector does not always ensure the same expression level between TALENs and marker proteins. For this reason, we redesigned the TALEN plasmid to contain the fluorescent protein within the same open reading frame as the TALEN, separated by a 2A peptide sequence. 2A peptide regions are known to act either as an unstable "self-cleaving" peptide, or possibly as a region that forces the ribosome to skip the incorporation of some amino acids during mRNA translation, resulting in two separate proteins from a single mRNA. This method appears to be more reliable than our attempts to use internal ribosome entry site (commonly known as IRES) in *Ciona*.

Among the known 2A peptide sequences (Kim et al. 2011) we have used the sequence GSGEGRGSLLTCGDVEENPGP in our TALEN constructions. The stop codon of the TALEN open reading frame is replaced with the 2A peptide sequence, immediately followed by an in-frame mCherry open reading frame. Using this construct, we can be confident that the TALEN is present in the cells, which exhibit red fluorescence derived from mCherry at expression levels that can be deduced from the intensity of red fluorescence (Fig. 12.1). This is currently our preferred

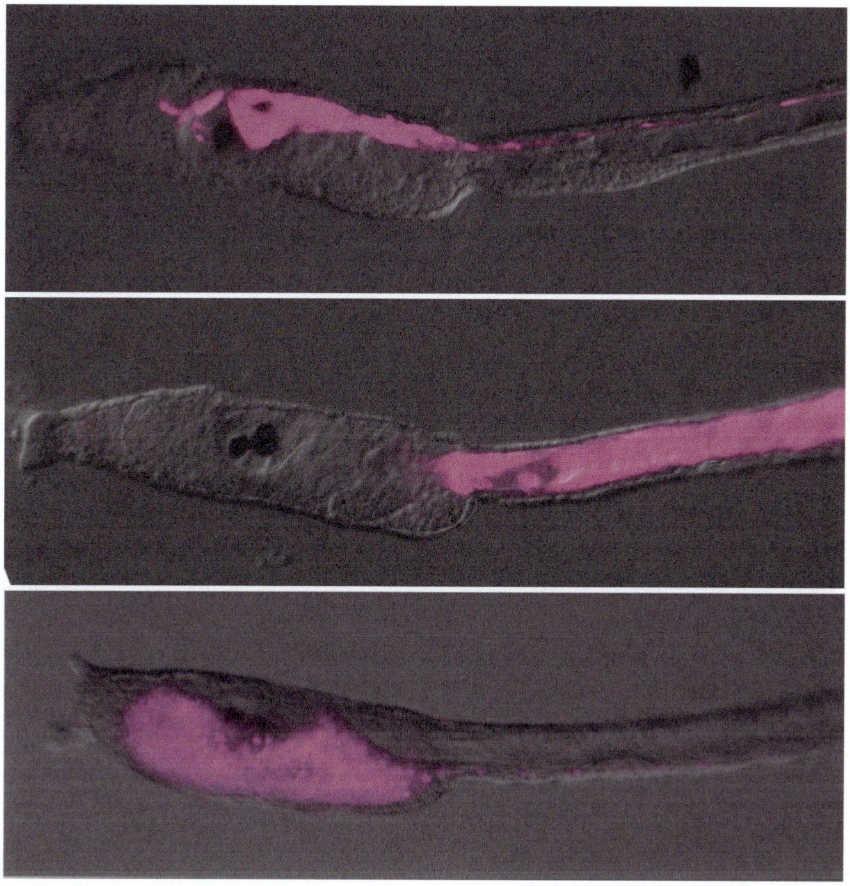

Fig. 12.1 Tissue-specific expression of transcriptional activator-like effector nucleases (TALENs) monitored by mCherry fluorescence. Representative larvae electroporated with *Nut* > *TALENs::2A::mCherry* (*top*), *TnI* > *TALENs::2A::mCherry* (*middle*), and *Titf1* > *TALENs::2A::mCherry* (*bottom*) that express mCherry in the central nervous system, muscle, and endoderm respectively

method of introducing TALENs and it can be used for ubiquitous knockouts using ubiquitous promoters, mRNA microinjections or for tissue-/cell-specific knockouts using appropriate expression plasmids. It has recently been shown using this method that a TALEN pair can be used to mutate an AP2 binding site within the epidermal enhancer of the gene encoding cellulose synthase (Sasakura et al. 2016).

This research demonstrates that TALENs can be reliably used to mutate a specific binding site of a transcription factor, thus allowing us to observe the real function of these binding sites in the context of the chromosome, which was difficult using conventional methods in *Ciona*.

12.3 Promoter-Switching System for Tissue-Specific TALEN Expression

Conditional knockouts allows us to knockout target genes in specific cells, tissues or organs. This analysis is particularly necessary for addressing the function of genes that are expressed in plural tissues and/or developmental timings. In a conventional mutant where its causative gene is disrupted in all of the cells, the relationships between a phenotype and an expression domain is not always easy to know. Conditional knockout can disrupt the gene in only one of its expression domains; we can easily know the causative expression of the gene for a phenotype. TALENs and CRISPR/Cas9 have provided easy conditional knockout systems in the *Ciona* community by expressing the components with appropriate *cis*-regulatory elements that can drive them in specific cells, tissues, and organs within a limited time (Tolkin and Christiaen 2016; Yoshida et al. 2017a). To facilitate this analysis, we established a method that can easily switch the promoter-enhancer elements in the TALEN expression vector.

The mutation efficiency of newly constructed TALEN pairs should be determined before functional analysis, because constructed TALENs are not always active enough to mutate genes efficiently. For this purpose, the TALENs are first assembled in a vector harboring the regulatory sequence of *EF1α*, which can express downstream genes in a ubiquitous manner (Fig. 12.2) (Treen et al. 2014; Sasakura et al. 2010).

When a TALEN pair with a good mutation rate is constructed, the *EF1α* regulatory element is replaced with appropriate *cis* elements for functional analyses of genes. In our vector system, the *EF1α*-regulatory element can be removed by digesting the vector with the *Not*I restriction enzyme followed by gel extraction of a TALEN-containing vector fragment (Fig. 12.2). After removal of the *EF1α* element, another *cis*-regulatory element can be inserted into the *Not*I site using an in vitro recombination system such as the In-Fusion HD Cloning Kit (Clontech). The insert fragments are required to have 15 bp extensions that are complementary to the ends of the *Not*I-digested TALEN vector (Fig. 12.2). The recombination-based method allows us to easily insert any sequence as the driver for TALEN expression. Once the insert fragment is prepared, we can use this fragment for any constructed TALENs, because the ends of *Not*I-digested TALEN vector are invariant. These properties facilitate tissue-specific knockout analysis using TALENs in ascidians.

12.4 Application of TALEN-Based Knockout

Transcription activator-like effector nuclease (TALEN)-based knockout is a powerful method of addressing the function of genes-of-interest. Here, we introduce an application of TALENs to ascidian research. This is TALEN-based knockout of the cellulose synthase gene (*CesA*), which improves in-situ hybridization analysis in ascidian larvae by reducing nonspecific, background staining in the tunic.

Whole-mount in-situ hybridization (WISH) is an essential method of visualizing gene expression, and is routinely performed in ascidian research. However, there is a difficulty in performing WISH in ascidian larvae because of the presence of the tunic, the thin noncellular film surrounding the body. Because the tunic causes a high background signal, expression of a gene is not easy to see at the larval stage by WISH. Cellulose is the major component in the tunic, whose synthesis requires *CesA* that encodes cellulose synthase and is expressed in the epidermis (Nakashima et al. 2004).

An effective way of avoiding the high background staining in the tunic is blocking cellulose formation by the disruption of *CesA* function. It has been shown that knockdown of *CesA* by morpholino antisense oligonucleotides (MOs) can be utilized for detecting gene expression at the endodermal strand, a very thin tissue that tends to give faint WISH signals (Kawai et al. 2015). However, this method requires purchasing MOs and the microinjection of MOs into eggs. Because MOs are expensive and microinjection is laborious, this may not be desirable for many researchers. To circumvent this issue, we constructed a TALEN pair designed to target *CesA*. The constructed *CesA*-TALEN pair showed a good mutation rate. To prevent the cellulose synthesis in the epidermis, *CesA*-TALENs are expressed using the regulatory sequence of *Friend of GATA* (*fog*) gene, which can drive gene expression in the animal hemisphere (Rothbächer et al. 2007).

We confirmed that cellulose formation is severely reduced in *CesA*-TALENs-introduced larvae by staining cellulose (Fig. 12.3). When we performed WISH using *CesA*-TALENs-introduced larvae, we successfully obtained clear signals with less background staining than control samples (Fig. 12.3).

An alternative way of obtaining cellulose-absent larvae is using *CesA*-mutant animals.

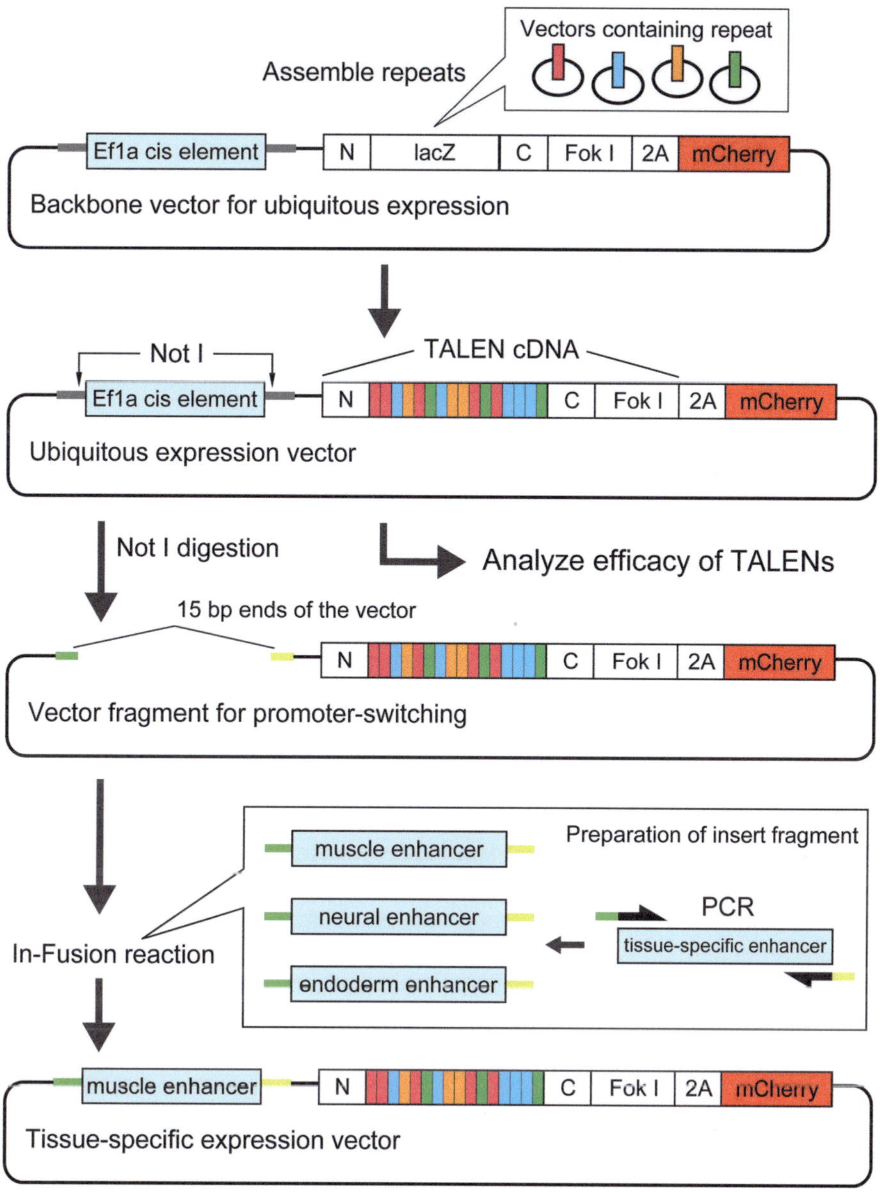

Fig. 12.2 Our TALEN vector systems for gene knockouts in *Ciona intestinalis*. *C* cDNA encoding carboxyl terminal region of TALE domain, *FokI* cDNA encoding the nuclease domain of FokI, *LacZ* a portion of *LacZ* cDNA, *N* cDNA encoding amino terminal region of TALE domain, *2A* cDNA encoding 2A peptide, *NotI* *Not*I restriction site. Note that the enhancers contain promoter elements

Ciona intestinalis mutant line for *CesA*, named *swimming juvenile* (Sasakura et al. 2005), has been established and is currently available through National BioResource Project (NBRP), Japan (http://marinebio.nbrp.jp/ciona/top/top.jsp). Using *CesA*-mutant larvae, we can obtain clear signals through WISH analysis as when using *CesA*-knockout animals (Fig. 12.3).

The advantages of the mutant over CesA–TALEN-introduced animals are as follows. First, CesA-mutant larvae completely lose cellulose in a nonmosaic fashion, whereas CesA–

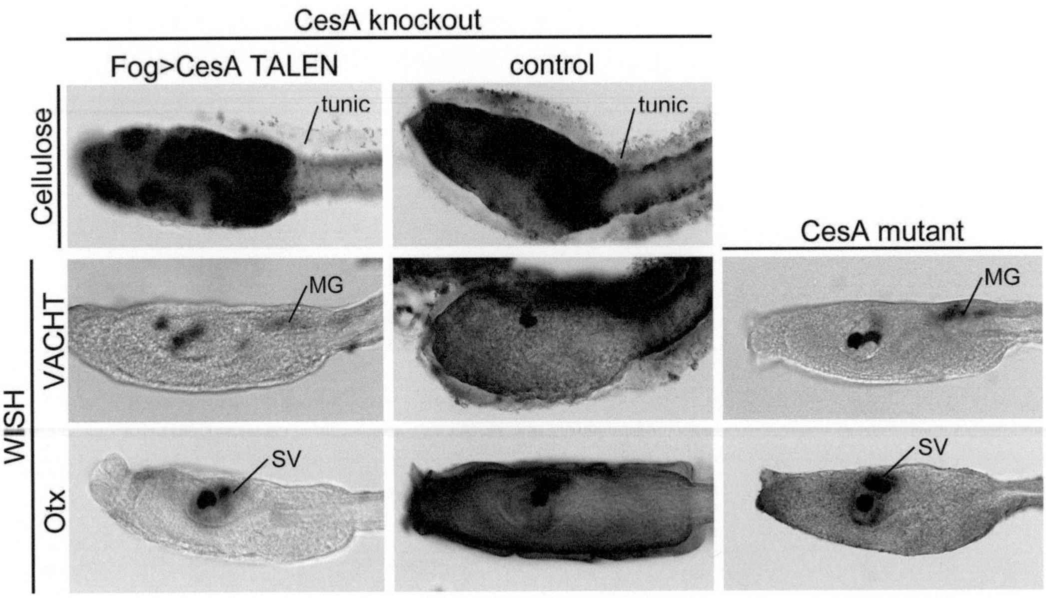

Fig. 12.3 Knockout of cellulose synthase gene (*CecA*) by TALENs. *Top panels* show cellulose staining in *CesA* knockout (*left*) and control (*right*) larvae. Cellulose in the tunic is hardly detectable in *CesA*-knockout larva. *Middle panels* show expression of *VACHT* detected by whole-mount in-situ hybridization (WISH) in *CesA* knockout (*left*), control (*center*) and *CesA* mutant (*right*) larvae. *Bottom panels* show expression of *Otx* detected by WISH in *CesA* knockout (*left*), control (*center*), and *CesA* mutant (*right*) larvae, *MG* motor ganglion, *SV* sensory vesicle

TALEN-introduced animals possibly retain some cells with functional CesA. Second, CesA mutants can be obtained without electroporation; because electroporation requires dechorionation (see Chap. 5), the resultant larvae usually lose their left–right axis (Shimeld and Levin 2006; Yoshida and Saiga 2008). CesA-mutant larvae possess a normal left–right axis, and the larvae are suitable for analyzing genes responsible for the left–right axis. CesA mutants are limited in that collecting many eggs from transgenic lines is not easy. For this reason, using mutants would not be good for functional analyses of genes through electroporation or microinjection. Therefore, TALEN-based knockout of *CesA* is a preferable way of obtaining many specimens for WISH analysis.

Using this method, we can prevent cellulose formation by both electroporations and microinjections in parallel with modifying functions of genes-of-interest, by mixing *CesA* TALEN vectors or mRNAs with other molecules that can disrupt functions of the genes.

12.5 Methods

12.5.1 TALEN Construction

The methods of assembling TALENs are not described here; please refer to protocols provided by each TALEN kit. We recommend using platinum TALENs (Sakuma et al. 2013a) for knockouts in *Ciona*. The TALEN repeat arrays can be assembled in the backbone vector for ubiquitous expression as described above (Fig. 12.2). Vector information is available at http://marinebio.nbrp.jp/ciona/forwardToKnockOutAction.do.

12.5.2 Checking the Efficiency of TALENs

The efficacy of constructed TALENs can be analyzed by expressing TALENs ubiquitously using *EF1α*-regulatory element. A brief procedure of our protocol for estimating TALEN activity is as follows:

1. After assembly of TALEN repeat arrays into the backbone vector with *EF1α*-promoter, incubate appropriate *E. coli* clones in 3 mL of L-broth medium. Midi- or maxiprep-scale culture is not necessary.
2. Extract plasmid DNAs using a commercially available miniprep kit. We use Nucleospin Plasmid Easypure kit (Macherey-Nagel). Elute DNA with 50 μl of buffer. Digest extracted DNAs (4 μl) by *Nru*I restriction enzyme to confirm the assembly. In the platinum TALEN kit, the NG module that recognizes T contains an *NruI* site; therefore, different TALEN assemblies usually have different restriction patterns.
3. Prepare DNA solution for the electroporation by mixing 35 μl each of left and right TALEN expression vectors and 10 μl of TE buffer.
4. Electroporation of a TALEN pair into eggs (see Chap. 5 for detail). We use a cuvette 4 mm wide containing 80 μl of DNA solution, 420 μl of mannitol/seawater solution, and 300 μl of eggs in mannitol/seawater solution for each electroporation.
5. When electroporated animals develop to the larval stage, pick up about 50–100 animals that exhibit ubiquitous mCherry fluorescence with a fluorescent microscope. It is not necessary to avoid collecting embryos that show abnormal morphology when estimating mutation rates.
6. Extract genomic DNA from the collected embryos in bulk. We use the Wizard genomic DNA purification kit (Promega) for isolating genomic DNA.
7. Amplify DNA fragments 200–300 bp long that contain the target site of a TALEN pair by polymerase chain reaction (PCR). The presence of mutations in PCR fragments is detected by means of surveyor nucleases such as Cel-I (Sakuma et al. 2013b) or electrophoresis with 15% polyacrylamide gel (PAGE) (Ota et al. 2013). We recommend PAGE because PCR fragments can be analyzed directly with clearer resolution. To estimate mutation rates, clone and sequence PCR fragments to determine the frequency of mutated clones.
8. Mutation frequency indicates the probability of a copy of the target genetic locus having a mutation. In our experience, a TALEN pair with a mutation efficiency of approximately 60% is sufficient for analyzing gene functions.

12.5.3 Analyzing Gene Functions by Tissue-Specific TALENs

Replace the *EF1α*-promoter in the TALEN expression vectors with the required *cis*-regulatory sequence using the method described above (Sect. 12.3). A detailed method of electroporation is described in Chap. 5. We use 30 μg each of plasmid DNAs harboring paired left and right TALEN cDNAs for each electroporation, when we use a cuvette 4 mm wide. In the case of simultaneous knockout of multiple genes or expressing TALENs in distinct tissues (electroporation of more than two kinds of vectors), reducing the amount (e.g., 20 μg each) of plasmid DNAs is preferred to avoid developmental defects. For control experiments, a two-fold number of expression vectors of only one side of each TALEN pair (left or right) are electroporated. Activity of TALENs expressed in this tissue-specific manner can be confirmed by the same procedure as described above.

When TALENs are expressed in tissues with small cell numbers (e.g., larval muscle), cleavage arresting at an appropriate stage with cytochalasin B improves detection of mutations. When we detect mutations in the muscle, we arrest cleavage at the 8- to 16-cell stage. The collection of embryos and following procedures are the same as those described in Sect. 12.5.2.

12.5.4 Microinjection of TALEN mRNAs

In *Ciona*, germ cell mutagenesis is achieved by introducing TALEN mRNAs into eggs (Yoshida et al. 2014). The introduction of TALEN mRNAs by microinjection is also recommended for ubiquitous knockout of target genes, because expressing TALENs by *EF1α*-promoter tends to cause developmental defects. Preparation for TALEN mRNAs is as follows:

1. For in vitro synthesis of mRNAs, TALEN repeats are assembled into a vector that contains partial cDNA encoding the N and C terminals of TALENs that are flanked by untranslated regions of a tubulin gene from the ascidian *Halocynthia roretzi*. This TALEN-compatible vector is based on the pBS-HTB vector (Akanuma et al. 2002).
2. We synthesize mRNAs with MegaScript T3 kit (Ambion) and Cap structure analog (New England Biolabs) according to the manufacturers' protocols.
3. The recommended final concentration of TALEN mRNAs in the injection solution is 100 ng/μl for each of the left and right TALENs.

For microinjection of mRNAs into eggs, see Chap. 2. We recommend analyzing the efficacy of TALENs expressed by mRNA injection first.

12.5.5 Germ Cell Mutagenesis via Primordial Germ Cell Removal

Although the above-described method can introduce mutations into germ-line cells in addition to somatic cells, the mutation frequencies of germ cells are variable among specimens and are sometimes very low. On the other hand, the mutation frequency of somatic cells is constantly high in animals injected with TALEN mRNA (Yoshida et al. 2014). In *Ciona*, primordial germ cells (PGCs) can be regenerated from somatic cells when the original PGCs, which are specified by maternal factors, are removed (Takamura et al. 2002). Therefore, mutagenesis in germ-line cells with higher frequency can be achieved by conversion of mutated somatic cells to germ cells via PGC regeneration. Here, we briefly describe our latest method for germ-line mutagenesis using PGC regeneration (Yoshida et al. 2017b):

1. Construct a TALEN pair and analyze its activity (see Sects. 12.5.1 and 12.5.2).
2. Replace the *EF1α*-promoter in the TALEN expression vectors with *TnI* (muscle) promoter (see Sect. 12.3). Although promoters for *Nut* (neural tissues) and *CesA* (epidermis) can also be used for this method, the *TnI* promoter shows the most reproducible result among them (Yoshida et al. 2017b).
3. Electroporation of TALEN expression vectors. When animals reach the larval stage (approximately 20–24 h after fertilization at 18 °C), at which PGCs are located in the posterior ventral side of the tail, cut off about a half of the tail using a scalpel; then, the PGCs are removed.
4. Culture these larvae until they reach a reproductive stage, isolate genomic DNA from sperm collected from individual animals, and analyze mutations.

This PGC regeneration-mediated method has two major advantages. First, this method enables us to generate mutant lines of genes that are essential for development by expressing TALENs in tissues where the target genes are not expressed. Second, germ-line mutagenesis can be achieved by electroporation of TALEN expression vectors, whereas a *cis* element that drives gene expression in germ cells during embryogenesis has not been isolated. We believe that this method eases and facilitates germ-cell mutagenesis in *Ciona*.

References

Akanuma T, Hori S, Darras S, Nishida H (2002) Notch signaling is involved in nervous system formation in ascidian embryos. Dev Genes Evol 212:459–472

Boch J, Scholze H, Schornack S et al (2009) Breaking the code of DNA binding specificity of TAL-type III effectors. Science 326:1509–1512

Carroll D (2014) Genome engineering with targetable nucleases. Annu Rev Biochem 83:409–439

Ikuta T, Satoh N, Saiga H (2010) Limited functions of Hox genes in the larval development of the ascidian Ciona Intestinalis. Development 137:1505–1513

Kawai N, Ochiai H, Sakuma T et al (2012) Efficient targeted mutagenesis of the chordate Ciona intestinalis genome with zinc-finger nucleases. Develop Growth Differ 54:535–545

Kawai N, Ogura Y, Ikuta T et al (2015) Hox10-regulated endodermal cell migration is essential for development of the ascidian intestine. Dev Biol 403:43–56

Kay S, Bonas U (2009) How Xanthomonas type III effectors manipulate the host plant. Curr Opin Microbiol 12:37–43

Kim JH, Lee SR, Li LH et al (2011) High cleavage efficiency of a 2A peptide derived from porcine teschovirus-1 in human cell lines, zebrafish and mice. PLoS One 6:e18556

Moscou MJ, Bogdanove AJ (2009) A simple cipher governs DNA recognition by TAL effectors. Science 326:1501

Nakashima K, Yamada L, Satou Y, Azuma J, Satoh N (2004) The evolutionary origin of animal cellulose synthase. Dev Genes Evol 214:81–88

Ota S, Hisano Y, Muraki M et al (2013) Efficient identification of TALEN-mediated genome modifications using heteroduplex mobility assays. Genes Cells 18:450–458

Rothbächer U, Bertrand V, Lamy C, Lemaire P (2007) A combinatorial code of maternal GATA, Ets and beta-catenin-TCF transcription factors specifies and patterns the early ascidian ectoderm. Development 134:4023–4032

Sakuma T, Ochiai H, Kaneko T et al (2013a) Repeating pattern of non RVD variations in DNA-binding modules enhances TALEN activity. Sci Rep 3:3379

Sakuma T, Hosoi S, Woltjen K et al (2013b) Efficient TALEN construction and evaluation methods for human cell and animal applications. Genes Cells 18:315–326

Sasakura Y, Nakashima K, Awazu S et al (2005) Transposon-mediated insertional mutagenesis revealed the functions of animal cellulose synthase in the ascidian Ciona intestinalis. Proc Natl Acad Sci U S A 102:15134–15139

Sasakura Y, Suzuki MM, Hozumi A, Inaba K, Satoh N (2010) Maternal factor-mediated epigenetic gene silencing in the ascidian Ciona intestinalis. Mol Gen Genomics 283:99–110

Sasakura Y, Ogura Y, Treen N et al (2016) Transcriptional regulation of a horizontally transferred gene from bacterium to chordate. Proc Biol Sci 283:20161712

Shi W, Peyrot SM, Munro E, Levine M (2009) FGF3 in the floor plate directs notochord convergent extension in the Ciona tadpole. Development 136:23–28

Shimeld SM, Levin M (2006) Evidence for the regulation of left-right asymmetry in Ciona intestinalis by ion flux. Dev Dyn 235:1543–1553

Takamura K, Fujimura M, Yamaguchi Y (2002) Primordial germ cells originate from the endodermal strand cells in the ascidian Ciona intestinalis. Dev Genes Evol 212:11–18

Tolkin T, Christiaen L (2016) Rewiring of an ancestral Tbx1/10-Ebf-Mrf network for pharyngeal muscle specification in distinct embryonic lineages. Development 143:3852–3862

Treen N, Yoshida K, Sakuma T et al (2014) Tissue-specific and ubiquitous gene knockouts by TALEN electroporation provide new approaches to investigating gene function in Ciona. Development 141:481–487

Yoshida K, Saiga H (2008) Left-right asymmetric expression of Pitx is regulated by the asymmetric Nodal signaling through an intronic enhancer in Ciona intestinalis. Dev Genes Evol 218:353–360

Yoshida K, Treen N, Hozumi A, Sakuma T, Yamamoto T, Sasakura Y (2014) Germ cell mutations of the ascidian Ciona intestinalis with TALE nucleases. Genesis 52:431–439

Yoshida K, Nakahata A, Treen N, Sakuma T, Yamamoto T, Sasakura Y (2017a) Hox-mediated endodermal identity patterns pharyngeal muscle formation in the chordate pharynx. Development 144:1629–1634

Yoshida K, Hozumi A, Treen N et al (2017b) Germ cell regeneration-mediated, enhanced mutagenesis in the ascidian Ciona intestinalis reveals flexible germ cell formation from different somatic cells. Dev Biol 423:111–125

13 CRISPR Knockouts in *Ciona* Embryos

Shashank Gandhi, Florian Razy-Krajka, Lionel Christiaen, and Alberto Stolfi

Abstract

Clustered regularly interspaced short palindromic repeats (CRISPR)/Cas9 has emerged as a revolutionary tool for fast and efficient targeted gene knockouts and genome editing in almost any organism. The laboratory model tunicate *Ciona* is no exception. Here, we describe our latest protocol for the design, implementation, and evaluation of successful CRISPR/Cas9-mediated gene knockouts in somatic cells of electroporated *Ciona* embryos. Using commercially available reagents, publicly accessible plasmids, and free web-based software applications, any *Ciona* researcher can easily knock out any gene of interest in their favorite embryonic cell lineage.

Keywords

Genome editing · Targeted mutagenesis · Somatic gene knockout · sgRNAs · Tunicates · Chordates

13.1 Introduction

Developmental biologists have always been interested in targeted loss-of-function mutations to probe the role of specific genes in embryogenesis and regeneration. One approach towards this goal has been to engineer the sequence specificity of DNA-binding domains found in natural transcription factors. When these customized DNA-binding proteins are fused to DNA nuclease domains, they are capable of inducing site-specific double-stranded breaks (DSBs), resulting in mutations through improper repair of these breaks by nonhomologous end-joining (NHEJ). Among these engineered reagents are the zinc finger nucleases (ZFNs) (Beerli and Barbas 2002; Bibikova et al. 2003; Maeder et al. 2008) and transcription activator-like effector nucleases (TALENs) (Christian et al. 2010; Miller et al. 2011). Both ZFNs and TALENs have been used for targeted mutagenesis in *Ciona* embryos (Kawai et al. 2012; Treen et al. 2014; Yoshida et al. 2014).

Although these programmable nucleases made it possible to cause site-directed DSBs at any part

S. Gandhi
Division of Biology and Biological Engineering, California Institute of Technology, Pasadena, CA, USA

F. Razy-Krajka · L. Christiaen (✉)
Center for Developmental Genetics, Department of Biology, New York University, New York, NY, USA
e-mail: lc121@nyu.edu

A. Stolfi (✉)
School of Biological Sciences, Georgia Institute of Technology, Atlanta, GA, USA
e-mail: alberto.stolfi@biosci.gatech.edu

Y. Sasakura (ed.), *Transgenic Ascidians*, Advances in Experimental Medicine and Biology 1029,
https://doi.org/10.1007/978-981-10-7545-2_13

of the genome, even in a tissue- or cell lineage-specific manner, expensive and tedious cloning procedures posed a barrier to their widespread adoption and hampered their scaling for higher-throughput applications such as genome-wide reverse genetic screens. More recently, a targeted platform known as clustered regularly interspaced short palindromic repeats (CRISPR)/Cas9 was developed, based on the immune response mechanism of *Streptococcus* bacteria (Barrangou et al. 2007; Jinek et al. 2012; Cong et al. 2013; Jinek et al. 2013; Mali et al. 2013). In these bacteria, processed short CRISPR RNA sequences guide the Cas9 protein to specific target sites on foreign DNA. Cas9 is characterized by two signature nuclease domains, and interacts with a DNA sequence ("NGG" for *S. pyogenes* Cas9) known as the protospacer adjacent motif (PAM). Sequence-specific base-pairing between the Cas9-associated short RNAs and protospacer DNA sequence of 20 bp adjacent to the PAM then triggers the protein's nuclease activity, resulting in cleavage of both strands of the target sequence (Garneau et al. 2010; Deltcheva et al. 2011; Gasiunas et al. 2012; Anders et al. 2014; Jinek et al. 2014).

In its native context, two distinct short RNAs guide Cas9: CRISPR RNA (crRNA) and trans-activating crRNA (tracrRNA). However, a chimeric "single-guide RNA" (sgRNA) is sufficient to mimic the roles of these two components (Jinek et al. 2012). This small but profound improvement has helped launch CRISPR/Cas9 as a cheap, simple, and efficient system for targeted mutagenesis in a remarkably wide variety of organisms (Perry and Henry 2015; Iaffaldano et al. 2016; Long et al. 2016; Nomura et al. 2016; Nymark et al. 2016; Tian et al. 2016), and in tunicates (Sasaki et al. 2014; Stolfi et al. 2014; Abdul-Wajid et al. 2015; Cota and Davidson 2015; Gandhi et al. 2017; Segade et al. 2016; Tolkin and Christiaen 2016).

Modifications to the CRISPR/Cas9 system have allowed for further applications, such as targeted knock-ins (Wang et al. 2013), transcriptional activation or repression (Maeder et al. 2013; Perez-Pinera et al. 2013; Qi et al. 2013), chromatin modifications (Hilton et al. 2015), and the visualization of genome organization and dynamics (Chen et al. 2013), although these approaches have yet to be adapted to tunicates. Similarly, other CRISPR variants, such as CRISPR/Cpf1, have been developed for targeted mutagenesis in mammalians (Kleinstiver et al. 2015; Zetsche et al. 2015), but their effects have not yet been tested in *Ciona.*

In *Ciona,* the most widely used application of CRISPR to date is for targeted mutagenesis in somatic cells of transiently transfected (electroporated) embryos. In this method, in vitro-fertilized embryos are electroporated at the one-cell stage with plasmids that drive the zygotic expression of Cas9 protein and sgRNAs. Although sgRNAs are transcribed ubiquitously from a U6 small RNA promoter (Nishiyama and Fujiwara 2008), by RNA polymerase III (RNAPolIII), Cas9 can be expressed in a cell-specific manner by using a lineage-specific promoter. We use a humanized Cas9 flanked by nuclear localization signals (NLS::Cas9::NLS) (Chen et al. 2013; Stolfi et al. 2014), although other Cas9 variants have not been thoroughly evaluated in *Ciona.* Targeted mutations occur only when both Cas9 and the sgRNA are present, and can happen on different sister chromatids in different cells at different times. This means that each embryo is actually a mosaic composed of cells bearing a combination of wild-type and/or distinct mutant alleles. In spite of this mosaicism, somatic knockouts are a powerful means of dissecting the tissue-specific functions of a gene in development.

Here, we present our latest protocols for generating successful CRISPR/Cas9-mediated mutagenesis (hereinafter referred to as "CRISPR knockouts") in somatic cells of *Ciona* embryos, based on our published and unpublished reports (Stolfi et al. 2014; Gandhi et al. 2017). The aim of this chapter is to empower laboratories working on *Ciona* (and other tunicates) to harness the power of this simple but very effective tool. The protocols presented here only use widely available commercial reagents, and all plasmids can be ordered from Addgene (https://www.addgene.org/Lionel_Christiaen/).

13.2 sgRNA Design

Perhaps nothing is more important for successful CRISPR knockouts in *Ciona* than selecting the right sgRNAs, which vary widely in their ability to actually induce Cas9-mediated DSBs. We refer to this as sgRNA mutagenesis "activity" or, more precisely, efficacy. Some sgRNAs are highly active, whereas others may not yield detectable mutations. Predicting which sgRNAs cause either frequent or rare mutations is an arduous and potentially frustrating task. Many high-throughput studies have sought to create predictive algorithms to distinguish, a priori, "good" vs "bad" sgRNAs. A recent meta-study of these methods (Haeussler et al. 2016) concluded that most available algorithms do not accurately predict the activity of sgRNAs outside a narrow range of organisms, cell types, or experimental conditions. The authors recommended two such algorithms, depending on the method of sgRNA transcription (in vivo by RNA polymerase III, or in vitro by viral T7 RNA polymerase). This is because the efficacy of an sgRNA is probably contingent upon its expression level and stability, which varies depending on the methods used to transcribe it. According to their comparisons, Fusi/Doench is the more accurate predictive algorithm for in vivo-transcribed sgRNAs in metazoans including *Ciona* (Fusi et al. 2015; Doench et al. 2016), whereas CRISPRScan (Moreno-Mateos et al. 2015) is recommended for predicting the activity of T7-transcribed sgRNAs.

The CRISPOR portal incorporates these findings and features into a useful web-based CRISPR sgRNA design tool (http://crispor.tefor.net/) (Haeussler et al. 2016). The input is any sequence from the *Ciona* genome (three different assembly versions are supported), and the output is every valid sgRNA target, their scores by the various algorithms used to predict efficacy and specificity, and primer sequences for constructing an expression vector.

Important considerations for sgRNA design and selection include not only predicted cutting efficiency, but also off-target effects and possible escape by polymorphisms in the target sequence. Ideally, an sgRNA should match extensively only one site in the genome (the target site) and no other site, which could be potentially cleaved as a result. On the other hand, single nucleotide polymorphisms (SNPs) and other naturally occurring mutations can prevent sgRNA pairing with the intended target, precluding efficient cleavage by Cas9. Although the compact genome of *Ciona* depresses off-target effects, SNPs are extremely frequent in genetically diverse wild *Ciona* populations (Satou et al. 2012). CRISPOR v4.0 takes both off-targets and SNPs into account. Individual SNPs and sites of potential off-target effect are shown for each candidate sgRNA, which allows the user to choose whether the sgRNA is worth using or not.

Considerable attention must also be paid to selecting the location of the sgRNA target within a locus of interest. Our analysis of CRISPR knockouts in *Ciona* indicates that, as in other organisms, NHEJ repair of targets cleaved by Cas9 overwhelmingly favors short indels (Gandhi et al. 2017). If targeting coding a sequence, there is a 2-in-3 chance of the indel resulting in a frameshift, and likely premature stop codon. Conversely, there is a 1-in-3 chance of an in-frame indel being generated, which may or may not affect the function of the resulting protein. Bear in mind that, once an indel is generated, the sgRNA no longer matches the target site. This means that CRISPR/Cas9-generated mutations are all-or-nothing and irreversible. If deleting a few amino acid residues from the target region does not affect the function of the protein of interest, then one third of the alleles in the embryo are virtually wild-type, even assuming a 100% mutagenesis rate.

Although a short out-of-frame indel can result in a loss-of-function allele, in certain cases the truncated protein may act as a neomorphic variant, like a "dominant-negative." The further the target is from the translation start site, the higher the chance of a CRISPR/Cas9-generated indel resulting in a truncated protein. However, if the indel is too close to the translation start, translation initiation may simply shift to a downstream start codon, with little impact on the resulting protein function. Thus, selecting a good sgRNA also depends on finding this "sweet spot,", which varies from protein to protein.

An effective strategy to circumvent all these potential pitfalls is to use two or more highly active sgRNAs in combination. This increases the odds of generating at least one out-of-frame indel, and the large deletions spanning multiple targets have been consistently observed in *Ciona* embryos (Gandhi et al. 2017), the largest deletion reported being ~13 kb (Abdul-Wajid et al. 2015).

13.3 sgRNA Expression Cassette Construction by One-Step Overlap PCR

CRISPOR returns a list of sgRNA targets and their relevant efficacy and specificity scores and information. A link is provided for each target to a page that lists the oligonucleotide sequences that need to be ordered to construct the sgRNA expression vector according to a variety of strategies. For *Ciona,* the relevant primers are for One-Step Overlap PCR (OSO-PCR) (Urban et al. 1997), which allows for the rapid synthesis of a U6 > sgRNA cassette in a single PCR reaction (Gandhi et al. 2017). The target-specific sequence (the "protospacer") of any sgRNA cassette is only 19 bp. Thus, in OSO-PCR, limiting amounts of unique overlap primers generate a protospacer "bridge" between universal U6 promoter and sgRNA scaffold sequences, which are amplified from separate template molecules. In *Ciona,* a modified sgRNA^{F+E} scaffold is used to increase stability and decrease premature termination of transcription (Orioli et al. 2011; Chen et al. 2013; Stolfi et al. 2014).

sgRNA expression cassettes can then be electroporated directly into *Ciona* embryos as unpurified PCR products for in vivo transcription, or further processed/purified for cloning into plasmid for long-term storage/propagation. We can reliably detect mutagenesis activity of sgRNAs transcribed in embryos electroporated with as little as 20 µl of unpurified OSO-PCR reaction per 700 µl electroporation volume (see peakshift assay, Sect. 13.7). This enables fast testing of a large number of candidate sgRNAs.

The step-by-step protocol (adapted from Gandhi et al. 2017) is described below:

1. If selecting targets using CRISPOR, choose those with high Fusi/Doench scores (>60) and no known SNPs or off-targets. Click on the "PCR primers" link underneath the target sequence and the pre-designed primers for OSO-PCR can be found, ready to be ordered from your preferred oligonucleotide vendor. With oligos in hand, skip ahead to step 5.

   If targets and design primers have to be identified manually, look for candidate targets of N(19) + PAM ("NGG") sequence.

```
              target N(19)    PAM
...TCAACCCAACTGAGGGTTGGACAACAGGAGGAGCAACAGT...
```

2. Add a "**G**" to 5′ end of target sequence, to obtain a G+(N)19 sequence. The initial "G" is important for transcription start by PolIII.

GCTGAGGGTTGGACAACAGG

3. Append "**GTTTAAGAGCTATGCTGGAAACAG**" to the 3′ end of the G + N(19) sequence. This is now the forward primer used to amplify the sgRNA scaffold part of the cassette ("OSO forward" primer):

GCTGAGGGTTGGACAACAGG**GTTTAAGAGCTATGCTGGAAACAG**

4. Copy reverse complement of G + N(19), append "**ATCTATACCATCGGATGCC TTC**" to the 3′ end of this. This is the reverse primer for amplifying the U6 promoter part of the cassette

   ("OSO reverse" primer):

CCTGTTGTCCAACCCTCAGC**ATCTATACCATCGGATGCCTTC**

---

5. Set up the following PCR reaction. Template plasmids are available from Addgene (https://www.addgene.org/Lionel_Christiaen/):
   For a 50-μl reaction:
   1.5 μl 10 mM dNTPs
   1 μl 50 mM MgSO4
   10 μl 10 × Pfx buffer*
   1 μl U6 > XX plasmid at 15 ng/μl
   1 μl X > sgRNA(F + E) plasmid at 15 ng/μl
   1.5 μl 20 μM U6 forward primer (5′-TGGCGGGTGTATTAAACCAC -3′)
   1.5 μl 20 μM sgRNA reverse primer (5′-GGATTTCCTTACGCGAAATACG -3′)
   1 μl 2 μM** OSO forward primer (designed in step 3, or obtained from CRISPOR)
   1 μl 2 μM** OSO reverse primer (designed in step 4, or obtained from CRISPOR)
   30 μl H_2O
   0.5 μl Pfx platinum
   PCR program:

   94° - 3′
   94° - 30″ |
   50° - 30″ | ×30
   68° - 3′ |
   68° - 5′

   *Final concentration is 2×, not 1×
   **The 1:10 dilution of your custom overlap target-specific primers forces the "fusion" of the entire cassette later in the reaction, as these primers are depleted from the solution

6. Check 2 μl of the PCR reaction on a gel. There should be a strong band at ~1.2 kbp. If the band is only 1 kbp, the fusion did not occur. In our hands, the success rate is 94%.

13.4 Cloning OSO-PCR Cassette Using In-Fusion

Although OSO-PCR cassettes can be directly tested in *Ciona* by co-electroporation with Cas9 expression plasmid, they can also be processed for cloning into an empty plasmid vector. This allows for their replication and long-term propagation in *E. coli* cells, and the preparation of pure, highly concentrated sgRNA expression vector plasmid DNA for electroporations. We recommend using the In-Fusion restriction enzyme-free cloning system from Clontech/Takara (https://www.clontech.com/), although restriction enzyme cloning and other systems can be used as well.

The step-by-step protocol is described below:

1. Set up a "boost" PCR reaction to add 15-nt overhangs to the ends of the cassette required for cloning into the empty vector:
   For a 50-μl reaction:
   1.5 μl 10 mM dNTPs
   1 μl 50 mM MgSO4
   10 μl ×10 Pfx buffer
   1 μl OSO-PCR reaction
   1.5 μl 20 μM In-Fusion forward primer (5′-ATTAATTAAGGCGCGCCTGGCGGG TGTATTAAACCAC-3′)
   1.5 μl 20 μM In-Fusion reverse primer (5′-CGCTCAGCTGGAATTCAAAA AAAGCAC -3′)
   35 μl H2O
   0.5 μl Pfx platinum
   PCR program:

94° - 2′
94° - 30″ |
55° - 30″ | ×15
68° - 3′ |
68° - 5′

2. Add 2 μl *DpnI* enzyme to the reaction and incubate for 2 h at 37 °C. This digests any remaining template plasmid.
3. Gel-purify boost PCR band, elute in 50 μl of water.
4. Set up In-Fusion reaction and incubate at 50 °C for 20 min:
   4 μl boost PCR, gel-purified
   4 μl* vector (e.g., any of the Christiaen Lab Cas9 vectors) cut with AscI + EcoRI, gel-purified
   2 μl 5× In-Fusion
   *According to the manufacturer, use 50–200 ng total of linearized vector
5. Transform 1 μl in 25 μl of Stellar competent *E. coli* cells, which come with the In-Fusion kit, and plate on an LB ampicillin agar plate.
6. Pick and grow at least four colonies, and screen for positive clones by colony PCR directly on cultured *E. coli* cells using the U6 forward primer (5′- TGGCGGGTGTATT AAACCAC -3′) and the In-Fusion reverse primer. The correct band should be ~1 kb in length.

13.5 Conventional sgRNA Expression Vector Assembly

sgRNA expression vectors can also be directly assembled in plasmid form by traditional ligation of annealed oligonucleotides into linearized vector. Our initial sgRNA vectors were constructed this way and this T4-ligase-based method is indeed a faster and more reliable approach for obtaining sgRNA expression plasmids. The obvious downside is that colony selection and plasmid preparation must be performed before testing sgRNA efficacy, which is notoriously difficult to predict a priori. As a result, we do not recommend the following method to assemble untested sgRNAs. However, this is a suitable approach to recreate expression vectors for sgRNAs that have already been tested and validated.

The step-by-step protocol (adapted from Stolfi et al. 2014) is described below:

1. Given the same N(19) + PAM ("NGG") target sequence that was provided as an example for OSO-PCR design.

```
                  target N(19)    PAM
...TCAACCCAACTGAGGGTTGGACAACAGGAGGAGCAACAGT...
```

2. Add a "**G**" to the 5′ end of the target sequence, to obtain a G+(N)19 sequence.

GCTGAGGGTTGGACAACAGG

3. Append "**AGAT**" to the 5′ end of the G + N(19) sequence. This is now the sense oligonucleotide to be ordered.

AGATGCTGAGGGTTGGACAACAGG

4. Copy reverse complement of G + N(19), append "**AAAC**" to the 5′ end of this now. This is the antisense oligonucleotide:

AAACCCTGTTGTCCAACCCTCAGC

5. Anneal the oligonucleotides at 10 μM by boiling for 5 min in 10 mM Tris pH 7.5, 50 mM NaCl, and then cool naturally to room temperature.

```
sense    AGATGCTGAGGGTTGGACAACAGC
             ||||||||||||||||||||
             CGACTCCCAACCTGTTGTCGCAAA   antisense
```

6. Dilute the annealed oligos 1:1000 and ligate this into U6 > sgRNA(F + E) linearized with *BsaI*.

```
5'...ATCCGATGGTAT AGATGCTGAGGGTTGGACAACAGC GTTTAAG...3'
3'...TAGGCTACCATATCTA CGACTCCCAACCTGTTGTCGCAAA TTC...5'
```

1 μl annealed oligonucleotides 1:1000
1 μl U6 > sgRNA(F + E), cut with BsaI, gel-purified
2.5 μl water
5 μl 2× rapid ligation buffer (Promega)
0.5 μl T4 DNA ligase (1.5 units)
Incubate at room temperature for 15 min.

7. Transform this ligation into *E. coli* cells, and screen colonies by PCR using U6 forward primer and the antisense oligonucleotide detailed above as a reverse primer.

13.6 Assaying CRISPR Knockouts

Either in plasmid or unpurified, PCR product format, sgRNA expression constructs should be assayed for their ability to cause on-target CRISPR knockouts. We have encountered a wide range of mutagenesis efficacies, from 0% to >60%, estimated by next-generation sequencing (Gandhi et al. 2017). Thus, it is advised that one test 4–8 candidate sgRNAs per target to identify the most effective ones to use in further experiments.

It is not absolutely necessary to use an sgRNA expression *plasmid* to assay its efficacy. We have verified highly active sgRNAs expressed from unpurified OSO-PCR products electroporated into *Ciona* embryos. This has allowed us to quickly test the efficacies of large numbers of sgRNAs, either by target sequence analysis or by phenotypic assay (Gandhi et al. 2017). Typically, 15–45 μl of unpurified products can be added to a single 700 μl electroporation solution, together with the Cas9 vector. However, the linear nature of the PCR product, and the reagents present in the reaction may interfere with normal development. Therefore, our current strategy is to assay sgRNA efficacy using OSO-PCR products, but then clone those products that prove most effective into a plasmid for use for publication-quality experiments.

There are different methods of estimating sgRNA efficacies in a quantitative manner. A very basic approach consists of amplifying target regions by PCR and cloning these products into a plasmid vector, then sequencing a handful of clones and counting the number of mutant clones (Sasaki et al. 2014; Stolfi et al. 2014). However, this approach is very time-consuming, labor-intensive, and not accurate, as a very large number of clones would need to be sequenced to approach a reliable sample size.

Efficacies of sgRNAs have also been measured in *Ciona* by Cel-I nuclease assay (Sasaki et al. 2014) or Thermo Fisher Scientific GeneArt Genomic Cleavage Detection kit (Stolfi et al. 2014). These methods depend on nucleases that recognize and cleave DNA bulges resulting from hybridization of DNA strands bearing distinct indels. The result is smaller "cleavage bands" that can be measured by fluorescence intensity on an agarose gel. However, the nuclease also cleaves bulges resulting from single-nucleotide mismatches, which is extremely problematic when using this assay on animals from a highly polymorphic population, as we do for *Ciona*.

More recently, we have employed next-generation sequencing to calculate the ratio of mutant and wild-type sequences amplified by PCR (Gandhi et al. 2017). This approach allowed us to assay the efficacies of over 80 sgRNAs in parallel, by pooling PCR products amplified from embryos electroporated with different sgRNA vectors. However, the cost and depth of this method of sequencing would not be justified if only a handful sgRNAs were being measured at a time. Therefore, we only recommend the next-generation sequencing route for large-scale assays (>100 sgRNAs).

13.7 Sanger Sequencing-Based "peakshift" Assay for sgRNA Activity

Currently, our recommended approach for estimating the efficacies of a few sgRNAs at a time is to use Sanger sequencing of target sequence PCR products. This is a relatively simple and cost-effective method that returns highly consistent, fairly quantitative estimates of sgRNA efficacy. Unlike next-generation sequencing, Sanger sequencing cannot resolve the sequences of individual molecules, but rather returns a composite of all the molecules sequenced in the reaction. Normally, the sequence is readable because all the molecules are identical. However, when there are many products bearing short indels because of CRISPR, the peaks in a typical Sanger sequencing trace appear mixed, with signal for more than one nucleotide base at the same position in the sequence (Fig. 13.1). This "peakshift" can be quantified by algorithms such as the ab1 Peak Reporter by Thermo Fisher Scientific (https://apps.thermofisher.com/ab1peakreporter/) (Roy and Schreiber 2014). We have shown a nearly linear correlation between CRISPR knockout peakshifts measured by ab1 Peak Reporter and frequency of a loss-of-function phenotype in F0 (Gandhi et al. 2017). This suggests that the sgRNAs that produce the highest peakshifts might be the most effective at generating loss-of-function alleles, which is ultimately the goal of CRISPR knockout experiments.

Up to three sgRNA cassettes targeting different genes have been electroporated in the same embryos and assayed in this manner, and their efficacies do not seem to be hampered by this multiplexing (A.S., unpublished observation). However, care must be taken not to test targets that are on the same chromosome, as large deletions or chromosomal breaks may occur as a result. What follows is a protocol for electroporating a given sgRNA construct (plasmid or OSO-PCR) and assaying its mutagenesis efficacy by peakshift.

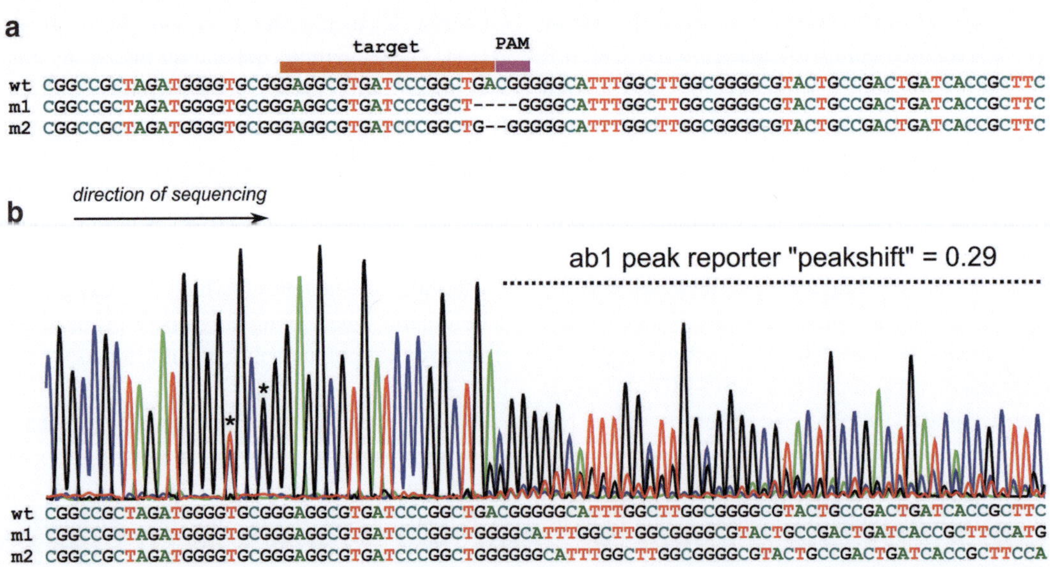

Fig. 13.1 Clustered regularly interspaced short palindromic repeats (CRISPR) indels. (**a**) Wild-type ("wt") target sequence aligned with two CRISPR knockout mutant sequences ("m1" and "m2") generated by imprecise repair of CRISPR/Cas9-mediated double-stranded breaks. Alignment shows gaps (–) in place of missing nucleotides in target or protospacer adjacent motif (PAM) sequence. (**b**) When sequenced by Sanger sequencing, pools of wild-type and mutant sequences produce a "peakshift," which can be quantified by the ab1 Peak Reporter web app (see text for details). Below, the same sequences in (**a**) aligned without gaps, showing the cause of the overlapping peaks seen in the peakshift area. *Asterisks* denote naturally-occurring single-nucleotide polymorphisms

The step-by-step protocol (adapted from Gandhi et al. 2017) is described below:

1. Following the standard electroporation protocol (Christiaen et al. 2009), prepare an electroporation mix:
   25–50 μg sgRNA plasmid or 15–45 μl unpurified sgRNA OSO-PCR
   25 μg Eef1a1 > nls::Cas9::nls
   400 μl 0.96 M D-Mannitol
   Water to 500 μl
   This solution is then mixed with 200 μl sea water containing fertilized *Ciona* eggs for electroporation.
2. Grow embryos at 18–24 °C until hatching. Collect hatched larvae and extract genomic DNA using the QIAamp DNA Micro Kit (Qiagen) following a modified protocol.
   Modifications to manufacturer's protocol:
   Lyse embryos in 180 μl buffer ATL + 5 μl proteinase K for 30 min
   Use carrier RNA (as supplied in the kit)
   Elute DNA in 20 μl of water
3. Measure the extracted DNA using a spectrophotometer. Prepare the following PCR reaction to amplify the target sequence. For best results, the aim should be to design primers to amplify a fragment 300–1500 bp long, with the target site(s) at least 150 bp away from either end of the fragment. We prefer Pfx platinum from Thermo Fisher Scientific, but any proof-reading polymerase should suffice.
   For a 50-μl reaction:
   1.5 μl 10 mM dNTPs
   1 μl 50 mM MgSO4
   10 μl 10× Pfx buffer
   1 μl genomic DNA at 200 ng/μl
   1.5 μl 20 μM forward primer
   1.5 μl 20 μM reverse primer
   35 μl H2O
   0.5 μl Pfx platinum
   "Touchdown" PCR program:

   94° - 3′
   94° - 30″ |
   60° - 30″ | ×5
   68° - 5′ |
   94° - 30″ |
   58° - 30″ | ×5
   68° - 5′ |
   94° - 30″ |
   55° - 30″ | ×15
   68° - 5′ |
   68° - 10′

4. Column- or gel-purify the resulting PCR product, and send off for Sanger sequencing. The primers used for sequencing can be the same as those used for PCR, provided the target is at least 150 bp and at most 500 bp away from the primer. This ensures large enough stretches of "normal" and "shifted" peaks for a proper quantification by ab1 Peak Reporter. The orientation of sequencing does not matter, but it is critically important to avoid sequencing reads that may encounter naturally occurring indels before the target site, which can cause a natural peakshift and mask the effect of CRISPR. Several internal primers may have to be designed and tested specifically for sequencing, if the PCR primers are not suitable.
5. The resulting .ab1 sequencing files are then uploaded to Thermo Fisher Scientific's ab1 Peak Reporter (https://apps.thermofisher.com/ab1peakreporter/), which may require registering/logging into the Thermo Fisher website. The program returns a .csv file, which can be opened in Microsoft Excel and saved as an .xlsx file.
6. The data should first be filtered to only display the values at each peak called. This is because the data contain signal reads at every position measured by the instrument, including in between peaks (in between individual basepairs in the sequence). To do this, create a filter for the "BaseCall" column (column B) and exclude "-". This leaves only the peaks, represented by "calls" indicating G, A, T, C, or N.
7. After filtering this way, the target sequence and PAM can be searched for in column B, displayed as 5′ to 3′ from top to bottom (Fig. 13.2). Once the target sequence is found, color-coding it may help to keep track of the position in the file.

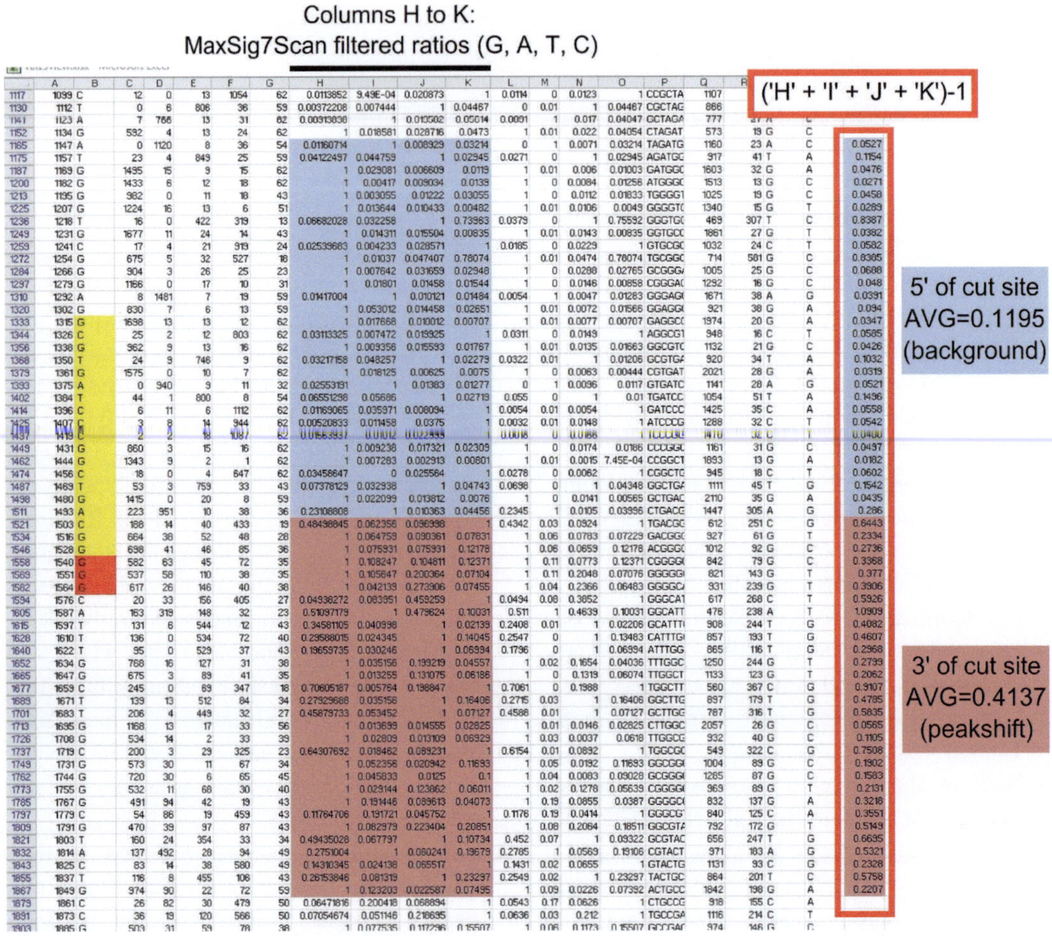

Fig. 13.2 ab1 Peak Reporter spreadsheet. Annotated example of an excel spreadsheet generated by the ab1 Peak Reporter web app. Each row represents a called peak, or nucleotide, of the sequence, from 5′ to 3′ (top to bottom respectively). Cells of interest are color=coded or outlined manually. In *yellow*, the sgRNA target, and in *red*, the PAM. In *light blue*, the MaxSig7Scan ratios for 30 nucleotides upstream of the Cas9 cut site, and in *pink*, the MaxSig7Scan ratios of 30 nucleotides downstream of the Cas9 cut site. Cas9 tends to cut in the target, ~3 basepairs from the PAM. Outlined in the *red box*: the sum of secondary MaxSig7Scan ratios for each nucleotide, using the formula indicated. The average of these values after the Cas9 cut site represents the "peakshift," the amount of secondary peak calling due to the presence of sequences with short indels in the target. The average of the value before the Cas9 cut site is the background signal. See text for details

8. In column U, calculate the sum of the secondary peaks by adding the values in columns H–K ("MaxSig7Scan Filtered Ratios") and subtracting 1. Subtracting 1 is to remove the contribution of the primary peak, which is always 1, regardless of its actual identity.
9. To obtain a quantitative estimate of the peakshift resulting from mutant reads, calculate the average value in column U, over 30 positions downstream of (3′ to) the Cas9 cleavage site, usually around the third basepair in the target from the PAM. To get a sense of the secondary signal background of the read, calculate the average in column U over 30 positions upstream of the cleavage site. By subtracting this background average from the peakshift average, a corrected peakshift value can be obtained.

Bear in mind that the peakshift can be suppressed by sequence homogeneity near the target site. Because CRISPR knockouts are usually short indels, shifting peaks of the same identity are not detected. For instance, a 1-bp deletion in the sequence GGGGAAAA only produces secondary peaks at one position, whereas a 1-bp deletion in the sequence GAGAGAGA results in secondary peaks at all positions.

13.8 Conclusion

As more *Ciona* research groups adopt CRISPR, more data will emerge on the best practices to ensure optimal CRISPR activity, including sgRNA efficacy prediction. We hope that the above protocols will speed up this adoption and bring about exciting improvements to CRISPR knockout strategies in *Ciona.*

Acknowledgments Research in the laboratory of L.C. is supported by R01 awards HL108643 and GM096032 from the NIH/NHLBI and NIH/NIGMS respectively; and by grant 15CVD01 from the Leducq Foundation. A.S. is supported by R00 award HD084814 from the NIH/NICHD.

References

Abdul-Wajid S, Morales-Diaz H, Khairallah SM, Smith WC (2015) T-type Calcium Channel regulation of neural tube closure and EphrinA/EPHA expression. Cell Rep 13:829–839

Anders C, Niewoehner O, Duerst A, Jinek M (2014) Structural basis of PAM-dependent target DNA recognition by the Cas9 endonuclease. Nature 513:569–573

Barrangou R, Fremaux C, Deveau H, Richards M, Boyaval P et al (2007) CRISPR provides acquired resistance against viruses in prokaryotes. Science 315:1709–1712

Beerli RR, Barbas CF (2002) Engineering polydactyl zinc-finger transcription factors. Nat Biotechnol 20:135–141

Bibikova M, Beumer K, Trautman JK, Carroll D (2003) Enhancing gene targeting with designed zinc finger nucleases. Science 300:764–764

Chen B, Gilbert LA, Cimini BA, Schnitzbauer J, Zhang W et al (2013) Dynamic imaging of genomic loci in living human cells by an optimized CRISPR/Cas system. Cell 155:1479–1491

Christiaen L, Wagner E, Shi W, Levine M (2009) Electroporation of transgenic DNAs in the sea squirt *Ciona*. Cold Spring Harbor protocols 2009**:** pdb. prot5345

Christian M, Cermak T, Doyle EL, Schmidt C, Zhang F et al (2010) Targeting DNA double-strand breaks with TAL effector nucleases. Genetics 186:757–761

Cong L, Ran FA, Cox D, Lin S, Barretto R et al (2013) Multiplex genome engineering using CRISPR/Cas systems. Science 339:819–823

Cota CD, Davidson B (2015) Mitotic membrane turnover coordinates differential induction of the heart progenitor lineage. Dev Cell 34:505–519

Deltcheva E, Chylinski K, Sharma CM, Gonzales K, Chao Y et al (2011) CRISPR RNA maturation by trans-encoded small RNA and host factor RNase III. Nature 471:602–607

Doench JG, Fusi N, Sullender M, Hegde M, Vaimberg EW et al (2016) Optimized sgRNA design to maximize activity and minimize off-target effects of CRISPR-Cas9. Nat Biotechnol 34:184

Fusi, N., I. Smith, J. Doench and J. Listgarten, 2015 In Silico Predictive Modeling of CRISPR/Cas9 guide efficiency. bioRxiv: 021568

Gandhi S, Haeussler M, Razy-Krajka F, Christiaen L, Stolfi A (2017) Evaluation and rational design of guide RNAs for efficient CRISPR/Cas9-mediated mutagenesis in Ciona. Dev Biol 425:8–20

Garneau JE, Dupuis M-È, Villion M, Romero DA, Barrangou R et al (2010) The CRISPR/Cas bacterial immune system cleaves bacteriophage and plasmid DNA. Nature 468:67–71

Gasiunas G, Barrangou R, Horvath P, Siksnys V (2012) Cas9–crRNA ribonucleoprotein complex mediates specific DNA cleavage for adaptive immunity in bacteria. Proc Natl Acad Sci 109:E2579–E2586

Haeussler M, Schönig K, Eckert H, Eschstruth A, Mianné J et al (2016) Evaluation of off-target and on-target scoring algorithms and integration into the guide RNA selection tool CRISPOR. Genome Biol 17:148

Hilton IB, D'Ippolito AM, Vockley CM, Thakore PI, Crawford GE et al (2015) Epigenome editing by a CRISPR-Cas9-based acetyltransferase activates genes from promoters and enhancers. Nat Biotechnol 33:510–517

Iaffaldano B, Zhang Y, Cornish K (2016) CRISPR/Cas9 genome editing of rubber producing dandelion Taraxacum kok-saghyz using agrobacterium rhizogenes without selection. Ind Crop Prod 89:356–362

Jinek M, Chylinski K, Fonfara I, Hauer M, Doudna JA et al (2012) A programmable dual-RNA–guided DNA endonuclease in adaptive bacterial immunity. Science 337:816–821

Jinek M, East A, Cheng A, Lin S, Ma E et al (2013) RNA-programmed genome editing in human cells. elife 2:e00471

Jinek M, Jiang F, Taylor DW, Sternberg SH, Kaya E et al (2014) Structures of Cas9 endonucleases reveal

RNA-mediated conformational activation. Science 343:1247997

Kawai N, Ochiai H, Sakuma T, Yamada L, Sawada H et al (2012) Efficient targeted mutagenesis of the chordate Ciona intestinalis genome with zinc-finger nucleases. Develop Growth Differ 54:535–545

Kleinstiver BP, Prew MS, Tsai SQ, Topkar VV, Nguyen NT et al (2015) Engineered CRISPR-Cas9 nucleases with altered PAM specificities. Nature 523: 481–485

Long S, Wang Q, Sibley LD (2016) Analysis of noncanonical calcium-dependent protein kinases in toxoplasma gondii by targeted gene deletion using CRISPR/Cas9. Infect Immun 84:1262–1273

Maeder ML, Thibodeau-Beganny S, Osiak A, Wright DA, Anthony RM et al (2008) Rapid "open-source" engineering of customized zinc-finger nucleases for highly efficient gene modification. Mol Cell 31: 294–301

Maeder ML, Linder SJ, Cascio VM, Fu Y, Ho QH et al (2013) CRISPR RNA-guided activation of endogenous human genes. Nat Methods 10:977–979

Mali P, Yang L, Esvelt KM, Aach J, Guell M et al (2013) RNA-guided human genome engineering via Cas9. Science 339:823–826

Miller JC, Tan S, Qiao G, Barlow KA, Wang J et al (2011) A TALE nuclease architecture for efficient genome editing. Nat Biotechnol 29:143–148

Moreno-Mateos MA, Vejnar CE, Beaudoin J-D, Fernandez JP, Mis EK et al (2015) CRISPRscan: designing highly efficient sgRNAs for CRISPR-Cas9 targeting in vivo. Nat Meth 12:982–988

Nishiyama A, Fujiwara S (2008) RNA interference by expressing short hairpin RNA in the Ciona intestinalis embryo. Develop Growth Differ 50:521–529

Nomura T, Sakurai T, Osakabe Y, Osakabe K, Sakakibara H (2016) Efficient and heritable targeted mutagenesis in mosses using the CRISPR/Cas9 system. Plant Cell Physiol 57:2600–2610

Nymark M, Sharma AK, Sparstad T, Bones AM, Winge P (2016) A CRISPR/Cas9 system adapted for gene editing in marine algae. Sci Rep 6

Orioli A, Pascali C, Quartararo J, Diebel KW, Praz V et al (2011) Widespread occurrence of non-canonical transcription termination by human RNA polymerase III. Nucleic Acids Res 39:5499–5512

Perez-Pinera P, Kocak DD, Vockley CM, Adler AF, Kabadi AM et al (2013) RNA-guided gene activation by CRISPR-Cas9-based transcription factors. Nat Methods 10:973–976

Perry KJ, Henry JQ (2015) CRISPR/Cas9-mediated genome modification in the mollusc, Crepidula Fornicata. Genesis 53:237–244

Qi LS, Larson MH, Gilbert LA, Doudna JA, Weissman JS et al (2013) Repurposing CRISPR as an RNA-guided platform for sequence-specific control of gene expression. Cell 152:1173–1183

Roy S, Schreiber E (2014) Detecting and quantifying low level gene variants in sanger sequencing traces using the ab1 peak reporter tool. J Biomol Techn JBT 25:S13

Sasaki H, Yoshida K, Hozumi A, Sasakura Y (2014) CRISPR/Cas9-mediated gene knockout in the ascidian Ciona intestinalis. Develop Growth Differ 56:499–510

Satou Y, Shin-i T, Kohara Y, Satoh N, Chiba S (2012) A genomic overview of short genetic variations in a basal chordate, Ciona Intestinalis. BMC Genomics 13:208

Segade F, Cota C, Famiglietti A, Cha A, Davidson B (2016) Fibronectin contributes to notochord intercalation in the invertebrate chordate, Ciona Intestinalis. EvoDevo 7:21

Stolfi A, Gandhi S, Salek F, Christiaen L (2014) Tissue-specific genome editing in Ciona embryos by CRISPR/Cas9. Development 141:4115–4120

Tian S, Jiang L, Gao Q, Zhang J, Zong M et al (2016) Efficient CRISPR/Cas9-based gene knockout in watermelon. Plant Cell Rep:1–8

Tolkin T, Christiaen L (2016) Rewiring of an ancestral Tbx1/10-Ebf-Mrf network for pharyngeal muscle specification in distinct embryonic lineages. Development 143:3852–3862

Treen N, Yoshida K, Sakuma T, Sasaki H, Kawai N et al (2014) Tissue-specific and ubiquitous gene knockouts by TALEN electroporation provide new approaches to investigating gene function in *Ciona*. Development 141:481–487

Urban A, Neukirchen S, Jaeger K-E (1997) A rapid and efficient method for site-directed mutagenesis using one-step overlap extension PCR. Nucleic Acids Res 25:2227–2228

Wang H, Yang H, Shivalila CS, Dawlaty MM, Cheng AW et al (2013) One-step generation of mice carrying mutations in multiple genes by CRISPR/Cas-mediated genome engineering. Cell 153:910–918

Yoshida K, Treen N, Hozumi A, Sakuma T, Yamamoto T et al (2014) Germ cell mutations of the ascidian Ciona Intestinalis with TALE nucleases. Genesis 52:431–439

Zetsche B, Gootenberg JS, Abudayyeh OO, Slaymaker IM, Makarova KS et al (2015) Cpf1 is a single RNA-guided endonuclease of a class 2 CRISPR-Cas system. Cell 163:759–771

Transgenic Techniques for Investigating Cell Biology During Development

14

Christina D. Cota

Abstract

Ascidians are increasingly being used as a system for investigating cell biology during development. The extreme genetic and cellular simplicity of ascidian embryos in combination with superior experimental tractability make this an ideal system for in vivo analysis of dynamic cellular processes. Transgenic approaches to cellular and sub-cellular analysis of ascidian development have begun to yield new insights into the mechanisms regulating developmental signaling and morphogenesis. This chapter focuses on the targeted expression of fusion proteins in ascidian embryos and how this technique is being deployed to garner new insights into the cell biology of development.

Keywords

Ascidians · *Ciona intestinalis* · *Phallusia mammillata* · *Halocynthia roretzi* · Cardiac induction · Collective cell migration · Cell cycle progression · Spindle positioning · Fusion proteins

14.1 The Study of Developmental Cell Biology in Ascidians

The development of multicellular organisms requires the precise coordination of processes within and between cells. However, high-resolution in vivo analysis of the coordinated cellular processes underlying development has been hindered by large cell numbers and technical challenges associated with manipulating and monitoring cells in developmental model systems. Ex-vivo cell culture systems have allowed researchers to circumvent some of these challenges, reducing cellular complexity and facilitating in vivo imaging. Yet, ex-vivo models fail to recapitulate the diversity of cellular interactions present in animal systems.

Ascidians have emerged as powerful systems for the in vivo analysis of cellular mechanisms underlying development. Development of transparent ascidian embryos is rapid. Larval development proceeds through a well-characterized series of stereotypic cell divisions that culminate within ~12 h post-fertilization. At the end of this period, each transparent free-swimming ascidian larva consists of fewer than 3,000 cells (Satoh 1994). In addition to their extreme cellular simplicity, ascidian genomes are compact, which greatly facilitates the identification of *cis*-regulatory elements controlling gene expression (Satoh et al. 2003). Moreover, ascidian genomes contain single orthologs to most vertebrate genes (Dehal

C. D. Cota (✉)
Swarthmore College, Swarthmore, PA, USA
e-mail: ccota1@swarthmore.edu

Y. Sasakura (ed.), *Transgenic Ascidians*, Advances in Experimental Medicine and Biology 1029,
https://doi.org/10.1007/978-981-10-7545-2_14

2002). This comes in stark contrast to vertebrate genomes, which possess multiple paralogous copies of genes as a result of genome duplication. Together, the extreme cellular and genetic simplicity of ascidian embryos greatly facilitates high resolution in vivo analysis of intracellular dynamics driving critical signaling and transcriptional decisions during development.

Ascidian embryos are particularly amenable to transgenic manipulations. Established injection and electroporation protocols allow researchers to rapidly and reliably generate thousands of transgenic embryos for experimental analysis (Hikosaka et al. 1993; Corbo et al. 1997; Zeller 2004). The ease and efficiency of generating transient transgenic embryos greatly facilitates high-throughput in vivo analysis of cellular processes. Genes of interest may be expressed in embryos ubiquitously or under the control of well-characterized, lineage-specific drivers (*cis*-regulatory enhancer + promoter) (Stolfi and Christiaen 2012). Transgenically expressed proteins can be used to monitor gene expression, label cells and organelles, or perturb gene function.

A wide variety of transgenic approaches to manipulating gene function have been adapted for use in ascidian embryos. Classically, these transgenic manipulations of gene function have relied heavily on the expression of constitutively active or dominant negative constructs (Stolfi and Christiaen 2012). More recently, however, knockdown and genome editing technologies have been adapted for use in ascidian systems, including: RNA interference (RNAi) (Nishiyama and Fujiwara 2008), zinc finger nucleases (ZFNs) (Kawai et al. 2012), transcription activator-like effector nucleases (TALENs) (Treen et al. 2014; Yoshida et al. 2014) and most recently clustered regularly interspaced short palindromic repeats (CRISPR/Cas9) (Sasaki et al. 2014; Stolfi et al. 2014). The application of techniques in ascidian embryos has already been surveyed in the opening chapters of this volume. Thus, this chapter focuses specifically on transgenic tools used to visualize and perturb fundamental cellular processes in ascidian embryos.

The adaptation of fluorescent proteins for use in ascidians has revolutionized the study of cellular processes in these animals. In the two decades that have passed since it was first demonstrated that heterologous expression of green fluorescent protein (GFP) from the jellyfish *Aequorea victoria* could be used as a non-invasive means of monitoring gene expression in vivo (Chalfie et al. 1994), strategies utilizing fluorescent proteins have been developed to visualize subcellular structures and organelles, visualize protein interactions, and monitor the distribution and trafficking of protein in cells (Miyawaki 2011). The relative ease of generating transgenic ascidian embryos coupled with their developmental robustness towards large quantities of exogenous DNA facilitates the expression of fluorescent proteins in ascidian embryos. Fluorescent proteins may be expressed as standalone proteins or as chimeric fusion proteins, in which the sequence encoding the fluorophore has been fused to a sequence encoding a protein of interest. The later approach represents a powerful strategy for monitoring cellular processes in vivo. Moreover, multiple fusion proteins can be expressed simultaneously in a single embryo to concomitantly label multiple cell populations or subcellular structures. Co-expression of fusion proteins can be used to assay the relative distribution of proteins (co-localization; Denker et al. 2013; Sehring et al. 2014; Gline et al. 2015; Newman-Smith et al. 2015) or identify protein interactions (fluorescence resonance energy transfer; Cooley et al. 2011). Additionally, both free fluorophores and fusion proteins can be expressed in embryos in conjunction with perturbation constructs (e.g., dominant-negative forms, CRISPR/Cas9 gene-targeting plasmids) to assess the impacts of these constructs on a cell or process of interest (Stolfi and Christiaen 2012).

14.2 Codon Optimization of Sequences Encoding Fluorescent Proteins

Commercially available fluorescent protein sequences have been widely used by ascidian biologists for visualizing cells and proteins in developing embryos. These fluorescent proteins are easily obtained and available with a range of spectral properties (Day and Davidson 2009). Moreover, sequences encoding many commercially available fluorescent proteins have been modified to improve their photostability, increase the rate of protein maturation and folding, or prevent oligomerization of fluorophores in cells (Shaner et al. 2004; Zeller et al. 2006; Shaner et al. 2008). The sequences encoding several commonly used, commercially available fluorescent proteins (e.g., eGFP, mCherry) have also been human codon-optimized to improve the efficiency of translation in mammalian cells (Yang et al. 1996; Snapp 2009). Despite having been selected for optimal performance in mammalian cells, even the human codon-optimized fluorescent proteins can be readily expressed and detected in ascidian systems (Zeller et al. 2006; Passamaneck et al. 2006). However, Zeller et al. demonstrated that commercially available eGFP, codon-optimized for expression in human cells, is sub-optimal for use in *Ciona intestinalis* embryos. Indeed, they showed that codon-optimized eGFP took longer to reach detectable levels in *Ciona* embryos than wild-type *Aequorea*-derived GFP. The Zeller group therefore generated a set of fluorescent protein sequences that have been codon-optimized for use in *Ciona* (green [GFPci], cyan [CFPci], yellow [YFPci], red [RFPci]; http://www.bio.sdsu.edu/faculty/zeller/Vector_Sequences.html) (Zeller et al. 2006). The *Ciona* codon-optimized, GFP-derived fluorescent proteins may be detected as many as 90 min earlier than their unoptimized counterparts (CFPci vs CFP) and may result in ≤50% increase in detectable fluorescence (GFPci vs GFP) making them ideal for experiments that require robust, early fluorescence (Zeller et al. 2006).

14.3 Fusion Proteins for Visualization of Cellular Processes in Ascidian Embryos

Fluorophore-tagged fusion proteins enable in vivo analysis of cellular proteins, structures, and organelles. Transgenic embryos, both live and fixed, can be imaged with sub-cellular resolution using standard confocal microscopy (Negishi et al. 2013; McDougall et al. 2015). A growing collection of fluorescent protein fusions (FP fusions) have been generated that label specific cellular structures or compartments in ascidian embryos (Table 14.1). High-resolution in vivo analysis of the spatiotemporal distribution of fusion proteins has provided mechanistic insight into a wide range of dynamic cellular processes. In the following sections, we discuss several examples, including cell division, spindle-positioning, morphogenesis, and cell signaling.

14.3.1 Cell Cycle Progression

Fusion proteins have been employed extensively to monitor the spatial and temporal dynamics of cell-cycle progression in ascidian embryos. Activation of the zygotic genome is strongly associated with shifts in the duration of cell-cycle phases. Dumollard et al. generated transgenic *Phallusia mammillata* embryos expressing a fluorophore-tagged histone, H2B::mRFP, to examine cell-cycle progression during zygotic genome activation in ascidians (Dumollard et al. 2013). Fluorophore-tagged histones are incorporated into DNA-bound nucleosomes, efficiently labeling chromatin in both interphase and mitotic cells and providing an efficient way of monitoring cell-cycle progression in vivo (Kanda et al. 1998). Using this approach, they were able to identify two distinct cell cycle phases in early ascidian embryos modulated by β-catenin-dependent transcription. Intriguingly, β-catenin-dependent activation of the zygotic endomesodermal gene regulatory network at the 16-cell stage was associated with a shortened interphase. In contrast, subsequent β-catenin-dependent transcription in the endoderm just prior to gastrulation was associated with prolonged interphase (Dumollard et al. 2013). These results highlighted specific regulatory circuits capable of modifying cell-cycle progression. This study also suggested that modulation of cell-cycle progression in the early embryo might be driven by transcriptional programs associated with cell fate induction.

Table 14.1 Ascidian cell imaging tool kit

Structure/molecule	Notes	Species[a]	Promoters	References
Cytoskeletal				
Microtubules				
Ensconsin (MAP7)::3xGFP/RFP	Full-length human E-Map-115 (Ensconsin; GenBank: LT734165.1)	*Ci*	FOG	Roure et al. (2007)
			Brachyury	Dong et al. (2011)
			Mesp	Cooley et al. (2011)
MAP7::eGFP/RFP/mCherry	Mouse MAP7 N terminal fragment (GenBank: BC052637.1)	*Pm*		Prodon et al. (2010)
Ensconsin::3xGFP	Mouse MAP7 N terminal fragment (GenBank: BC052637.1)	*Pm*		Negishi and Yasuo (2015)
Microtubule-binding				
EB3::GFP	Mouse EB3	*Pm*		Negishi and Yasuo (2015)
Actin				
GFP/mCherry::hActin	Full-length human β-actin	*Ci*	Brachyury	Dong et al. (2011)
LifeAct::meGFP	First 17 aa of Abp140 (*Saccharomyces cerevisiae)*	*Ci*	Brachyury	Dong et al. (2011) and Sehring et al. (2014)
LifeAct::GFP/RFP	First 17 aa of Abp140 (*Saccharomyces cerevisiae)*	*Pm*		McDougall et al. (2015)
mCherry::UtrCH	CH domain from *Ciona* utrophin	*Ci*	Brachyury	Dong et al. (2011)
Utrophin::GFP	First 783 bp of human utrophin (GenBank: NM_007124)	*Ci*	Mesp	Cooley et al. (2011)
ABD::Venus	ABD from ABP-120	*Pm*		Prodon et al. (2010)
Actin-binding				
Cofilin::mCherry	*Ciona*	*Ci*	FOG	Sehring et al. (2014)
α-actinin::mCherry	*Ciona*	*Ci*	FOG	Sehring et al. (2014)
mCherry::tropomyosin	*Ciona*	*Ci*	FOG	Sehring et al. (2014)
EB1::mCherry	*Ciona*	*Ci*	FOG	Sehring et al. (2014)
IQGAP::mCherry	*Ciona*	*Ci*	FOG	Sehring et al. (2014)
Anillin::mCherry	*Ciona*	*Ci*	FOG	Sehring et al. (2014)
Septin-2::mCherry	*Ciona*	*Ci*	FOG	Sehring et al. (2014)
Myosin				
mCherry::MLC	*Ciona* MLC	*Ci*	Brachyury	Dong et al. (2011)
iMyo::YFP[b]	YFP-tagged intrabody that recognizes nonmuscle myosin II A	*Ci*	FOG	Hashimoto et al. (2015)
Chromatin				
H2B::Venus/CFP/mCherry/RFP/YFP	*Human histone 2B (GenBank:* X00088)	*Ci*	FOG	Roure et al. (2007)
			Mesp	Stolfi et al. (2010)
			Zip	Kawai et al. (2015)
			Tensin	Kourakis et al. (2014)
			Brachyury	Kourakis et al. (2014)
			FoxB	Imai et al. (2009)
			FoxF	Gline et al. (2015)
			Snail	Imai et al. (2009)
			Etr	Veeman et al. (2010)
			Mnx	Navarrete and Levine (2016)

(continued)

Table 14.1 (continued)

Structure/molecule	Notes	Species[a]	Promoters	References
H2B::RFP1	Mouse histone H2B (HH2B) from lung tumor-BI105582)	*Pm*		Prodon et al. (2010)
Plasma membrane				
GAP43::Venus/CFP	*Mouse Gap-43*	*Ci*	FOG	Roure et al. (2007)
			Epi1	Ogura et al. (2011)
GAP43::GFP	First 20 amino acids of Ciona GAP-43	*Ci*	Bra	Dong et al. (2009)
PH::eGFP/Tomato	PIP2 binding domain from human PLCδ1	*Pm*		Prodon et al. (2010)
YFP/Kaede/Venus/ mCherry::CAAX		*Ci*	Epi1	Dong et al. (2009)
			Zip	Kawai et al. (2015)
			FoxB	Navarrete and Levine (2016)
memb::GFP	First 20aa of *Ciona* GAP-43	*Ci*	Brachyury	Dong et al. (2009)
			Mesp	Cooley et al. (2011)
Caveolin::GFP/RFP	*Full-length Ciona* caveolin *(CavA;* Ciinte.g00012490)	*Ci*	Mesp	Cota and Davidson (2015)
Endosomes				
Caveolin::GFP/RFP	*Full-length Ciona* caveolin *(CavA;* Ciinte.g00012490)	*Ci*	Mesp	Cota and Davidson (2015)
Adhesions				
DE-cadherin::mCherry	Open reading frame and the 3′ untranslated region for *Drosophila melanogaster* E-cadherin	*Ci*	FOG	Roure et al. (2007)
			Brachyury	Dong et al. (2009)
ZO1::3×GFP/tGFP	*Ciona* ZO-1 (Cicl035p23)	*Ci*	FOG	Hashimoto et al. (2015)
				Denker et al. (2013)
		Pm		Sherrard et al. (2010)
GFP::Talin	*Ciona* talin (Ciinte.g00005295)	*Ci*	Mesp	Norton et al. (2013)
Golgi				
ST::GFP	Membrane signal anchor sequence from Rat sialyltransferase	*Ci*	EpiB	Zeller et al. (2006)
Endoplasmic reticulum				
KDELR::GFP/RFP	*Ciona* KDELR (ci0100149851)	*Ci*	Brachyury	Oda-Ishii et al. (2010)
			Mesp	Gline et al. (2015)
Centrosomes/spindle poles			Mesp	
Aurora::Venus/mCherry	Full-length *Ciona* aurora kinase (Aur; Ciinte.g00008046)	*Ci*	FOG	Roure et al. (2007)
	Full-length *Ciona* aurora kinase (Ciinte.g00008046; transcript model KH.C7.219.v1.A.SL1-1)	*Pm*		Hebras and McDougall (2012)
Venus/mCherry::First 632 amino acids of Girdin	First 632 amino acids of *Ciona* Girdin (Ciinte.g00002344; gene model KH2012:KH.C11.506)	*Pm*	FOG	McDougall et al. (2015)
Plk1::mCherry/Venus	Full-length *Ciona* PLK1 (Ciinte. g00002774; transcript model KH. C12.238.v1.A.SL1-1)	*Pm*	Brachyury	Hebras and McDougall (2012)
Midbody				
Aurora::Venus/mCherry	Full-length *Ciona* aurora kinase (Ciinte.g00008046; transcript model KH.C7.219.v1.A.SL1-1)	*Pm*		Hebras and McDougall (2012)

(continued)

Table 14.1 (continued)

Structure/molecule	Notes	Species[a]	Promoters	References
Venus:TPX2	Full-length *Ciona* Tpx2 (Ciinte. g00005340; transcript model KH. C3.181.v1.A.SL1-1)	*Pm*		Hebras and McDougall (2012)
Incep:mCherry	Full-length *Ciona* INCEP (Ciinte. g00004061)	*Pm*		Hebras and McDougall (2012)
Plk1:mCherry/Venus	Full-length *Ciona* PLK1 (Ciinte. g00002774; transcript model KH. C12.238.v1.A.SL1-1)	*Pm*		Hebras and McDougall (2012)

Fluorescent fusion proteins and bioprobes used to visualize subcellular dynamics in ascidian embryos (adapted from Negishi et al. 2013; McDougall et al. 2015). *Ciona intestinalis* (*Ci*) and *Phallusia mammillata* (*Pm*). Species, gene model, transcript model, and/or GenBank identifiers for source sequences are indicated where reported

CH calponin homology, *ABD* actin-binding domain, *MLC* myosin II light chain, *YFP* yellow fluorescent protein

[a]"Species" indicates species of ascidian in which a probe was used

[b]iMyo:YFP is an intrabody that has been used to recognize activated myosin in *Ciona intestinalis* (Hashimoto et al. 2015)

Transgenically expressed fluorescence ubiquitination cell cycle indicator (FUCCI) probes represent an alternative method of monitoring cell-cycle phasing in living cells using a two-color fusion protein system (Sakaue-Sawano et al. 2008). Labelled Cdt1 protein with a peak expression during G1 and its inhibitor, geminin, whose own expression peaks during S and G2 phases, are simultaneously expressed in cells. Time-lapse imaging is then used to monitor the fluorescence of the two FUCCI probes and thereby monitor cell cycle progression. The simultaneous expression of these two probes enables more precise discrimination between cells in G1 and S/G2/M phases than is possible relying only on markers of chromatin confirmation alone (Sakaue-Sawano et al. 2008). Ogura et al. employed FUCCI technology to examine cell-cycle progression in the epidermis of *Ciona intestinalis* embryos during neurulation. Using this approach, they were able to identify a prolonged G2 phase in epidermal cells associated with neural tube closure/epidermal fusion. They were also able to demonstrate that precocious division of the epidermis resulting from ectopic expression of the cell cycle regulator, Cdc25, disrupted neural tube closure. In addition, live imaging of the epidermal divisions revealed that resumption of cell cycle progression is staggered along the anterior–posterior (AP) axis, such that cells posterior to the embryo divide before those anterior. The staggered division of epidermal cells correlate with AP zippering of the neural tube. Epidermal division that occurs at this time is oriented along the AP axis contributing to elongation of the embryo along this axis (Ogura et al. 2011).

14.3.2 Spindle Positioning

Imaging of transgenic ascidian embryos has also garnered insights into the mechanisms driving spindle positioning in the early embryo. Ascidian embryos display a unique invariant cleavage pattern prior to gastrulation that is dependent upon the precise orientation of the cell division plane, including some instances of unequal cleavage. Work in *Halocynthia roretzi* embryos initially showed that the orientation and positioning of the cleavage plane during these unequal cleavages is mediated by an electron-dense, multilayered structure termed the centrosome-attracting body (CAB) (Hibino et al. 1998; Nishikata et al. 1999; Iseto and Nishida 1999). Subsequent work has shown the CAB to be highly conserved amongst ascidian species (Brown and Swalla 2007; Hibino et al. 1998; Nishikata et al. 1999; Iseto and Nishida 1999; Sardet et al. 2003; Patalano et al. 2006; Gyoja 2006). Prodon et al. employed 4D live imaging to examine CAB-mediated spindle dynamics during unequal cleavage in transgenic *Phallusia mammillata* embryos expressing a GFP-tagged microtubule binding protein (MAP7::EGFP). During the interphase, the nucleus and associated centrosomes migrate toward the CAB. As the spindle begins to form during mitotic entry, the nearest, medial spindle pole is attracted to the CAB, where it remains tethered through cytokinesis. Finally, tethering of the lateral spindle pole further rotates the mitotic spindle during anaphase. Blastomere isolation experiments demonstrated that tethering is dependent upon cell–cell contacts. Further analysis of furrow formation following orientation of the mitotic spindle in transgenic embryos expressing either a membrane-tethered GFP (PH::EGFP) or a Venus-tagged actin-binding domain (ABD::Venus) revealed an unusual process of asymmetric furrow formation in these cells during cytokinesis. In particular, lateral spindle attachment to cell–cell contact appears to inhibit initiation of the cytokinetic furrow at the site of attachment (Prodon et al. 2010).

14.3.3 Morphogenesis of the Ascidian Neural Tube

The ascidian neural tube arises from a flat sheet of ~40 neural progenitor cells termed the neural plate. Primary invagination of the neural plate occurs during gastrulation. (Nicol and Meinertzhagen 1988b; Nicol and Meinertzhagen 1988a). Live imaging of primary neural plate invagination in transgenic embryos expressing H2B::YFP revealed a two-step process in which the medial cells of the neural plate undergo a single oriented division aligning the daughter cells along the AP axis. These cells then intercalate via a convergent extension to form the ventral floorplate of the neural tube, a single linear row of cells extending along the AP axis of the embryo. Subsequently, the lateral cells of the neural plate undergo concomitant lateral stacking, forming two lateral rows of cells flanking the ventral floorplate. These cellular rearrangements are dependent on fibroblast growth factor (FGF) and nodal signaling, respectively (Navarrete and Levine 2016).

A subsequent secondary invagination completes the closure of the neural tube. Secondary invagination of the posterior neural tube occurs via unidirectional zippering, which proceeds in a posterior to anterior direction and completes formation of the nerve chord. The cellular mechanism underlying neuroepidermal zippering was recently described by Hashimoto et al. (2015). Using transgenic embryos expressing a GFP-labeled marker of tight junctions (ZO1::3×GFP) in the neural plate and overlying epidermis in conjunction with live imaging, they observed a striking pattern of junctional contractions occurring ahead of the closing “zipper” of the neural tube. Using an embryo expressing a YFP-conjugated intrabody to monitor activated myosin (iMYO::YFP), they found that junctional contractions were associated with a transient increase in activated myosin. Further investigation of the actomyosin contractions during this process using a combination of transgenic manipulations and cytoskeletal inhibitors indicated that myosin is activated in cells by ROCK/Rho. Finally, laser ablation of junctions revealed that the transient activation of myosin ahead of zippering generated increased the tension across tight junctions required to pull the neural plate closed at the midline. Junctional exchange ahead of cell detachment serves to dissipate this increased tension and promote zipper progression.

14.3.4 Cell Migration

The migration of two bilaterally positioned pairs of cardiopharyngeal progenitors (trunk ventral cells; TVCs) during development of the ascidian heart represents an elegant, highly simplified model for investigating collective cell migration. TVCs migrate as polarized pairs of cells, tethered to one another by cadherin-rich cell–cell adhesions (Norton et al. 2013; Gline et al. 2015). High-resolution analysis of transgenic TVCs expressing a GFP-tagged moesin fusion protein (GFP::Moesin) revealed a pronounced leader–trailer polarity, in which the leader cells exhibit robust enrichment of lateral actin-rich protrusions at the leading edge, whereas the trailer cell displays a concomitant constriction of the trailing edge (Christiaen et al. 2008; Gline et al. 2015).

Analysis of transgenic TVCs in which mosaic incorporation of transgenes resulted in the selective inhibition of cell migration in either the leader or the trailer revealed that individual TVCs are capable of migrating alone. However, using the green-to-red photoconvertible fluorophore, Kaede (*Mesp*>nls::Kaede), to label individual cardiopharyngeal founder cells and thereby distinguish between individual transgenic TVCs within pairs, Gline et al. (2015) found that the more anterior founder cell gave birth to the leader TVC in 95% of embryos examined. These results indicated that there is a strong positional bias that contributes to the establishment of leader trailer polarity.

Although TVC pairs are capable of migrating independently, robust migration is dependent on responses to signals from surrounding tissues. The precise extrinsic cues necessary for robust directional migration of TVC pairs have not been identified. However, using live imaging to assess TVC migration (*Mesp*>GFP, *FoxF*>H2B::mCherry), it was shown that extrinsic cues have an impact on migration by altering leader–trailer cell polarity.

By selectively inhibiting secretion in tissues surrounding the TVCs during migration via targeted expression of a dominant negative form of the GTPase, Sar1, in the mesenchyme (*Twist*>Sar1^{T35N}), the notochord (*Brachyury*>Sar1^{T35N}), the epidermis (*EphrinB*>Sar1^{T35N}), and the endoderm (*Nkx2–1*>Sar1^{T35N}), Gline et al. found that polarization of protrusions within the leader–trailer cell pair requires secretion from the endoderm. In contrast, cues secreted from the mesenchyme are required to reinforce trailer cell identity, whereas cues secreted by the epidermis are required to maintain adhesion between leader and trailer cells (Gline et al. 2015).

14.3.5 Signal Processing

The cardiopharyngeal lineage in *Ciona* can be traced back to a multipotent pre-cardiac founder cell lineage established through exclusive expression of the bHLH transcription factor, Mesp (Satou 2004). At the onset of neurulation, each founder cell divides asymmetrically to produce a larger anterior tail muscle cell and a smaller heart progenitor cell (TVC). FGF activates mitogen activated protein kinase signaling (MAPK) in the TVCs, initiating a well-characterized transcriptional program that induces heart progenitor fate (Evans-Anderson and Christiaen 2016; Cota et al. 2017). During heart progenitor induction, founder cells are exposed uniformly to FGF (Cooley et al. 2011). Analysis of dividing founder cells with transgenic labeling of membrane (GPI::RFP) and filamentous actin enrichment (utrophin::GFP) in transgenic embryos revealed polarized actin-enriched membrane protrusions that correlated with localized inductive signaling along the nascent TVC membrane. Manipulations of actin dynamics including targeted expression of a constitutively active form of CDC42 (CDC42^{Q61L}), co-expression of a truncated form of a downstream actin nucleator (Wasp$^{\Delta VCA}$), along with pharmacological inhibition of actin-related protein (ARP)-mediated actin nucleation (Ck-666), indicated that localized protrusion was required for localized heart progenitor induction (Cooley et al. 2011).

A subsequent study suggested that membrane protrusion drives polarized founder cell adhesion to the underlying extracellular matrix during heart progenitor induction (Norton et al. 2013). Analysis of focal adhesion dynamics in transgenic founder cells expressing GFP::Talin revealed that matrix adhesion is maintained along the presumptive TVC membrane during founder cell division. Targeted perturbations of adhesion through manipulations of Rap GTPase activity and co-expression of specific Integrin receptor chains (Rap^{S17N}, Rap^{G12V} or RapGAP, Integrin-β1, Integrin-β2, Integrin-α11) revealed that localized mitotic adhesion is both necessary and sufficient for polarized induction (Norton et al. 2013).

The most recent study that looked at pre-cardiac founder cells revealed that transgenically labeled FGF receptor (FGFR::Venus) is localized to the adherent membrane during founder cell division. This study also suggested that FGFRs are trafficked in adherent membrane domains enriched with Caveolin (caveolin::GFP/RFP). Targeted disruption and restoration of these adhesive domains using dominant negative constructs

(Caveolin^{P132L}, Rap^{S17N}), CRISPR knockdown (*Mesp*>nls::Cas9:nls, *U6*>CavA.39gRNA), and targeted co-expression of Integrin β-chains (*Mesp*>Integrin-β2 or *Mesp*>Integrin-β1) showed that they are both necessary and sufficient for polarized FGFR trafficking and during founder cell mitosis and polarized TVC induction immediately after division (Cota and Davidson 2015).

Taken together, these results suggest that polarized adhesive interactions might be able to alter mitotic membrane trafficking. In this way, matrix adhesion can compartmentalize the cellular response to widespread inductive FGF signals.

14.4 Conclusions

Over the past decade, ascidians have emerged as a powerful model in which to study key aspects of cell biology. For this purpose, researchers have adapted a set of molecular tools for labeling cells and intracellular structures that can be used in conjunction with established methods of manipulating gene expression and function to explore the contribution of dynamic cellular processes to development (Table 14.1). However, the repertoire of fusion proteins available for labeling cellular structures in transgenic ascidians remains limited in comparison to more established model systems. Thus, further adaptation of this system requires development of additional cellular reporters.

Most notably absent from the ascidian "toolkit" are markers of specific membrane compartments within the endolysomal system, which are critical for future analysis of protein trafficking and degradation in cells. Continuing efforts are also focused on further optimizing protocols for live imaging. Although high-resolution live imaging has been successfully employed to examine cellular processes in early ascidian embryos (Ogura et al. 2011; Dumollard et al. 2013; Negishi and Yasuo 2015), the success of this approach has been limited at later stages of development, particularly in cells positioned deep within embryos.

The continued development of improved methods for transgenically manipulating the optically transparent ascidian, *Phallusia mammillata*, holds great promise in overcoming this limitation (Negishi et al. 2013; McDougall et al. 2014; McDougall et al. 2015). Moreover, newly developed SNAP, CLIP, and HALO peptide-

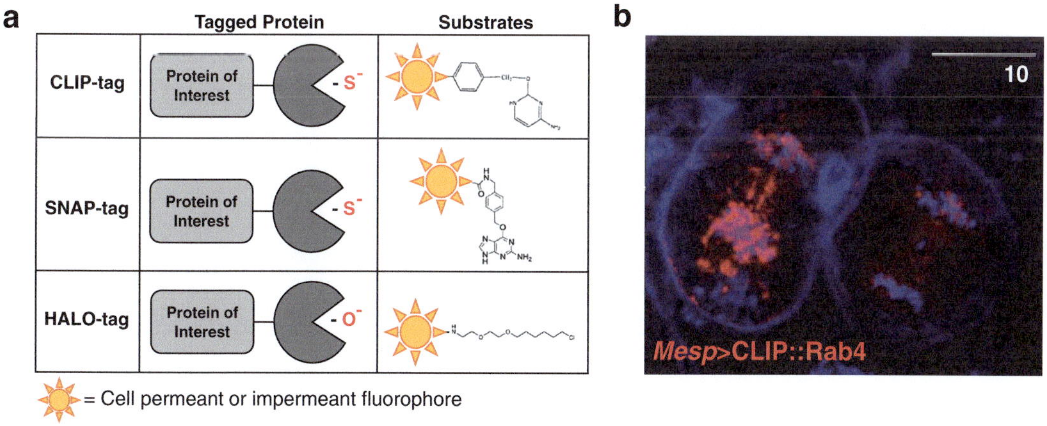

Fig. 14.1 CLIP, SNAP, and HALO tags for covalent labeling of ascidian protein. Genetically encoded CLIP tag (New England BioLabs, NEB), SNAP tag (NEB) and HALO tag (Promega) schematically depicted in (**a**) can be used in conjunction with commercially available fluorophore-conjugated substrates to covalently label and visualize proteins in living or fixed transgenic ascidian embryos. (**b**) Ventral projection of a pair of mitotic precardiac founder cells in a *Ciona* embryo expressing an endosome-associated CLIP tag fusion protein under the control of the cardiopharyngeal lineage specific promoter, Mesp (*Mesp*>CLIP::Rab4). The embryo was fixed and then incubated with the CLIP-Cell TMR-Star (red; NEB) substrate before imaging using confocal microscopy. Phalloidin was used to detect F-actin and DRAQ5 was used to detect DNA (*blue*). Anterior is to the right

tagging strategies that allow covalent chemical labeling of proteins in vivo hold some potential for achieving brighter, more photo-resistant labeling of proteins to facilitate live imaging (Fig. 14.1) (Keppler et al. 2003; Gautier et al. 2008; England et al. 2015).

Despite these challenges, research into this system has already begun to provide novel insights into the cellular mechanisms underlying development. In particular, ascidian researchers have begun to integrate transgenic approaches to in vivo analysis of cellular processes with existing methods for manipulating gene function. This integrative approach to investigating cellular processes during ascidian development has proven to be a powerful strategy for identifying the molecular mechanisms underlying complex developmental processes, including morphogenesis, cell migration, and cell signaling.

Acknowledgments The author would like to thank Dr. Brad Davidson (Swarthmore College) for comments on the manuscript. Work in the Davidson laboratory is supported by Swarthmore College and the NIH (1R01HL091027, R15 HD080525-01). CDC is supported by an American Heart Association Postdoctoral Award (16POST27250075).

References

Brown FD, Swalla BJ (2007) Vasa expression in a colonial ascidian, Botrylloides violaceus. Evol Dev 9:165–177. https://doi.org/10.1111/j.1525-142X.2007.00147.x

Chalfie M, Tu Y, Euskirchen G et al (1994) Green fluorescent protein as a marker for gene expression. Science (New York, NY) 263:802–805. https://doi.org/10.1126/science.8303295

Christiaen L, Davidson B, Kawashima T et al (2008) The transcription/migration interface in heart precursors of Ciona intestinalis. Science New York NY 320:1349–1352. https://doi.org/10.1126/science.1158170

Cooley J, Whitaker S, Sweeney S et al (2011) Cytoskeletal polarity mediates localized induction of the heart progenitor lineage. Nat Cell Biol 13:952–957. https://doi.org/10.1038/ncb2291

Corbo JC, Levine M, Zeller RW (1997) Characterization of a notochord-specific enhancer from the Brachyury promoter region of the ascidian, Ciona intestinalis. Development 124:589–602

Cota CD, Davidson B (2015) Mitotic membrane turnover coordinates differential induction of the heart progenitor lineage. Dev Cell 34:505–519. https://doi.org/10.1016/j.devcel.2015.07.001

Cota CD, Palmquist K, Davidson B (2017) Heart development in Ciona. In: Reference module in life sciences. Elsevier, January. ISBN 9780128096338. http://dx.doi.org /https://doi.org/10.1016/B978-0-12-809633-8.12148-X

Day RN, Davidson MW (2009) The fluorescent protein palette: tools for cellular imaging. Chem Soc Rev 38:2887–2921. https://doi.org/10.1039/b901966a

Dehal P (2002) The draft genome of Ciona intestinalis: insights into chordate and vertebrate origins. Science (New York, NY) 298:2157–2167. https://doi.org/10.1126/science.1080049

Denker E, Bocina I, Jiang D (2013) Tubulogenesis in a simple cell cord requires the formation of bi-apical cells through two discrete Par domains. Development 140:2985–2996. https://doi.org/10.1242/dev.092387

Dong B, Horie T, Denker E et al (2009) Tube formation by complex cellular processes in Ciona Intestinalis notochord. Dev Biol 330:237–249. https://doi.org/10.1016/j.ydbio.2009.03.015

Dong B, Deng W, Jiang D (2011) Distinct cytoskeleton populations and extensive crosstalk control Ciona notochord tubulogenesis. Development 138:1631–1641. https://doi.org/10.1242/dev.057208

Dumollard R, Hebras C, Besnardeau L, McDougall A (2013) Beta-catenin patterns the cell cycle during maternal-to-zygotic transition in urochordate embryos. Dev Biol 384:331–342. https://doi.org/10.1016/j.ydbio.2013.10.007

England CG, Luo H, Cai W (2015) HaloTag technology: a versatile platform for biomedical applications. Bioconjug Chem 26:975–986. https://doi.org/10.1021/acs.bioconjchem.5b00191

Evans-Anderson H, Christiaen L (2016) Ciona as a simple chordate model for heart development and regeneration. JCDD 3:25. https://doi.org/10.3390/jcdd3030025

Gautier A, Juillerat A, Heinis C et al (2008) An engineered protein tag for multiprotein labeling in living cells. Chem Biol 15:128–136. https://doi.org/10.1016/j.chembiol.2008.01.007

Gline S, Kaplan N, Bernadskaya Y et al (2015) Surrounding tissues canalize motile cardiopharyngeal progenitors towards collective polarity and directed migration. Development 142:544–554. https://doi.org/10.1242/dev.115444

Gyoja F (2006) Expression of a muscle determinant gene, macho-1, in the anural ascidian Molgula tectiformis. Dev Genes Evol 216:285–289. https://doi.org/10.1007/s00427-005-0056-1

Hashimoto H, Robin FB, Sherrard KM, Munro EM (2015) Sequential contraction and exchange of apical junctions drives zippering and neural tube closure in a simple chordate. Dev Cell 32:241–255. https://doi.org/10.1016/j.devcel.2014.12.017

Hebras C, McDougall A (2012) Urochordate ascidians possess a single isoform of aurora kinase that localizes to the midbody via TPX2 in eggs and cleavage stage embryos. PloS One 7(9):e45431

Hibino T, Nishikata T, Nishida H (1998) Centrosome-attracting body: a novel structure closely related to unequal cleavages in the ascidian embryo. Develop Growth Differ 40:85–95

Hikosaka A, Satoh N, Makabe KW (1993) Regulated spatial expression of fusion gene constructs with the 5′ upstream region of Halocynthia roretzi muscle actin gene in Ciona savignyi embryos. Rouxs Arch Dev Biol 203:104–112. https://doi.org/10.1007/BF00539896

Imai KS, Stolfi A, Levine M, Satou Y (2009) Gene regulatory networks underlying the compartmentalization of the Ciona central nervous system. Development 136:285–293. https://doi.org/10.1242/dev.026419

Iseto T, Nishida H (1999) Ultrastructural studies on the centrosome-attracting body: electron-dense matrix and its role in unequal cleavages in ascidian embryos. Develop Growth Differ 41:601–609

Kanda T, Sullivan KF, Wahl GM (1998) Histone-GFP fusion protein enables sensitive analysis of chromosome dynamics in living mammalian cells. Curr Biol 8:377–385

Kawai N, Ochiai H, Sakuma T et al (2012) Efficient targeted mutagenesis of the chordate Ciona Intestinalis genome with zinc-finger nucleases. Develop Growth Differ 54:535–545. https://doi.org/10.1111/j.1440-169X.2012.01355.x

Kawai N, Ogura Y, Ikuta T et al (2015) Hox10-regulated endodermal cell migration is essential for development of the ascidian intestine. Dev Biol 403:43–56. https://doi.org/10.1016/j.ydbio.2015.03.018

Keppler A, Gendreizig S, Gronemeyer T et al (2003) A general method for the covalent labeling of fusion proteins with small molecules in vivo. Nat Biotechnol 21:86–89. https://doi.org/10.1038/nbt765

Kourakis MJ, Reeves W, Newman-Smith E et al (2014) A one-dimensional model of PCP signaling: polarized cell behavior in the notochord of the ascidian Ciona. Dev Biol 395:120–130. https://doi.org/10.1016/j.ydbio.2014.08.023

McDougall A, Lee KW-M, Dumollard R (2014) Microinjection and 4D fluorescence imaging in the eggs and embryos of the ascidian Phallusia mammillata. Methods Mol Biol 1128:175–185. https://doi.org/10.1007/978-1-62703-974-1_11

McDougall A, Chenevert J, Pruliere G et al (2015) Centrosomes and spindles in ascidian embryos and eggs. Methods Cell Biol 129:317–339. https://doi.org/10.1016/bs.mcb.2015.03.006

Miyawaki A (2011) Proteins on the move: insights gained from fluorescent protein technologies. Nat Rev Mol Cell Biol 12:656–668. https://doi.org/10.1038/nrm3199

Navarrete IA, Levine M (2016) Nodal and FGF coordinate ascidian neural tube morphogenesis. Development 143:4665–4675. https://doi.org/10.1242/dev.144733

Negishi T, Yasuo H (2015) Distinct modes of mitotic spindle orientation align cells in the dorsal midline of ascidian embryos. Dev Biol 408:66–78. https://doi.org/10.1016/j.ydbio.2015.09.019

Negishi T, McDougall A, Yasuo H (2013) Practical tips for imaging ascidian embryos. Develop Growth Differ 55:446–453. https://doi.org/10.1111/dgd.12059

Newman-Smith E, Kourakis MJ, Reeves W et al (2015) Reciprocal and dynamic polarization of planar cell polarity core components and myosin. Elife 4:e05361. https://doi.org/10.7554/eLife.05361

Nicol D, Meinertzhagen IA (1988a) Development of the central nervous system of the larva of the ascidian, Ciona intestinalis L. II. Neural plate morphogenesis and cell lineages during neurulation. Dev Biol 130:737–766

Nicol D, Meinertzhagen IA (1988b) Development of the central nervous system of the larva of the ascidian, Ciona intestinalis L: I. The early lineages of the neural plate. Dev Biol 130:721–736

Nishikata T, Hibino T, Nishida H (1999) The centrosome-attracting body, microtubule system, and posterior egg cytoplasm are involved in positioning of cleavage planes in the ascidian embryo. Dev Biol 209:72–85. https://doi.org/10.1006/dbio.1999.9244

Nishiyama A, Fujiwara S (2008) RNA interference by expressing short hairpin RNA in the Ciona intestinalis embryo. Develop Growth Differ 50:521–529. https://doi.org/10.1111/j.1440-169X.2008.01039.x

Norton J, Cooley J, Islam AFMT et al (2013) Matrix adhesion polarizes heart progenitor induction in the invertebrate chordate Ciona intestinalis. Development 140:1301–1311. https://doi.org/10.1242/dev.085548

Oda-Ishii I, Ishii Y, Mikawa T (2010) Eph regulates Dorsoventral asymmetry of the notochord plate and convergent extension-mediated notochord formation. PLoS One 5:e13689. https://doi.org/10.1371/journal.pone.0013689.s007

Ogura Y, Sakaue-Sawano A, Nakagawa M et al (2011) Coordination of mitosis and morphogenesis: role of a prolonged G2 phase during chordate neurulation. Development 138:577–587. https://doi.org/10.1242/dev.053132

Passamaneck YJ, Di Gregorio A, Papaioannou VE, Hadjantonakis A-K (2006) Live imaging of fluorescent proteins in chordate embryos: from ascidians to mice. Microsc Res Tech 69:160–167. https://doi.org/10.1002/jemt.20284

Patalano S, Pruliere G, Prodon F et al (2006) The aPKC-PAR-6-PAR-3 cell polarity complex localizes to the centrosome attracting body, a macroscopic cortical structure responsible for asymmetric divisions in the early ascidian embryo. J Cell Sci 119:1592–1603. https://doi.org/10.1242/jcs.02873

Prodon F, Chenevert J, Hebras C et al (2010) Dual mechanism controls asymmetric spindle position in ascidian germ cell precursors. Development 137:2011–2021. https://doi.org/10.1242/dev.047845

Roure A, Rothbächer U, Robin F et al (2007) A multicassette Gateway vector set for high throughput and comparative analyses in Ciona and vertebrate embryos. PLoS One 2:e916. https://doi.org/10.1371/journal.pone.0000916

Sakaue-Sawano A, Kurokawa H, Morimura T et al (2008) Visualizing spatiotemporal dynamics of multicellular

cell-cycle progression. Cell 132:487–498. https://doi.org/10.1016/j.cell.2007.12.033
Sardet C, Nishida H, Prodon F, Sawada K (2003) Maternal mRNAs of PEM and macho 1, the ascidian muscle determinant, associate and move with a rough endoplasmic reticulum network in the egg cortex. Development 130:5839–5849. https://doi.org/10.1242/dev.00805
Sasaki H, Yoshida K, Hozumi A, Sasakura Y (2014) CRISPR/Cas9-mediated gene knockout in the ascidian Ciona intestinalis. Develop Growth Differ 56:499–510. https://doi.org/10.1111/dgd.12149
Satoh N (1994) Developmental biology of ascidians. Cambridge University Press, Cambridge
Satoh N, Satou Y, Davidson B, Levine M (2003) Ciona intestinalis: an emerging model for whole-genome analyses
Satou Y (2004) The ascidian Mesp gene specifics heart precursor cells. Development 131:2533–2541. https://doi.org/10.1242/dev.01145
Sehring IM, Dong B, Denker E et al (2014) An equatorial contractile mechanism drives cell elongation but not cell division. PLoS Biol 12:e1001781. https://doi.org/10.1371/journal.pbio.1001781
Shaner NC, Campbell RE, Steinbach PA et al (2004) Improved monomeric red, orange and yellow fluorescent proteins derived from Discosoma sp. red fluorescent protein. Nat Biotechnol 22:1567–1572. https://doi.org/10.1038/nbt1037
Shaner NC, Lin MZ, McKeown MR et al (2008) Improving the photostability of bright monomeric orange and red fluorescent proteins. Nat Meth 5:545–551. https://doi.org/10.1038/nmeth.1209
Sherrard K, Robin F, Lemaire P, Munro E (2010) Sequential activation of apical and basolateral contractility drives ascidian endoderm invagination. Curr Biol CB 20:1499–1510. https://doi.org/10.1016/j.cub.2010.06.075
Snapp EL (2009) Fluorescent proteins: a cell biologist's user guide. Trends Cell Biol 19:649–655. https://doi.org/10.1016/j.tcb.2009.08.002
Stolfi A, Christiaen L (2012) Genetic and genomic toolbox of the chordate Ciona intestinalis. Genetics 192:55–66. https://doi.org/10.1534/genetics.112.140590
Stolfi A, Gainous TB, Young JJ et al (2010) Early chordate origins of the vertebrate second heart field. Science (New York, NY) 329:565–568. https://doi.org/10.1126/science.1190181
Stolfi A, Gandhi S, Salek F, Christiaen L (2014) Tissue-specific genome editing in Ciona embryos by CRISPR/Cas9. Development 141:4115–4120. https://doi.org/10.1242/dev.114488
Treen N, Yoshida K, Sakuma T et al (2014) Tissue-specific and ubiquitous gene knockouts by TALEN electroporation provide new approaches to investigating gene function in Ciona. Development 141:481–487. https://doi.org/10.1242/dev.099572
Veeman MT, Newman-Smith E, El-Nachef D, Smith WC (2010) The ascidian mouth opening is derived from the anterior neuropore: reassessing the mouth/neural tube relationship in chordate evolution. Dev Biol 344:138–149. https://doi.org/10.1016/j.ydbio.2010.04.028
Yang TT, Cheng L, Kain SR (1996) Optimized codon usage and chromophore mutations provide enhanced sensitivity with the green fluorescent protein. Nucleic Acids Res 24:4592–4593
Yoshida K, Treen N, Hozumi A et al (2014) Germ cell mutations of the ascidian Ciona Intestinalis with TALE nucleases. Genesis 52:431–439. https://doi.org/10.1002/dvg.22770
Zeller RW (2004) Generation and use of transgenic ascidian embryos. In: Methods in cell biology. Elsevier, pp 713–730
Zeller RW, Weldon DS, Pellatiro MA, Cone AC (2006) Optimized green fluorescent protein variants provide improved single cell resolution of transgene expression in ascidian embryos. Dev Dyn 235:456–467. https://doi.org/10.1002/dvdy.20644

Cellular Processes of Notochord Formation

15

William C. Smith

Abstract

This review covers recent advances in our understanding of the cell biology and morphogenesis of the ascidian notochord. In its development, the ascidian notochord undergoes a rapid series of cellular and morphogenic events that transform a group of 40 loosely packed cells in the neurula embryo into a tubular column with central lumen in the larva. The ascidian notochord has been a subject of intensive research in recent years, and particular focus in this review will be on events associated with the development and function of polarized cell properties, and the mechanism of lumen formation.

Keywords

Ascidian · *Ciona* · Notochord · Planar cell polarity · Morphogenesis · Lumenization

Note on Species Names It has recently been brought to the attention of the ascidian research community that there has been confusion regarding the names of commonly used experimental species. In particular, there has been widespread misidentification of the species *Ciona robusta* as *Ciona intestinalis* (Brunetti et al. 2015; Pennati et al. 2015). In fact, the natural range of *C. robusta* is much broader, and it has been used experimentally much more frequently than *C. intestinalis*. When it is known, the correct *Ciona* species name is used here (*C. intestinalis* or *C. robusta*, or *Ciona savignyi*), although the manuscripts referenced may use a different name. Where it is unclear whether *C. robusta* or *C. intestinalis* was used, the animals are simply referred to as *Ciona*.

15.1 Introduction

The notochord is one of the most prominent and widely studied organs in the tunicate tadpole larva. Structurally, it forms a central rod from the posterior tip of the tail, immediately ventral to the caudal nerve cord, and runs anterior to the start of the trunk, terminating under the motor ganglion (Fig. 15.1a). The ascidian notochord is largely passive—it is neither instructive on the development of surrounding tissues, as are vertebrate and cephalochordate notochords (Placzek 1995; Shimeld 1999), nor is it contractile like the cephalochordate notochord (Flood et al. 1969; Urano et al. 2003).

Despite its passivity, tunicate notochord development is surprisingly dynamic. The notochord

W. C. Smith (✉)
Department of Molecular, Cell and Developmental Biology and Neuroscience Research Institute, University of California, Santa Barbara, Santa Barbara, CA, USA
e-mail: w_smith@lifesci.ucsb.edu

Y. Sasakura (ed.), *Transgenic Ascidians*, Advances in Experimental Medicine and Biology 1029,
https://doi.org/10.1007/978-981-10-7545-2_15

cells, from their initial specification in the blastula embryo, through the formation of the mature lumenized structural column, display a wide repertoire of behaviors, starting with polarized motility and intercalation to form a single-file column, followed by a dramatic actin-driven change in cell shape as the notochord elongates, and then a transformation and sub-cellular partitioning to epithelial-like secretory cells (Fig. 15.1). This is followed by a new round of cell motility as the cells flatten and extend along the inner surface of an encapsulating extracellular sheath of matrix proteins, resulting in a tubular structure with a central lumen. Finally, the notochord is only a transient structure and is quickly retracted once metamorphosis is initiated, and then consumed via an apoptotic pathway.

Although these processes are common to morphogenesis throughout the metazoans, the tunicate notochord is a particularly tractable experimental model. In the ascidian class of the tunicates, which includes the widely studied spe-

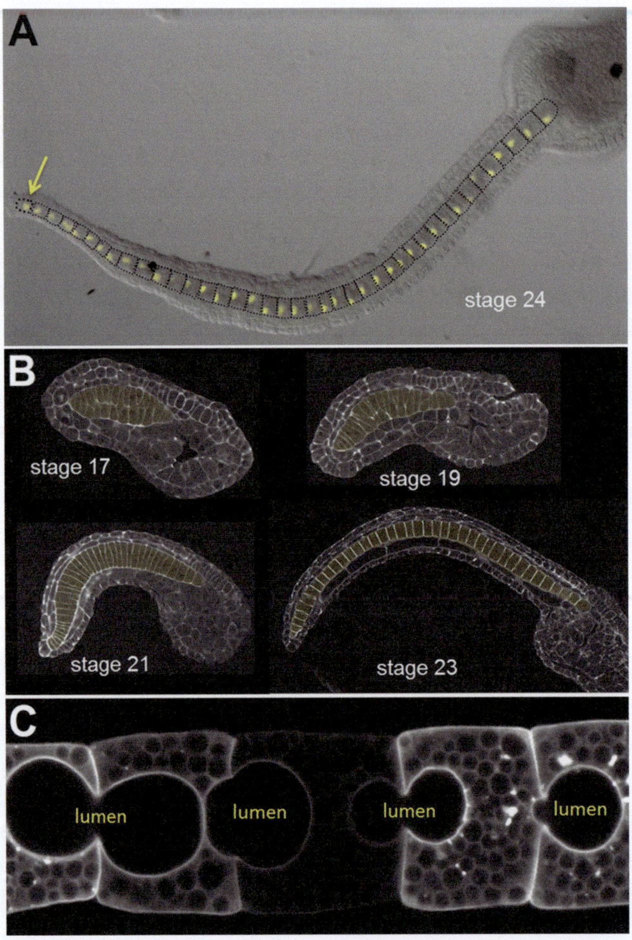

Fig. 15.1 (**a**) *Ciona robusta* late tailbud stage embryo (stage 24). The notochord is outlined and the notochord nuclei are labeled with a transgenic construct containing the *cis*-regulatory element from the *brachyury* gene (Corbo et al. 1997) driving nuclear localized green fluorescent protein. Note that the nuclei are posterior in all notochord cells with the exception of the posterior-most cell (*arrow*). (**b**) Stages of notochord intercalation and elongation in *Ciona robusta* embryos. Notochords are highlighted in *yellow*. Images in **b** courtesy of Michael Veeman. Staging according to Hotta et al. (2007). Anterior is to the right for all images. (**c**) Lumen pockets in five notochord adjacent cells the process of fusing. Notochord cells were visualized by expression of green fluorescent protein driven by the brachyury promoter (For details, see Jiang and Smith 2007)

cies *Ciona intestinalis*, *Ciona savignyi*, *Ciona robusta*, and *Halocynthia roretzi*, the notochord consists of only 40 cells. Moreover, all 40 cells are present by the end of gastrulation, meaning that all of the above morphogenetic events occur in the absence of cell division. Additionally, the small size, cellular simplicity, and transparency of tunicate embryos make them ideal for live imaging, without which the dynamic nature of notochord morphogenesis would never have been appreciated.

This series of reviews is dedicated to the topic of transgenesis in ascidian development, and the dynamic visualization of morphogenesis would not have been possible without transient and stable transgenesis methods for the introduction of fluorescent and epitope-tagged reporter proteins, and the collection of characterized notochord-specific *cis*-regulatory drivers.

Over the past decade, a number of reviews dedicated in full, or in part, to the tunicate notochord have been published (Jiang and Smith 2007; Lemaire et al. 2008; Ogura and Sasakura 2013; Passamaneck and Di Gregorio 2005; Satoh 2014). The purpose of this review is not to repeat what has already been the subject of reviews, but rather to emphasize recent reports and highlight unresolved issues pertaining to notochord morphogenesis.

15.2 Induction of the Notochord

This review focuses primarily on notochord morphogenesis, particularly at the later stages of notochord development. The initial induction of the notochord occurs in the blastula embryo via a fibroblast growth factor-dependent pathway, and has been the subject of several reviews (Jiang and Smith 2007; Lemaire et al. 2008; Satoh 2014; Kumano and Nishida 2007). The notochord arises from two of the four lineages established at the eight-cell stage, which are A, a, B, and b (Nishida 1987). The anterior 32 notochord cells are derived from the A lineage, and notochord fate becomes restricted to the A7.3 and A7.7 blastomere pairs at the 64-cell stage. Each of these four blastomeres divides three more times to give 32 cells by the neurula stage. The posterior eight cells are derived from the B lineage. Their induction occurs later, starting at the 64-cell stage, and becomes restricted to the B8.6 blastomere pairs at the 110-cell stage (*H. roretzi*) (Nishida 1987) or 112-cell stage (*Ciona*) (Lemaire 2009), and follows an inductive pathway distinct from the primary notochord cells (Hudson and Yasuo 2006). This early division of the anterior and posterior notochord lineages has served as a useful experimental tool, allowing the researcher to target misexpression and knockdown constructs and reagents into either anterior or posterior notochord (e.g., (Kourakis et al. 2014)). It should also be noted that the anterior and posterior lineages have been found to differentially express certain genes (Reeves et al. 2014), and to behave differently, for example, in their mechanism of intercalation and response to intercalation disruptors (Jiang et al. 2005; Segade et al. 2016; Veeman et al. 2008).

15.3 Medial Intercalation

During the neurula and early tailbud stages the notochord takes on its elongated form as the presumptive notochord field transitions from a disk of mesenchyme-like cells into a rod, one cell in diameter (Fig. 15.1b). This process, medial intercalation, involves the interdigitation of the cells from the right and left sides as they extend actin-rich protrusions toward the midline of the field (Jiang et al. 2005; Munro and Odell 2002a; Munro and Odell 2002b). For the most part, the interdigitation results in a column with cells deriving alternately from the left and right sides of the field, although this is not invariant (Carlson et al. 2015). Because of the initial disk shape of the field, the anterior and posterior ends complete their intercalation first. This, along with other factors, such as asymmetric cell division, contribute to the final tapered shape of the notochord (Veeman and Smith 2013).

Examination of the actin protrusions of the intercalating cells shows that they are biased toward the midline (Munro and Odell 2002b). The molecular mechanisms responsible for this bias are not fully resolved, but involve the planar cell polarity (PCP) pathway. The PCP pathway was first described in *Drosophila*, and was so-named because it mediates polarity across the plane of epithelial cells (i.e., at a right angle to apical/basal polarity) (Gubb and Garcia-Bellido 1982; Wong and Adler 1993). The PCP pathway is important for developmental processes in a number of organs and tissues in chordates, including convergent extension in the axial mesoderm, closure of the neural tube, neurite extension, and morphogenesis of kidney tubules and the inner ear (McGreevy et al. 2015; Schnell and Carroll 2014; Wallingford 2012; Wallingford et al. 2000). Key players mediating cell–cell communication in the PCP signaling pathway (called the core PCP components), include the transmembrane proteins Frizzled (Fz), Flamingo (Fmi; also known as Starry Night) and Strabismus (Stbm; also known as Van Gogh), and the cytoplasmic proteins Prickle (Pk), Dishevelled (Dsh; also abbreviated as Dvl), and Diego (Axelrod 2009; Goodrich and Strutt 2011; Gray et al. 2011; Seifert and Mlodzik 2007). In models of PCP signaling, the asymmetric distribution of the core components across the cell is thought to lead to the polarized activation of Dsh, which then acts through RhoA/Rac to bring about polarized cytoskeletal rearrangements (Schlessinger et al. 2009; Winter et al. 2001). In the *C. savignyi* mutant *aimless* (*aim*), which has a null mutation in *pk*, the medial bias of the notochord cell protrusions during intercalation is lost (although there is not an overall reduction in protrusions), and medial convergence is greatly retarded, although not completely absent (Jiang et al. 2005).

Despite the importance of the PCP pathway for notochord morphogenesis, there is a fundamental issue that remains unresolved. The core PCP pathway responds to and transduces polarity signals, but the nature of the underlying polarizing signal itself is unknown. Some core PCP components (Fz and Dsh) are shared with the wnt/β-catenin pathway, and although in some cases Wnt ligands have been shown to be essential for PCP, only recently have Wnt ligands in *Drosophila* been shown to convey polarity information (Wu et al. 2013). In vertebrates and other chordates it remains controversial whether Wnt ligands play a "permissive" role versus an "instructive" role (Gros et al. 2009; Heisenberg et al. 2000; Kilian et al. 2003; Niwano et al. 2009; Shafer et al. 2011). One candidate polarizing factor in the ascidian notochord is wnt5. In *H. roretzi*, wnt5 is first expressed at the gastrula and neurula stages in the notochord and flanking muscle cells (Niwano et al. 2009). In *Ciona*, the expression is similar, although the notochord expression is very weak (Imai et al. 2004; our unpublished observations). However, in a knockdown of wnt5 with morpholino oligonucleotides, either in *H. roretzi* (Niwano et al. 2009), or *C. robusta* (our unpublished observations), wnt5 acts strictly cell-autonomously, and is therefore unlikely to have a long range/global instructive activity. A more relevant factor in directing medial convergence may be FGF3, which is produced in the floor plate of the developing neural tube. Thus, its expression is centered dorsal and medial to the intercalating notochord (Shi et al. 2009). Loss of FGF3 function disrupts intercalation, apparently via the non-MAPK pathway.

An alternative pathway that has been proposed to mediate global polarity upstream of the core PCP pathway is the Dachsous/Fat/Four-Jointed (Ds/Ft/Fj) pathway (Casal et al. 2006; Thomas and Strutt 2012). Ft and Ds are members of the cadherin family and signal via heterodimers formed by their extracellular domains, whereas Fj is a kinase that modulates Ds and Ft (Axelrod 2009; Thomas and Strutt 2012). Members of this family have been found in gradients across the polarizing axis (Brittle et al. 2012; Hale et al. 2015; Matakatsu and Blair 2004). Although tissue-wide polarity is disrupted by the loss of Ft, Ds or Fj activity in *Drosophila*, how this pathway interacts with the core PCP pathway is only beginning to be understood (Lawrence et al. 2007; Matis et al. 2014; Merkel et al. 2014; Olofsson et al. 2014). In vertebrates, the Ft/Ds/Fj pathway is also active, and mutants can cause polarity defects (Le Pabic et al. 2014; Mao et al.

2011; Saburi et al. 2012), although a possible role in notochord medial intercalation has not been explored. An analysis of the *Ciona* genome identified 15 members of the cadherin superfamily, with two orthologs of Ft (annotated as cadherin-related-10 and cadherin-related-6), and one of Ds (annotated *Ciona* Cadherin-related-7) (Noda and Satoh 2008). The two *Ciona* Ft genes are expressed in the notochord and CNS, whereas *Ciona* Ds is expressed exclusively in the CNS (Noda and Satoh 2008). Graded expression is not obvious in the expression of these genes, although expression levels are very low, making the expression patterns hard to visualize.

15.4 Other Factors Contributing to Medial Intercalation

As noted above, the notochord cells of *aim* mutants do show significant, although incomplete, medial convergence (Jiang et al. 2005). The residual convergence is not likely to be the product of residual PCP activity as the *aim* mutation appears to result in a complete loss of PCP signaling, indicated by the complete absence of medial bias to the actin protrusions, and of notochord cell elongation in the mediolateral (ML) axis (Jiang et al. 2005). Additionally, there is only a single *pk* gene in the *Ciona* genome. Thus, the PCP pathway is only part of the convergence mechanism.

Another mechanism that appears to act largely in parallel with the PCP pathway is boundary capture by the sheath of extracellular matrix (ECM) that surrounds the notochord. The ECM sheath contributes to medial convergence through a mechanism first described in amphibians in which adhering cells lose their motility, thereby being captured at the boundary (Keller et al. 2000). This boundary capture mechanism is evident in the *C. savignyi* laminin α3/4/5 null mutant *chongmague* (*chm*). Laminin α3/4/5 is one of the major components of the *Ciona* ECM. It is produced by notochord cells and is localized to the periphery of the notochord field by the early neurula stage (Veeman et al. 2008). In the absence of laminin α3/4/5, the notochord field does initially narrow medially, but the motile behavior of the cells persists and the cells ultimately end up loosely arranged in the tail, with many cells separating and migrating away from the field (Veeman et al. 2008).

In a double mutant of *aim* and *chm* the notochord field shows no apparent medial narrowing, although the cells remain highly motile, supporting the hypothesis that these two mechanisms act in parallel on medial convergence. Nevertheless, there is an interplay between the ECM and the PCP pathway. In *aim* mutants, the localization of laminin α3/4/5 initially appears normal, but early tailbud stage ECM deposition is observed inside the notochord field, not just at the periphery (Veeman et al. 2008). This abnormal ECM deposition is a possible explanation why notochord intercalation in *aim* embryos never progresses beyond a column two cells wide—the cells become captured by the surrounding ECM before they are able to complete intercalation.

Finally, knockdown of another ECM component, *fibronectin* (*fn*), in *Ciona* also leads to disruption in medial intercalation (Segade et al. 2016). However, *fn* knockdown does not appear to disrupt the integrity of the sheath. Rather, a defect was observed in the ability of cells to interdigitate, presumably due to a loss of adhesion disrupting the ability of opposing cells to crawl past one another.

15.5 Elongation Phase

At the end of medial intercalation (mid-tailbud stage) the notochord is a single cell-wide column of cells at the center of the tail. Initially, the notochord cells are exceptionally tall in the ML axis and narrow in the anterior/posterior (AP) axis, a configuration known as a stack-of-coins (Fig. 15.1b, stage 21). The forces that hold the cells in this constrained shape are not known. However, over the ensuing ~5 h, the shape of the notochord cells changes dramatically as they narrow in the ML axis and elongate in the AP axis, whereas cell volume remains constant (Miyamoto and Crowther 1985), resulting in the cells assuming a drum-like shape (Jiang and Smith 2007; Miyamoto and Crowther 1985). Driving the elongation of the cell is the formation of a circumferential contractile ring at the midpoint of the cell's AP axis, in addition to contractile blebs at the basal surface of the cells (i.e., where the cells contact the notochord sheath) (Fig. 15.2) (Sehring et al. 2014). The presence of the equatorial band is evident as a mid-point furrow of the cell, and is the focus for localization of actomyosin machinery including actin, myosin II, cofilin, tropomyosin, and talin. These same proteins were also observed to localize to the basal surface blebs. It is thought that as the blebs retract away from the basal surface, they contribute to elongation by creating a longitudinal force at the cell–cell interfaces, see Figure 3 in Sehring et al. (2014).

The nascent contractile ring can be seen forming when cells intercalate as patches of cortical actin on the anterior sides of the cells (Sehring et al. 2015). By the stack-of-coins phase, the equatorial ring is complete, but polarized anteriorly. Only once the cell elongates does the ring relocate to the midpoint. Its position at the midpoint appears to be dynamically maintained by the actin contractile machinery itself and the action of the planar cell polarity (PCP) pathway (Sehring et al. 2015).

Near the end of the elongation phase the formation of intercellular pockets (lumens) is visible, growing between the juxtaposed AP faces of the cells (Fig. 15.1c). In *Ciona* these lumens eventually fuse to form a contiguous cell-free space running the length of the notochord, surrounded by the now flattened, endothelial-like, notochord cells and constrained by the ECM sheath, making a tubular structure. The formation of a contiguous notochord lumen appears unique to tunicates, and is distinct from the intracellular vacuolization seen in vertebrate notochords (Ellis et al. 2013), although the two processes likely serve similar purposes of increasing the volume and rigidity of the notochord. Adding to the confusion, it should be noted that in some of the older literature, tunicate notochord lumens are called vacuoles, e.g., (Miyamoto and Crowther 1985), although, being extracellular, they are not true vacuoles.

Even though the final purpose of the notochord is to facilitate swimming, lumen formation also plays a role in notochord morphogenesis by contributing to elongation, although the most dramatic increase in notochord length, driven by the earlier contractile rings (Sehring et al. 2014), is observed well before the first appearance of the luminal pockets. There is also considerable variation between tunicate species regarding the extent of lumenization, with some species (e.g., *Styela plicata*) showing no apparent lumen formation, whereas others (e.g., *Clavelina hunstmani*) develop unfused lumens (Jiang and Smith 2007).

15.6 Cell Polarity Following Intercalation

The period between the end of intercalation and the onset of lumenization is characterized by a partitioning of the notochord cell surfaces into apical, basal, and lateral domains, similar to that of a typical vertebrate epithelial cell, but with an important twist: the notochord cell forms two distinct apical domains, one each at the centers of the anterior and posterior interfaces (Denker et al. 2013). The lumens initiate and grow from the two apical domains, which comprise approximately the central one-third of the interface surfaces, localize proteins active in the lumen formation, such as the ion transporter SLC26aα (Deng et al. 2013), and are surrounded by a ring of tight junctions (Denker et al. 2013) (Fig. 15.2). Surrounding the central apical domains at the cell–cell interfaces are domains that localize E-cadherin and that have been equated with the lateral domains of epithelial cells. Finally, the basal domain forms circumferentially around the drum-shaped cell where contact is made with the extracellular sheath.

Essential to the formation of the apical domains is the partitioning protein Par3, which is first observed at the end of intercalation as a broad disk at the AP interfaces and then progressively constricts to a tighter domain at the center, and which is necessary for subsequent localiza-

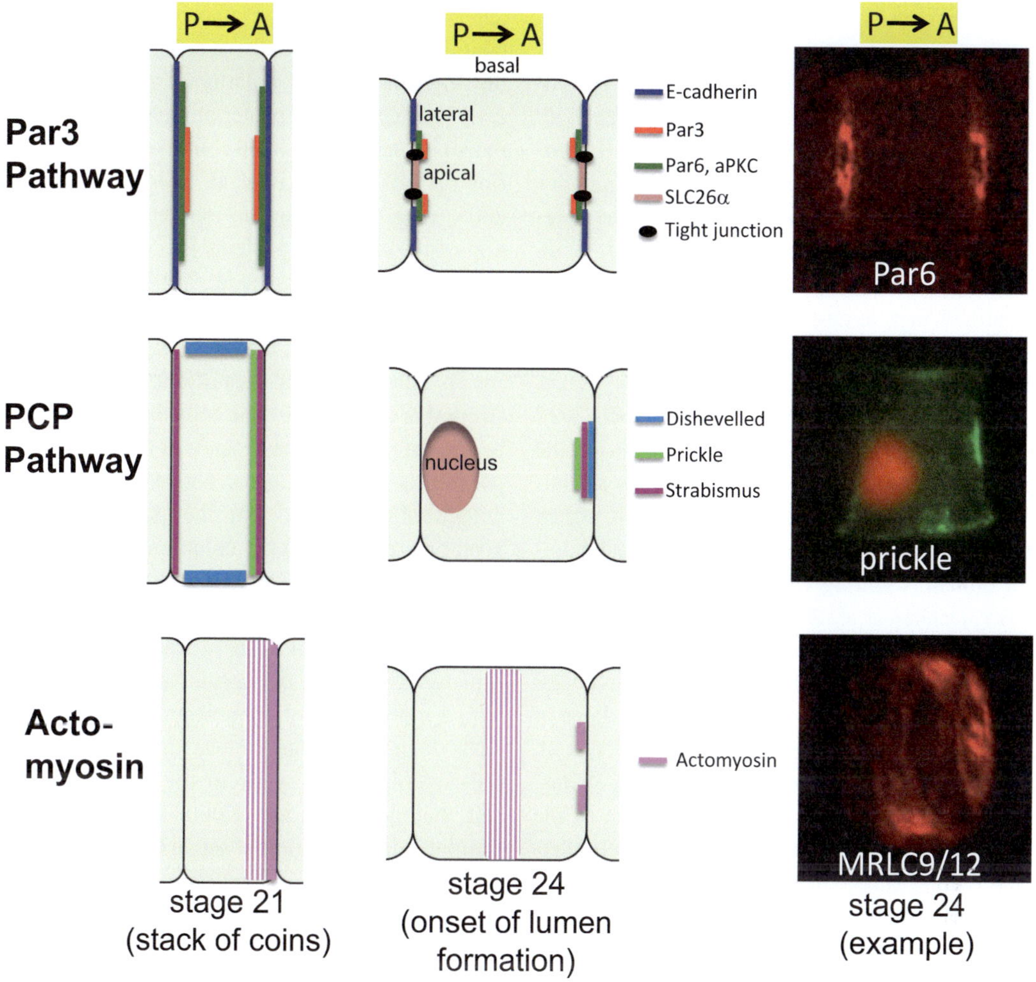

Fig. 15.2 Summary of cell polarities in the intercalated notochord at two stages of notochord development: the onset of elongation (stage 21) and the onset of lumen formation (stage 24). The anterior (*A*) to posterior (*P*) axis are as indicated. Par6 image courtesy of Di Jiang. Staging according to Hotta et al. (2007). The examples shown at stage 24 (*right column*) were generated by transient transgenesis of brachyury promoter constructs expressing either a Par6-green fluorescent protein fusion (*top*) (Denker et al. 2013), nuclear-localized red fluorescent protein and prickle-myc (*middle*) (Newman-Smith et al. 2015), or Venus-tagged myosin regulatory light chain 9/12 (*bottom*) (Newman-Smith et al. 2015)

tion of PAR6 and aPKC to the apical domains. The molecular signals that lie upstream of PAR3 localization to the AP interfaces, and the subsequent constriction of the apical disks toward the centers of the interfaces is not known, but experimental manipulation of embryos to block intercalation, leaving the notochord cells with multiple cell–cell contacts instead of the normal two, results in apical domains forming at all sites of cell–cell contact (Denker et al. 2013).

Cytoskeletal elements also figure in the epithelial partitioning of the cells and apical constriction. Microtubules are enriched at the apical domains, and although their disruption leads to loss of lumen formation, implying that they play a role in polarized secretion, this does not rule out a role in creating or maintaining the apical domain itself (Dong et al. 2011). The role of the actin cytoskeleton appears to be more direct, and disruption of the actin network leads to the failure of lumen formation (Dong et al. 2011). Furthermore, in addition to the circumferential actin/myosin II band found midway on the AP axis and that drives cell elongation (Sehring et al. 2014), it was also observed that the type II myosin *myosin10/11/14*, and the myosin regulatory light chain *MRLC9/12*, are localized to the *anterior* cortex of each cell (Newman-Smith et al. 2015) (Fig. 15.2). These two genes were first identified in a large-scale screen for notochord expressed genes (Satou et al. 2001), and together comprise a two-gene operon in *Ciona* (Satou et al. 2008). *Myosin10/11/14* and *MRLC9/12* localization at the cell–cell interface follows the pattern of apical/basal partitioning. They are initially found broadly across the anterior cortex and then resolve to a ring that surrounds the apical domain (Sehring et al. 2014; Sehring et al. 2015; Denker et al. 2013; Newman-Smith et al. 2015). This ring of myosin and its regulated contractility by the TGFβ-ROC pathway is essential for controlling the rate of expansion and shape of the early forming lumen pockets (Denker et al. 2015). By forming a contractile "purse-string" around the apical domain, the expansion of the apical domains laterally along the cell–cell interface is constrained, instead directing the force of lumen expansion along the AP axis (see Figure 3 in Denker et al. (2015).

In cells treated with myosin II inhibitor blebbistatin, the lumens expand abnormally along the cell–cell interface rather than along the AP axis. The contractile ring is dynamically regulated during lumen expansion, and following its initial constriction, the apical domain is then observed to expand laterally along with lumen volume. As described above, the myosin-containing contractile ring is only at the anterior cortex of each cell (Newman-Smith et al. 2015), whereas the constriction is evident at both the anterior and posterior apical domains. Perhaps the tight junctional ring around the apical domain transduces the contractile force from the anterior contracting ring to the posterior apical domain of the neighboring anterior cell.

The simultaneous bipolar localization of the apical/basal partitioning proteins PAR3 to the anterior and posterior interfaces and unipolar/anterior localization of myosin highlights the complexity of the notochord cells. In addition to the myosin, the AP polarity of the notochord cells is evident in the posterior localization of the nuclei, which is seen in all cells except the posterior tip cell (Kourakis et al. 2014; Jiang et al. 2005) (Fig. 15.1a), and in the anterior localization of the core PCP proteins Pk, Stbm, and Dsh (Fig. 15.2) (Kourakis et al. 2014; Jiang et al. 2005; Sehring et al. 2015).

In the *aim* mutant, the nuclei randomly polarize to either anterior or posterior to the cells (Kourakis et al. 2014). The cells do not appear to be apolar; rather, in the absence of PCP signaling, polarity is not coordinated between cells in the notochord. The fact that the polarity of the posterior-most cell in the column is not in register with the rest of the cells suggests that the cell-to-cell PCP signaling might be from posterior to anterior (Kourakis et al. 2014; Jiang et al. 2005). This was confirmed by mosaic knockdown of Pk with a morpholino oligonucleotide (MO), which leads to noncell autonomous disruption in the polarity of cells anterior to the knocked-down cell, but not posterior (Kourakis et al. 2014). Similarly, laser ablation of single GFP-expressing notochord cells of a stable transgenic resulted in the disruption of polarity in cells anterior to the ablation, but not posterior (Kourakis et al. 2014).

Both the mosaic MO knockdown and ablation studies also proved informative about the range of PCP signaling within the notochord. Noncell autonomous polarity disruption was never observed more than two cells from the targeted cell (and in most cases only a single cell away). Thus, the AP polarity of the column of cells that make up the notochord is very robust to localized disruption of polarity, suggesting that the tissue-wide AP polarity cue might be distributed along the length of the column. Consistent with this, proper AP polarity persists following bisection of the notochord (Kourakis et al. 2014). However, as with medial intercalation, the nature of the upstream polarity signal that the core PCP pathway follows in AP polarity remains unknown (Newman-Smith et al. 2015).

Although it is tempting to consider that the PAR and PCP pathways may function independently in the notochord cells, crosstalk between the PAR and PCP pathways is observed in other cell types, as reviewed in McCaffrey and Macara (2012), and the localization of core PCP proteins in notochord cells is similar to PAR3 and PAR6 in that they are initially observed broadly along the anterior cortex and then constrict toward the center of the cell–cell interface (Kourakis et al. 2014; Jiang et al. 2005) (Fig. 15.2), thus following the formation of the *anterior* apical domain. Nevertheless, we have observed that lumens do form in *aim* mutants, albeit abnormally shaped (our unpublished observations), suggesting that PCP signaling might not be required for apical domain formation. The interaction between the anteriorly localized core PCP proteins and the myosin complexes is equally complex, and neither appears to be strictly upstream of the other.

The polarization of Pk and MRLC9/12 to the anterior cortex of each cell occurs within the first hour following the completion of intercalation, before the cells begin AP elongation (between stages 21 and 22) (Hotta et al. 2007). By contrast, at these same developmental stages (21/22) Stbm is found at the anterior *and* posterior poles of the cells, and then over the next couple of hours resolves to be exclusively anterior (Jiang et al. 2005). On the other hand, Dsh is initially concentrated at the lateral edges, but later also localizes to the anterior cortex (Sehring et al. 2015). Thus, the core PCP protein assemblies appear to be still forming *after* observable polarization of myosin. Moreover, treatment of notochord cells either before or after the onset of Pk and myosin polarity with the actin depolymerizing drugs cytochalasin B or latrunculin, or blebbistatin, results in the loss of Pk and Stbm polarity. These treatments also cause the nuclei of these cells to detach from the posterior poles and drift toward the middle of the cells. However, in these same cells, myosin retains its anterior polarization (Jiang et al. 2005). Thus, myosin can anteriorly polarize, even when the polarization of Pk and Stbm is disrupted.

Although these observations indicate the independence of myosin localization from core PCP signaling, the converse does not appear to be true. It was observed that when embryos are washed of cytochalasin B, Pk and nuclei repolarize to the anterior and posterior poles respectively, but not in the presence of blebbistatin, supporting the role of myosin motor activity and an intact actin network in the polarization of Pk and Stbm.

Contrary to the above results, in which the core PCP protein localization was disrupted by cytoskeletal inhibiting drugs, the localization of myosin in the Pk null-mutation line *aim* (Jiang et al. 2005) suggests a more complex relationship. As described above, the polarity of the notochord cells in *aim* is not lost, only randomized with regard to their neighbors. In other words, in the absence of core PCP signaling, cells autonomously polarize, as has been seen in *Drosophila* eye and wing disks (Gubb and Garcia-Bellido 1982; Wong and Adler 1993; Maung and Jenny 2011). Moreover, myosin polarity in *aim*-mutant notochord cells could be predicted by the nuclear polarity within these cells.

Two cellular phenotypes were observed: in those cells with posterior nuclei (i.e., the wild-type condition), a normal polarization of myosin to the anterior pole was observed; however, in those cells with anterior nuclei (the inverse of the wild type) enrichment of myosin to neither the anterior nor the posterior pole was observed (Newman-Smith et al. 2015). One possibility is that these different cellular phenotypes are due to the interaction of cell autonomous polarity with the global polarity cue, which is hypothesized to act upstream of PCP signaling (Axelrod 2009; Wu and Mlodzik 2009). This signal may act on myosin polarity in a Pk-independent way. Thus, in wild-type cells, the core PCP pathway responds to a global polarity cue, perhaps via myosin polarization, and reinforces polarity by ensuring alignment between neighboring cells.

In the absence of Pk (and therefore of cell-to-cell polarity coordination), the autonomous polarity of the cells is randomized and can either be in phase with global polarity or in opposition. When they are in sync, we observe a surprisingly normal cell in terms of myosin polarity, but when they are not, myosin polarity is lost.

The subcellular redistribution of the actin/myosin cytoskeleton has been extensively studied as downstream targets of PCP signaling. However, these results question the view of a direct, one-directional relationship between the PCP pathway and actin/myosin, and indicate, at least in the *Ciona* notochord, that polarization of the core PCP proteins is a target of myosin polarization, rather than the other way around. Many questions remain unanswered, the most important being: what sets the overall AP polarity of the notochord, and are the cues different for the core PCP pathway and myosin?

15.7 Final Development and Fate of the Notochord

As the luminal pockets continue to expand in the *Ciona* notochord, a remarkable transformation occurs in the behavior and properties of the cells, resulting in formation of a contiguous lumen running the length of the notochord, and giving the notochord its final mature form, just as the larvae begin swimming (Dong et al. 2009). This process of tubule formation has been the subject of a recent in-depth review (Denker and Jiang 2012); therefore, it is only summarized here. Looking just at the lumens, one sees that the initial set of isolated luminal pockets at the centers of each cell–cell interface continue to enlarge and then tilt on the AP axis. Further expansion leads to tips of adjacent lumens touching and then fusing as the transiently drum-like cells that once spanned the circumference of the notochord flatten to one side of the sheath, while retracting from the opposing side (see Figure 1 in (Denker and Jiang 2012). During spreading, the cells continue to show bipolarized behavior, sending out tractive protrusions both anteriorly and posteriorly. The cells continue to spread along the periphery, transforming into endothelial-like cells surrounding the lumen with apical domains facing inward and basal domains tightly associated with the encapsulating notochord sheath.

With the exception of a few chordate clades—the tunicate group known as the larvaceans being one (along with handful of vertebrate species, such as the lamprey) —the notochord is strictly an embryonic and larval organ (Annona et al. 2015). In ascidians, the notochord, along with rest of the tail, is retracted at the onset of metamorphosis (Cloney 1978; Karaiskou et al. 2015; Nakayama-Ishimura et al. 2009). This retraction can happen in as little as 30 min (Matsunobu and Sasakura 2015). The remnant of the notochord, expressing markers of apoptosis (Chambon et al. 2002), can be observed as a collapsed and coiled structure in the body of the developing juvenile (Chambon et al. 2002; Chiba et al. 2009). Over the course of metamorphosis, the remnant grows progressively smaller as it is fully consumed, and thus leaves no descendants in the adult, unlike the vertebrate notochord, which contributes cells to the intervertebral disks (McCann and Seguin 2016). There appear to be numerous developmental cues at work throughout metamorphosis, but the rapid retraction of the tail appears to be triggered via stimulation of palp sensory neurons (Nakayama-Ishimura et al. 2009). The mechanisms effecting the tail retraction are not full known, but are likely to include the deflation of the notochord lumen. The recent observation of nerve endings terminating at the notochord sheath (Ryan et al. 2016) suggest a possible neural pathway linking larval settling and tail retraction.

Acknowledgements I thank Erin Newman-Smith, Matt Kourakis, and Di Jiang for their critical reading of this manuscript. This work is supported by award HD038701 from the National Institutes of Health.

References

Annona G, Holland ND, D'Aniello S (2015) Evolution of the notochord. EvoDevo 6:30

Axelrod JD (2009) Progress and challenges in understanding planar cell polarity signaling. Semin Cell Dev Biol 20(8):964–971

Brittle A, Thomas C, Strutt D (2012) Planar polarity specification through asymmetric subcellular localization of fat and Dachsous. Curr Biol CB 22(10):907–914

Brunetti R et al (2015) Morphological evidence that the molecularly determined Ciona intestinalis type A and type B are different species: Ciona robusta and Ciona intestinalis. J Zool Syst Evol Res 53(3):186–193

Carlson M, Reeves W, Veeman M (2015) Stochasticity and stereotypy in the Ciona notochord. Dev Biol 397(2):248–256

Casal J, Lawrence PA, Struhl G (2006) Two separate molecular systems, Dachsous/fat and starry night/friz-

zled, act independently to confer planar cell polarity. Development 133(22):4561–4572
Chambon JP et al (2002) Tail regression in Ciona intestinalis (Prochordate) involves a Caspase-dependent apoptosis event associated with ERK activation. Development 129(13):3105–3114
Chiba S et al (2009) Brachyury null mutant-induced defects in juvenile ascidian endodermal organs. Development 136(1):35–39
Cloney RA (1978) Ascidian metamorphosis: review and analysis. In: Chia FS, Rice ME (eds) Settlement and metamorphosis of marine larvae. Elsevier, Amsterdam, pp 255–282
Corbo JC, Levine M, Zeller RW (1997) Characterization of a notochord-specific enhancer from the Brachyury promoter region of the ascidian, *Ciona intestinalis*. Development 124(3):589–602
Deng W et al (2013) Anion translocation through an Slc26 transporter mediates lumen expansion during tubulogenesis. Proc Natl Acad Sci U S A 110(37):14972–14977
Denker E, Jiang D (2012) Ciona intestinalis notochord as a new model to investigate the cellular and molecular mechanisms of tubulogenesis. Semin Cell Dev Biol 23(3):308–319
Denker E, Bocina I, Jiang D (2013) Tubulogenesis in a simple cell cord requires the formation of bi-apical cells through two discrete Par domains. Development 140(14):2985–2996
Denker E et al (2015) Regulation by a TGFbeta-ROCK-actomyosin axis secures a non-linear lumen expansion that is essential for tubulogenesis. Development 142(9):1639–1650
Dong B et al (2009) Tube formation by complex cellular processes in Ciona Intestinalis notochord. Dev Biol 330(2):237–249
Dong B, Deng W, Jiang D (2011) Distinct cytoskeleton populations and extensive crosstalk control Ciona notochord tubulogenesis. Development 138(8):1631–1641
Ellis K, Hoffman BD, Bagnat M (2013) The vacuole within: how cellular organization dictates notochord function. BioArchitecture 3(3):64–68
Flood PR, Guthrie DM, Banks JR (1969) Paramyosin muscle in the notochord of amphioxus. Nature 222(5188):87–88
Goodrich LV, Strutt D (2011) Principles of planar polarity in animal development. Development 138(10):1877–1892
Gray RS, Roszko I, Solnica-Krezel L (2011) Planar cell polarity: coordinating morphogenetic cell behaviors with embryonic polarity. Dev Cell 21(1):120–133
Gros J, Serralbo O, Marcelle C (2009) WNT11 acts as a directional cue to organize the elongation of early muscle fibres. Nature 457(7229):589–593
Gubb D, Garcia-Bellido A (1982) A genetic analysis of the determination of cuticular polarity during development in Drosophila melanogaster. J Embryol Exp Morphol 68:37–57
Hale R et al (2015) Cellular interpretation of the long-range gradient of four-jointed activity in the drosophila wing. Elife 4
Heisenberg CP et al (2000) Silberblick/Wnt11 mediates convergent extension movements during zebrafish gastrulation. Nature 405(6782):76–81
Hotta K et al (2007) A web-based interactive developmental table for the ascidian Ciona intestinalis, including 3D real-image embryo reconstructions: I. From fertilized egg to hatching larva. Develop Dynam 236(7):1790–1805
Hudson C, Yasuo H (2006) A signalling relay involving nodal and Delta ligands acts during secondary notochord induction in Ciona embryos. Development 133(15):2855–2864
Imai KS et al (2004) Gene expression profiles of transcription factors and signaling molecules in the ascidian embryo: towards a comprehensive understanding of gene networks. Development 131(16):4047–4058
Jiang D, Smith WC (2007) Ascidian notochord morphogenesis. Dev Dyn 236(7):1748–1757
Jiang D, Munro EM, Smith WC (2005) Ascidian prickle regulates both mediolateral and anterior-posterior cell polarity of notochord cells. Curr Biol 15(1):79–85
Karaiskou A et al (2015) Metamorphosis in solitary ascidians. Genesis 53(1):34–47
Keller R et al (2000) Mechanisms of convergence and extension by cell intercalation. Philos Trans R Soc Lond Ser B Biol Sci 355(1399):897–922
Kilian B et al (2003) The role of Ppt/Wnt5 in regulating cell shape and movement during zebrafish gastrulation. Mech Dev 120(4):467–476
Kourakis MJ et al (2014) A one-dimensional model of PCP signaling: polarized cell behavior in the notochord of the ascidian Ciona. Dev Biol 395(1):120–130
Kumano G, Nishida H (2007) Ascidian embryonic development: an emerging model system for the study of cell fate specification in chordates. Dev Dyn 236(7):1732–1747
Lawrence PA, Struhl G, Casal J (2007) Planar cell polarity: one or two pathways? Nat Rev Genet 8(7):555–563
Le Pabic P, Ng C, Schilling TF (2014) Fat-Dachsous signaling coordinates cartilage differentiation and polarity during craniofacial development. PLoS Genet 10(10):e1004726
Lemaire P (2009) Unfolding a chordate developmental program, one cell at a time: invariant cell lineages, short-range inductions and evolutionary plasticity in ascidians. Dev Biol 332(1):48–60
Lemaire P, Smith WC, Nishida H (2008) Ascidians and the plasticity of the chordate developmental program. Curr Biol CB 18(14):R620–R631
Mao Y et al (2011) Characterization of a Dchs1 mutant mouse reveals requirements for Dchs1-Fat4 signaling during mammalian development. Development 138(5):947–957
Matakatsu H, Blair SS (2004) Interactions between Fat and Dachsous and the regulation of planar cell polarity in the Drosophila wing. Development 131(15):3785–3794

Matis M et al (2014) Microtubules provide directional information for core PCP function. elife 3:e02893
Matsunobu S, Sasakura Y (2015) Time course for tail regression during metamorphosis of the ascidian Ciona intestinalis. Dev Biol 405(1):71–81
Maung SM, Jenny A (2011) Planar cell polarity in drosophila. Organogenesis 7(3):165–179
McCaffrey LM, Macara IG (2012) Signaling pathways in cell polarity. Cold Spring Harb Perspect Biol 4(6)
McCann MR, Seguin CA (2016) Notochord cells in intervertebral disc development and degeneration. J Dev Biol 4(1):1–18
McGreevy EM et al (2015) Shroom3 functions downstream of planar cell polarity to regulate myosin II distribution and cellular organization during neural tube closure. Biol Open 4(2):186–196
Merkel M et al (2014) The balance of prickle/spiny-legs isoforms controls the amount of coupling between core and fat PCP systems. Curr Biol 24(18):2111–2123
Miyamoto DM, Crowther RJ (1985) Formation of the notochord in living ascidian embryos. J Embryol Exp Morphol 86:1–17
Munro EM, Odell G (2002a) Morphogenetic pattern formation during ascidian notochord formation is regulative and highly robust. Development 129(1):1–12
Munro EM, Odell GM (2002b) Polarized basolateral cell motility underlies invagination and convergent extension of the ascidian notochord. Development 129(1):13–24
Nakayama-Ishimura A et al (2009) Delineating metamorphic pathways in the ascidian Ciona intestinalis. Dev Biol 326(2):357–367
Newman-Smith E et al (2015) Reciprocal and dynamic polarization of planar cell polarity core components and myosin. elife 4:e05361
Nishida H (1987) Cell lineage analysis in ascidian embryos by intracellular injection of a tracer enzyme. III. Up to the tissue restricted stage. Dev Biol 121(2):526–541
Niwano T et al (2009) Wnt5 is required for notochord cell intercalation in the ascidian Halocynthia Roretzi. Biol Cell 101(11):645–659
Noda T, Satoh N (2008) A comprehensive survey of cadherin superfamily gene expression patterns in Ciona intestinalis. Gene Expr Patterns 8(5):349–356
Ogura Y, Sasakura Y (2013) Ascidians as excellent models for studying cellular events in the chordate body plan. Biol Bull 224(3):227–236
Olofsson J et al (2014) Prickle/spiny-legs isoforms control the polarity of the apical microtubule network in planar cell polarity. Development 141(14):2866–2874
Passamaneck YJ, Di Gregorio A (2005) Ciona Intestinalis: chordate development made simple. Dev Dyn 233(1):1–19
Pennati R et al (2015) Morphological differences between larvae of the Ciona intestinalis species complex: hints for a valid taxonomic definition of distinct species. PLoS One 10(5):e0122879
Placzek M (1995) The role of the notochord and floor plate in inductive interactions. Curr Opin Genet Dev 5(4):499–506
Reeves W, Thayer R, Veeman M (2014) Anterior-posterior regionalized gene expression in the Ciona notochord. Dev Dyn 243(4):612–620
Ryan K, Lu Z, Meinertzhagen IA (2016) The CNS connectome of a tadpole larva of Ciona intestinalis (L.) highlights sidedness in the brain of a chordate sibling. elife 5
Saburi S et al (2012) Functional interactions between fat family cadherins in tissue morphogenesis and planar polarity. Development 139(10):1806–1820
Satoh N (2014) Developmental genomics of ascidians. Wiley Blackwell, Hoboken, xi, 201 pages
Satou Y et al (2001) Gene expression profiles in Ciona intestinalis tailbud embryos. Development 128(15):2893–2904
Satou Y et al (2008) Improved genome assembly and evidence-based global gene model set for the chordate Ciona intestinalis: new insight into intron and operon populations. Genome Biol 9(10):R152
Schlessinger K, Hall A, Tolwinski N (2009) Wnt signaling pathways meet Rho GTPases. Genes Dev 23(3):265–277
Schnell U, Carroll TJ (2014) Planar cell polarity of the kidney. Exp Cell Res 343:258
Segade F et al (2016) Fibronectin contributes to notochord intercalation in the invertebrate chordate, Ciona intestinalis. EvoDevo 7(1):21
Sehring IM et al (2014) An equatorial contractile mechanism drives cell elongation but not cell division. PLoS Biol 12(2):e1001781
Sehring IM et al (2015) Assembly and positioning of actomyosin rings by contractility and planar cell polarity. elife 4:e09206
Seifert JR, Mlodzik M (2007) Frizzled/PCP signalling: a conserved mechanism regulating cell polarity and directed motility. Nat Rev Genet 8(2):126–138
Shafer B et al (2011) Vangl2 promotes Wnt/planar cell polarity-like signaling by antagonizing Dvl1-mediated feedback inhibition in growth cone guidance. Dev Cell 20(2):177–191
Shi W et al (2009) FGF3 in the floor plate directs notochord convergent extension in the Ciona tadpole. Development 136(1):23–28
Shimeld SM (1999) The evolution of the hedgehog gene family in chordates: insights from amphioxus hedgehog. Dev Genes Evol 209(1):40–47
Thomas C, Strutt D (2012) The roles of the cadherins Fat and Dachsous in planar polarity specification in Drosophila. Develop Dynam 241(1):27–39
Urano A et al (2003) Expression of muscle-related genes and two MyoD genes during amphioxus notochord development. Evol Dev 5(5):447–458
Veeman MT, Smith WC (2013) Whole-organ cell shape analysis reveals the developmental basis of ascidian notochord taper. Dev Biol 373(2):281–289

Veeman MT et al (2008) Chongmague reveals an essential role for laminin-mediated boundary formation in chordate convergence and extension movements. Development 135(1):33–41

Wallingford JB (2012) Planar cell polarity and the developmental control of cell behavior in vertebrate embryos. Annu Rev Cell Dev Biol 28:627–653

Wallingford JB et al (2000) Dishevelled controls cell polarity during Xenopus gastrulation. Nature 405(6782):81–85

Winter CG et al (2001) Drosophila Rho-associated kinase (Drok) links frizzled-mediated planar cell polarity signaling to the actin cytoskeleton. Cell 105(1):81–91

Wong LL, Adler PN (1993) Tissue polarity genes of Drosophila regulate the subcellular location for prehair initiation in pupal wing cells. J Cell Biol 123(1):209–221

Wu J, Mlodzik M (2009) A quest for the mechanism regulating global planar cell polarity of tissues. Trends Cell Biol 19(7):295–305

Wu J et al (2013) Wg and Wnt4 provide long-range directional input to planar cell polarity orientation in Drosophila. Nat Cell Biol 15(9):1045–1055

Check for updates

16 Morphology and Physiology of the Ascidian Nervous Systems and the Effectors

Atsuo Nishino

Abstract

Neurobiology in ascidians has made many advances. Ascidians have offered natural advantages to researchers, including fecundity, structural simplicity, invariant morphology, and fast and stereotyped developmental processes. The researchers have also accumulated on this animal a great deal of knowledge, genomic resources, and modern genetic techniques. A recent connectomic analysis has shown an ultimately resolved image of the larval nervous system, whereas recent applications of live imaging and optogenetics have clarified the functional organization of the juvenile nervous system. Progress in resources and techniques have provided convincing ways to deepen what we have wanted to know about the nervous systems of ascidians. Here, the research history and the current views regard ing ascidian nervous systems are summarized.

Keywords

Simple brain · Motor ganglion · Muscle · Motor control · Development of membrane excitability · Chordate evolution · Motor pattern generation · Metamorphosis · Connectome

A. Nishino (✉)
Department of Biology, Faculty of Agriculture and Life Science, Hirosaki University, Hirosaki, Aomori, Japan
e-mail: anishino@hirosaki-u.ac.jp

16.1 Overview

Nervous systems of ascidians have attracted much interest. In ascidians, two distinct nervous networks develop in tadpole larvae and in adults. Both of them are relatively simple, offering us a cellular-based understanding of the working principles of the systems. Also, both the larval and adult nervous systems have molecular developmental bases that are comparable to those of nervous systems in vertebrates, including humans.

Here, I summarize various aspects of research history in this field and current views on ascidian nervous systems. The field is now integrating knowledge from classic and recent sophisticated neuroanatomy, embryology, neurophysiology, and new technologies, including transgenics. Ascidian neurons, especially those of larvae, are derived from defined cell lineages through an invariable cleavage pattern, and we can therefore reproducibly identify most of the neurons. The lower redundancy of genes in the genome and of neurons in the nervous systems of the larva can offer the chance to tightly connect the operational logics of the genes, neurons, and whole body. Triggering metamorphosis is also a function of the larval nervous system and thus a target of the neurobiology in this animal. The adult central nervous system represents an example of wholly regenerative tissues; on the other hand, its structural correspondence can be pursued to our

Y. Sasakura (ed.), *Transgenic Ascidians*, Advances in Experimental Medicine and Biology 1029,
https://doi.org/10.1007/978-981-10-7545-2_16

brain (e.g., Dufour et al. 2006). I hope that this review is of help for the future understanding of such appealing ascidian nervous systems. This research field is now advancing and expanding rapidly; readers can refer to recent excellent reviews on structural, molecular developmental, and evolutionary perspectives of the nervous systems of ascidians (e.g., Satoh 2014, 2016; Hudson 2016).

16.2 Larval Nervous System

Ascidian larvae sense light, presumably gravity and mechanical stimuli, and possibly water pressure and temperature, and they generate swimming patterns in response to those sensory stimuli (Grave 1920; Mast 1921; Crisp and Ghobashy 1971; Kajiwara and Yoshida 1985; Svane and Young 1989; Nakagawa et al. 1999; Tsuda et al. 2003a, b; McHenry and Strother 2003; Sakurai et al. 2004; Zega et al. 2006; Nishino et al. 2011). The nervous system of ascidian tadpole-shaped larvae is composed of a few hundred neurons and a smaller number of non-neuronal cells, mostly ependymal cells lining the central cavity of the sensory vesicle and the caudal nerve cord. Their number, relative positions of subtypes, and connections among them are believed to be mostly invariant among individuals in a species-specific manner. Recently, those of a single larva of *Ciona intestinalis* were completely identified by means of 3D reconstruction of serial images of electron-microscopic ultrathin sections (ssEM studies) (Ryan et al. 2016). Also, developmental origins of many of the cells in nervous systems have been identified through careful observations and extensive labeling studies (Nicol and Meinertzhagen 1988a, 1988b; Okada et al. 1997; Cole and Meinertzhagen 2004; Taniguchi and Nishida 2004; Stolfi and Levine 2011; Nishitsuji et al. 2012; Oonuma et al. 2016). Based on these fascinating research foundations, the developmental process of the neurons identified can be reproducibly examined and confirmed for the "same" type of neurons over individuals (e.g., Imai et al. 2009; Stolfi and Levine 2011; Stolfi et al. 2015; Ohtsuka et al. 2014; Esposito et al. 2017). These characteristics indicate the advantage of this nervous system as a research model comparable to that of the nematode *Caenorhabditis elegans*, for which the developmental origins and connections of all the neurons have already been uncovered (e.g., Sulston et al. 1983; White et al. 1986).

Differing from the nervous system of *C. elegans*, on the other hand, the ascidian larval nervous system can be directly compared with those of vertebrates (e.g., Meinertzhagen and Okamura 2001). The tunicate clade including ascidians constitutes a sister group of vertebrates. The larvae of ascidians are tadpole-shaped and can freely swim in water as vertebrate fish do, despite their ultimately simplified body (Nishino et al. 2011). Studying the swimming mechanisms of ascidian larvae should provide insights into the ancient way of swimming that originated before the evolution of vertebrates.

The nervous system of ascidian larvae is divided into four subdomains (Fig. 16.1) (Katz 1983; Nicol and Meinertzhagen 1991; Takamura 1998; Horie et al. 2010; Takamura et al. 2010; Ryan et al. 2016). One subdomain is the peripheral nervous system (PNS), or peripheral sensory system, which includes sensory neurons in adhesive papillae and epidermal sensory neurons in the trunk and tail. Another is the brain vesicle (BV), or sensory vesicle, with an interneuron center in the posterior portion. The BV is followed by a bottleneck narrowing connected to another subdomain, the motor ganglion (MG). The MG includes cell bodies of five paired motor neurons that innervate tail muscle cells. The final one is the caudal nerve cord (CNC) composed of dorsal, ventral, and lateral rows of ependymal cells, and of recently discovered, countable number of neurons. The latter three subdomains constitute the central nervous system (CNS) of the larvae (Fig. 16.1). The regional correspondence of these subdomains with those in the vertebrate CNS and PNS has been discussed (e.g., Meinertzhagen and Okamura 2001; Moret et al. 2005a).

Ryan et al. (2016) reported that the *Ciona* larval nervous system is composed of about 180 neurons in the CNS and about 28 neurons in the

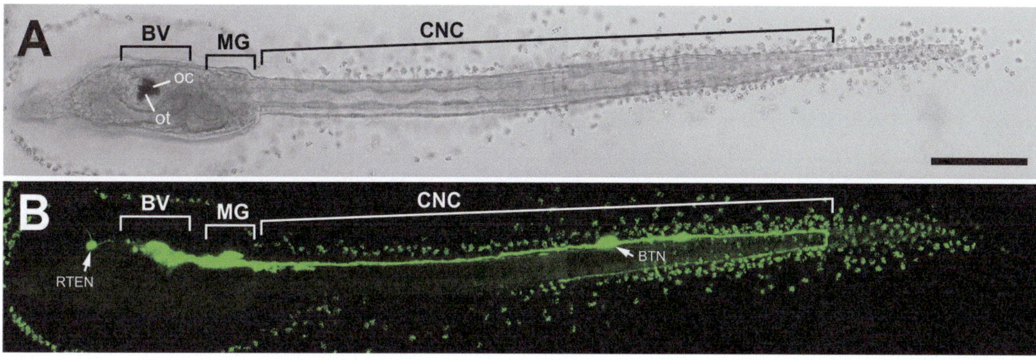

Fig. 16.1 Nervous systems of the *Ciona intestinalis* larva highlighted by mosaic expression of the enhanced green fluorescent protein (EGFP) under the control of a *cis*-regulatory sequence derived from the *synaptotagmin* gene of *Halocynthia*. This transgenic construct has been utilized to visualize single neurons in the larval central and peripheral nervous systems (see Okada et al. 2001, 2002; Katsuyama et al. 2002; Imai and Meinertzhagen 2007a, 2007b). (**a**) A transmission image of a *Ciona* larva. The central nervous system is in the dorsal side of the trunk and tail: brain vesicle (BV) containing otolith and ocellus pigment cells (*ot* and *oc* respectively), motor ganglion (MG) in the posterior trunk, and caudal nerve cord (CNC) extending along the notochord. (**b**) Mosaic EGFP expression here labels a rostral trunk epidermal neuron (RTEN), several neurons in the dorsal and posterior BV, motor neurons in the MG, and a bipolar tail neuron (BTN) of extensive anterior and posterior neurites. The posterior neurite extends to the posterior-most terminal muscle cell and turns ventrally and then elongates anteriorly. This larva was derived from an egg with an intact chorion into which the transgenic DNA construct was manually injected (see text). Scale bar = 100 μm

PNS. They identified 45 neuron subtypes according to the positions, shapes, and connections. Over the past 30 years, Meinertzhagen and colleagues have succeeded in clarifying the architecture and building-up process of the *Ciona* larval nervous system by means of scanning and transmission electron microscopy, digital reconstruction of serial-section images, and consecutive nuclear staining (Nicol and Meinertzhagen 1988a, 1988b, 1991; Cole and Meinertzhagen 2004; Ryan et al. 2016, 2017). They also performed mosaic labeling of larval neurons to determine the morphologies of neuron subtypes by electroporation-mediated transgenesis of the green fluorescent protein (GFP) gene under control of the synaptotagmin gene promotor/enhancer, as shown in Fig. 16.1 (Okada et al. 2001, 2002; Imai and Meinertzhagen 2007a, 2007b).

Detection of pan-neural or subtype-specific molecular markers using antibodies and in situ hybridization has also enhanced our knowledge of the larval nervous system. The whole image of the larval nervous system was first visualized by expression of a sodium channel gene (Okamura et al. 1994) and later by a monoclonal antibody that specifically labels neurons (Takamura 1998). Many molecular markers specific for neuronal subtypes have been utilized to analyze the network. The markers include genes for specific transcription factors, signaling molecules, vesicular transporters or synthetic enzymes for specific neurotransmitters, receptors for those, and functional molecules for specific sensory processing (Takamura et al. 2002; Moret et al. 2005b; Horie et al. 2005, 2008a, 2008b; Zega et al. 2008; Imai et al. 2009; Horie et al. 2010; Nishino et al. 2010, 2011; Stolfi and Levine 2011; Stolfi et al. 2015; Razy-Krajka et al. 2012).

Structural overviews of the above-stated four regions of the ascidian larval nervous system are given below.

16.2.1 Peripheral Nervous System

The peripheral nervous system of the ascidian larva includes papillar neurons in adhesive papillae, epidermal neurons (ENs) that each project a basal axon and an apical process into the larval tunic, and bipolar tail neurons (BTNs) with long bipolar neurites along the CNC and a cell body outside the CNC (Fig. 16.1) (Torrence and Cloney 1982, 1983; Katz 1983; Takamura 1998; Imai and Meinertzhagen 2007b; Horie et al. 2008a; Stolfi et al. 2015).

In general, ascidian larvae have three adhesive papillae: two dorsal papillae (called P1 on the right and P2 on the left) and one ventral papilla (P3). Imai and Meinertzhagen (2007b) proposed that each papilla harbors eight or more neuronal cells called anchor cells. The anchor cells extend an axon terminating on RTENs or on the posterior portion of the BV. They express vesicular glutamic acid transporter (VGluT), a reliable marker for neurons having glutamate for chemical transmissions, while acetylcholinesterase historeactivity was also reported in the papillae (Imai and Meinertzhagen 2007b; Horie et al. 2008a). The anchor cells are thought to function as both mechanosensors and chemosensors, although physiological studies have not yet proved the cues convincingly.

Epidermal neurons are classified into several subpopulations according to their position. The first subpopulation includes six or seven pairs of rostral trunk ENs (RTENs) regularly aligned on the dorsal and lateral surface of the trunk anterior to the BV in *Ciona* (Takamura 1998; Yokoyama et al. 2014). The second subpopulation consists of anterior and posterior apical trunk ENs (four ATENa and four ATENp cells) at the dorsal side of the anterior and posterior BV respectively. The third one includes dorsal and ventral caudal ENs (DCENs and VCENs) and an EN at the tail tip (TTN) (Yokoyama et al. 2014). DCENs and VCENs form eight sets and six sets of neighboring pairs respectively, on average, and are serially aligned along the dorsal and ventral midlines of the tail (Crowther and Whittaker 1994; Yokoyama et al. 2014). A similar distribution pattern of ENs has been reported in a distantly related ascidian species, *Halocynthia*, while VCENs are scarce in this species (Ohtsuka et al. 2001). These ENs each extend a long apical process to form a mesh-like network in the larval tunic, an extracellularly built, flat, fin-like structure (Crowther and Whittaker 1994; Ohtsuka et al. 2001; Pasini et al. 2006; Terakubo et al. 2010; Yokoyama et al. 2014). Although these ENs seem to express VGluT as anchor cells do (Horie et al. 2008a; Pasini et al. 2012), the physiological properties of the ENs have not yet been clarified. Functions as mechanosensors and/or chemosensors have been proposed, but the adequate stimuli remain to be elucidated.

Bipolar tail neurons are recently identified neurons in the larval tail (Imai and Meinertzhagen 2007b; Coric et al. 2008; Stolfi et al. 2015). They are derived from the posterior margin of the neural plate and migrate anteriorly during the tailbud stage. They extend anterior and posterior processes. The posterior (distal) process almost reaches the tip of the tail and often turns ventrally, whereas the anterior (proximal) one terminates on the posterior BV (Fig. 16.1) (Imai and Meinertzhagen 2007b; Ryan et al. 2017). The somata of BTNs are located in the middle of the larval tail, and they are associated with, but outside, the CNC (Stolfi et al. 2015). These neurons express an acid-sensing ion channel gene that encodes a possible molecular mediator for nociception and mechanosensation (Coric et al. 2008; Stolfi et al. 2015). As BTNs seemingly correspond to what were immunolabeled against γ-aminobutyric acid (GABA) in the *Ciona* tail (Brown et al. 2005), they are regarded as being GABAergic. Accordingly, BTNs express genes for glutamic acid decarboxylase (GAD), which is the enzyme to synthesize GABA (Zega et al. 2008; Stolfi et al. 2015), and for a homolog of vesicular inhibitory amino acid transporter (VIAAT), a factor with the activity of incorporating GABA or glycine into the synaptic vesicles (Horie et al. 2010). It should also be noted that some of the cells that Horie et al. (2008a) reported as being glutamatergic appear to share morphological traits with BTNs, and thus glutamatergic BTN-like cells may also be present.

16.2.2 Brain Vesicle

Ryan et al. (2016) reported that there are 143 neuronal cell bodies in the BV. The BV has a cavity inside, and a melanin-containing otolith protrudes from the ventromedial wall into the cavity. A single layer or a few layers of cells, mostly neurons, encircle the cavity; a single ocellus melanocyte and lens cells reside in the right side of the BV wall and directly face the cavity. These melanocytes have been proposed to be a phylogenetic counterpart of vertebrate neural crest cells (Abitua et al. 2012). Thirty or more photoreceptor cells (approximately 20 type I and 10 type II) are located in close vicinity to the ocellus pigment cell. The outer segments of type I photoreceptors are covered by the pigment cup, whereas those of type II face the cavity of the BV (Horie et al. 2008b). These photoreceptors are ciliary, not rhabdomeric, and express a type of opsin (Opsin-1 in *Ciona intestinalis*) and arrestin (Taniguchi and Nishida 2004; Horie et al. 2008b; Oonuma et al. 2016). Another group of six or seven photoreceptors expressing Opsin-1, type III photoreceptors, is located on the ventral wall of the BV and projects bulbous outer segments into the ventricle (Horie et al. 2008b; Ryan et al. 2016). Possible differences in the functions of these photoreceptor groups have not been determined. The melanin-containing otolith cell is associated with the fine processes of a pair of antennal mechanosensory neurons that are adjacent to the type III photoreceptors (Torrence 1986; Sakurai et al. 2004; Ryan et al. 2016). Some experiments, including laser ablation, suggested that the otolith and the associated antennal neurons constitute a gravity sensor of the larva (Tsuda et al. 2003a; Sakurai et al. 2004). Other than these, hypothetical pressure sensors called coronet cells also reside in the left wall of the BV (Eakin and Kuda 1971; Horie et al. 2008b; Ryan et al. 2016), although the result of an experimental trial to determine whether *Ciona* larvae can sense water pressure was negative (Tsuda et al. 2003a). About half of the coronet cells express a gene encoding tyrosine hydroxylase (TH), which catalyzes the formation of L-dihydroxyphenylalanine (L-DOPA) from tyrosine to be a rate-limiting factor to generate a catecholamine neurotransmitter, dopamine (DA) (Moret et al. 2005b). Ventrolateral TH-positive cells in the BV, including several coronet cells, do have DA immunoreactivity and expression of a gene for another enzyme, aromatic amino acid decarboxylase (AADC), which derives DA from L-DOPA (Moret et al. 2005b; Razy-Krajka et al. 2012). Pharmacological analyses in *Ciona* larvae have suggested that these DA cells might negatively regulate swimming activities (Razy-Krajka et al. 2012).

The photoreceptor cells and mechanoreceptive antennal neurons appear to express VGluT (Horie et al. 2008a). These neurons mostly send their axons to the posterior portion of the BV (Horie et al. 2008a, b), and connectome analysis has revealed that there are several interneuron groups that relay inputs from the sensory cells, including photoreceptors, antennae, epidermal sensory neurons, and BTNs (Stolfi et al. 2015; Ryan et al. 2016). One of the posterior BV neurons called eminens2 extends a descending axon to mediate sensory signaling from ENs to the MG, mainly to ddNs (see below), to possibly evoke a specific locomotor output (Ryan et al. 2017). These observations suggest that the posterior portion of the BV is the processing center of sensory inputs. In fact, a representative type of glutamate receptor is expressed in this region and is involved in the development of sensory cells and in the proper progression of metamorphosis (Hirai et al. 2017). Some subunits for the acetylcholine (ACh) receptor and GABA receptor are also known to be expressed in this posterior region of the BV (Zega et al. 2008; Nishino et al. 2011; our unpublished observation).

16.2.3 Motor Ganglion

According to Ryan et al. (2016), the MG of *Ciona intestinalis* contains 25 neurons including a pair of descending decussating neurons (ddNs) (Ryan et al. 2017). Results of transgenic studies using fluorescent protein genes suggested that this pair of ddNs might be vesicular acetylcholine transporter (VAChT)-positive, a reliable marker for cholinergic neurons, and paired-class homeobox gene *Dmbx*-positive (Takamura et al. 2010; Stolfi and Levine 2011). Recently, Ryan et al. (2017) showed that the network around this pair of interneurons is quite reminiscent of the Mauthner system, which is well known in relation to the startle response in vertebrate fish and tadpoles (Fetcho 1991; Eaton et al. 2001).

Five pairs of motor neurons that innervate tail muscle cells have been identified in *Ciona* (Imai and Meinertzhagen 2007a; Ryan et al. 2016, 2017). Two pairs of cholinergic neurons, which are *VAChT* and *Mnx* gene-positive, have been identified as "evident" motor neurons by their morphology of axons with button-like terminals, visualized by expressed transgenic fluorescent proteins (Takamura et al. 2010; Stolfi and Levine 2011; Hudson 2016). The anterior pair is *Nk6*/*Mnx* gene-positive A11.118, whereas the posterior pair is A10.57, expressing *Islet* and *Mnx* genes (Stolfi and Levine 2011). These pairs seem to correspond with MN1 and MN2 respectively, which were defined by Ryan et al. (2017). The former pair has larger frondose-like terminals on the anterior-most dorsal (D1) muscle cell and the anterior-most middle (M1) muscle cell (Fig. 16.2) (Imai and Meinertzhagen 2007a; Stolfi and Levine 2011; Nishino et al. 2011; Ryan et al. 2016, 2017). The MN2 pair has long axons down the tail, with formation of many neuromuscular junctions (Ryan et al. 2016). Ryan et al. (2017) reported the possibility that MN2 is involved in flicks of the tail, not the typical swimming of bilateral propagating waves, based on the dense innervation by ddNs. The other three pairs of motor neurons, MN3-5, are located slightly more dorsally than MN1-2 and have relatively thin and smooth axons with few varicosities descending down the CNC (Ryan et al. 2017). MN3-5, and MN2, have a long ipsilateral axon to form synaptic contacts on dorsal muscle cells (Fig. 16.2) (Nishino et al. 2011; Ryan et al. 2016). These cells may have synaptic contact with each other or other neurons in the CNC to also function as interneurons, not just as motor neurons (Takamura et al. 2010; Ryan et al. 2017).

Neuron types other than motor neurons have also been reported (Imai and Meinertzhagen 2007a; Ryan et al. 2017). These neurons include three pairs of MG interneurons (MGINs) and seven dorsally located ascending MG neurons (AMGs or contrapelo cells). MGINs have descending axons and have been proposed to mediate or suppress the signaling relay from ddNs to MNs (Ryan et al. 2017). AMGs may include two populations that extend presumed cholinergic and GABAergic axons to the BV respectively (Brown et al. 2005; Takamura et al. 2010). Some of the AMGs, including the cholinergic ones, have been proposed to transmit mechanosensory signals from the epidermal sensory neurons to ddNs (Takamura et al. 2010; Ryan et al. 2017).

16.2.4 Caudal Nerve Cord

It had once been assumed that the CNC does not contain any neurons and is merely composed of ependymal cells forming dorsal, ventral, and lateral walls facing the neural tube canal. We now know, however, that the dorsal nerve tube in the tail is not merely a scaffold for the motor nerves, but contains a countable number, nine or more, of neurons (Imai and Meinertzhagen 2007a; Horie et al. 2010; Takamura et al. 2010; Ryan et al. 2016, 2017). Two pairs of anterior caudal interneurons, ACINs, of *Ciona* larvae have anteriorly extending commissural axons and express VIAAT (Horie et al. 2010). Although the gene for the GABA synthesizing enzyme, *GAD*, does not appear to be expressed in this anterior caudal region (Zega et al. 2008), cells immunolabeled with a glycine antibody were found there (Nishino et al. 2010), and thus the ACINs are presumably glycinergic. These are derived from two siblings of A11.116 descendants and develop under the control of SoxB1 expression (Nishitsuji et al. 2012). Although the *Ciona* larva subjected to the ssEM study by Ryan et al. (2016, 2017) did not have two pairs of them, these ACINs have been proposed to receive a signal from ipsilateral excitatory axons at their cell body and to inhibit contralateral motor axons by releasing the inhibitory transmitter (Horie et al. 2010; Ryan et al. 2017). The motor neurons, and possibly other MG-descending neurons, express a glycine receptor gene, and the reciprocal inhibition between the left and right sides represents a network basis for alternating tail beats (Horie et al. 2010; Nishino et al. 2010).

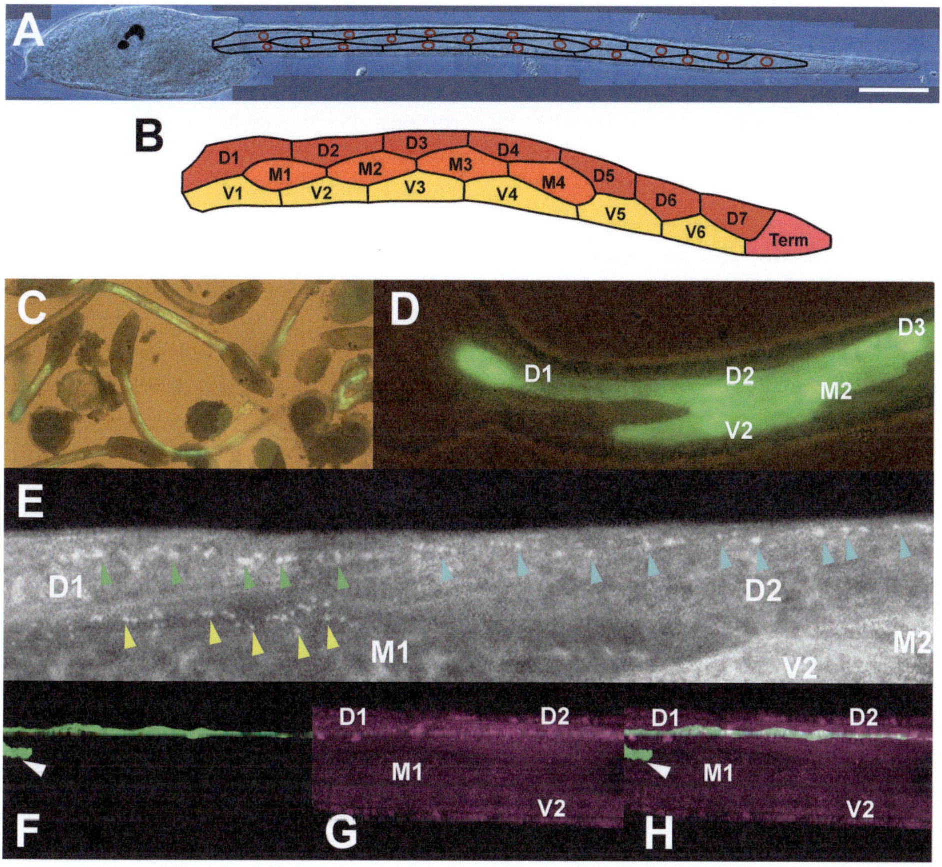

Fig. 16.2 Muscular system of a *Ciona intestinalis* larva. (**a**) Cell nuclei and margins of the single-cell-layered, left muscle band are indicated by *red* and *black* lines respectively. (**b**) The three-row arrangement of the 18 muscle cells is mirror-imaged between the sides and is invariant among larvae, so that the muscle cells can be identified over individuals, labeled here as D1, D2, etc. (**c**) Transgenic larvae whose muscle cells express enhanced green fluorescent protein (EGFP) by the control of a *Ci-rapsyn* gene *cis*-regulatory sequence. The transgene was delivered via electroporation by the procedures first introduced by Corbo et al. (1997). (**d**) Mosaic expression of EGFP from the same construct allows us to see the shape of the muscle cells. (**e**) Transgenic expression of EGFP-tagged Ci-nAChR-A1 in muscle cells driven by a *Halocynthia* muscle actin gene *cis*-regulatory sequence. Dorsal muscle cells (D1 and D2 here) and the M1 muscle cell form postsynaptic nAChR localization patches (*green*, *cyan*, and *yellow arrowheads* on D1, D2, and M1 respectively). (**f**–**h**) Simultaneous transgenic expression of EGFP-tagged Ci-synaptobrevin in motor neurons driven by the *Halocynthia synaptotagmin* gene *cis*-regulatory sequence (*green*, **f**) and mCherry-tagged Ci-nAChR-A1 in muscle cells (*magenta*, **g**). D1 and D2 have postsynaptic localization patches of nAChR, which are just along the motor axons (**h**). The M1 muscle cell is also innervated by a frondose motor terminal (*white arrowheads* in **f** and **h**), where no nAChR patches appear, probably because of the absence of transgene expression in M1 in this larva. Panels **a** and **b** are modified from a supplementary figure in Nishino et al. (2011). Scale bar = 100 μm

Pairs of midtail neurons, MTNs, with fusiform somata, are located posterior to ACINs and scattered along the CNC. Ryan et al. (2016, 2017) identified four of them by ssEM up to the anterior half of the tail, but there seem to be more of them posteriorly. Six pairs visualized by *VAChT* gene expression seem to correspond to MTNs (Horie et al. 2010; Takamura et al. 2010). Postsynaptic nAChR clusters on the dorsal muscle cells also seem to be formed in association with this cell type (Nishino et al. 2011), and these MTNs are therefore categorized as motor neurons (Ryan et al. 2017). Other than these, Ryan et al. (2016, 2017) found some interneurons in the anterior CNC, called posterior MG interneurons (PMGNs). BTNs have recently been identified in the tail, as mentioned above; their somata are located out of CNC and are a component of the PNS (Stolfi et al. 2015).

16.3 Larval Muscular System

The number of larval tail muscle cells is generally small, although the larvae of some species have hundreds of muscle cells (Jeffery and Swalla 1992). *Ciona* and *Halocynthia* larvae have 18 and 21 muscle cells respectively, on either side of the tail (Nishida and Satoh 1985; Nishino et al. 2011). The number and arrangement of muscle cells are left–right symmetrical and invariant in a species-specific manner (Fig. 16.2). The muscle cells in each side form a single-layered sheet and are arranged in three rows in *Ciona*: dorsal D1-7, middle M1-4, ventral V1-6, and posterior-most terminal cell in a total of 18 cells (Fig. 16.2) (Nishino et al. 2011). Despite this structural invariance and simplicity, the larval muscle band is not a mere gathering of muscle cells, but constitutes a system, as do our skeletal muscles and heart. The dynamics of the muscle bands is variable and graded. High-speed video imaging revealed that the tail of a *Ciona* larva produces repetitive antero-posteriorly-propagating and left–right alternating waves with magnitudes that vary stochastically and in response to light–dark conditions (Nishino et al. 2011). This implies that the muscle band represents a "soft" actuator that functions not in an all-or-none fashion, but in a graded manner. Although mature muscle cells can show regenerative Ca^{2+}-based action potentials, Ca^{2+} spikes, in experimental conditions, the action potential is not often seen in direct electrophysiological recordings (Ohmori and Sasaki 1977; Bone 1992; Nishino et al. 2011). The muscle cells on either side adhere to each other to form gap junctional communication (Bone 1992). This gap junctional communication considerably decreases the input resistance, making it difficult to evoke action potentials (Bone 1992). This property of the muscle band of tending not to make action potentials seems to provide a physiological basis for the graded dynamics.

Ascidian larval muscle responds to the neurotransmitter acetylcholine (ACh) (Ohmori and Sasaki 1977). We examined expression patterns in the larvae of all of the nicotinic ACh receptor (nAChR) subunit genes encoded in the *Ciona* genome, and we found that three subunit genes, called A1, B2/4, and BGDE3, and a gene for a clustering protein of muscle nAChR, rapsyn, are expressed in the muscle band (Fig. 16.2) (Nishino et al. 2011). At least two of these genes, A1 and rapsyn, are expressed more strongly in dorsal cells than in ventral or middle cells, whereas the expression of B2/4 and BGDE3 genes appears to be uniform among all of the muscle cells (see Fig. S4 in Nishino et al. 2011). Transgenic expression of fluorescent protein-tagged nAChR subunits and rapsyn protein allowed us to detect postsynaptic nAChR clusters in D1-7, M1, and posterior terminal muscle cells (Fig. 16.2) (Nishino et al. 2011; our unpublished observation). D1-7, M1, and probably posterior terminal muscle cells seem to receive ACh released from the axons of motor neurons. On the other hand, M2-4 and V1-6 cells harbor well-developed myofibrils and receive the activation signals via gap junctions. Because of the low input resistance of the muscle band, endplate potentials elicited by local nAChR activation do not evoke an action potential, but propagate over the surface of the muscle band with attenuation. The voltage-gated Ca_V1 channels and Ca^{2+}-induced Ca^{2+} release mechanisms of ryanodine receptors on the intercellular Ca^{2+} stores, both residing in the muscle band, would work as amplification components of the signal derived from the endoplate potential (Nakajo et al. 1999; Nishino et al. 2011). Considering different projection patterns of motor axons of MN1–5 and MTNs, spatiotemporally regulated ACh inputs into the muscle bands may achieve graded and propagating patterns, neither all-or-none nor simultaneous patterns, of flexions on the muscle band (Nishino et al. 2011; Ryan et al. 2017). *Halocynthia* larvae are known to have a different number, three pairs, of motor neurons and a different pattern of projections of their axons; the middle motoneuron pair called Moto-b innervates the anterior-most ventral muscle cell (Okada et al. 2002). These morphological variations may explain the differences between the swimming performance of the larvae of this species and that of the larvae of *Ciona*. The glycine receptor gene is expressed in the muscle bands of *Ciona* (Nishino et al. 2010). Although it is unknown whether the gene product functions there, inhibitory inputs into the muscle band through this receptor channel may further support the fine regulation of graded flexions.

16.4 Excitability of Neurons and Muscle Cells

Independently of the research on the morphology of ascidian nervous systems, there has been important research history on electrophysiology that has focused on the excitability of ascidian larval cells. This research was pioneered by the past Dr. Kunitaro Takahashi (Moody and Okamura 2013). Neurons and muscle cells are excitable cells, and each cell type has a specific excitation pattern that is represented by the combined functions of Na^+, Ca^{2+}, K^+, and other ion channels. The excitability of each cell type emerges in the course of development, and ascidian embryos have provided an ideal model for study. Takahashi et al. (1971) reported a beautiful sequence of four developmental stages in the changes in the membrane excitability of muscle cell lineage: a naïve state of excitability in the egg and cleavage stage; emergence of signs of excitability in the gastrula; occurrence in the tailbud stage of sustained all-or-none fashioned action potential (AP); and strengthening of currents to repolarize the membrane potential to sharpen the AP in hatchlings. As the muscle develops, the permeability of Na^+ declines and mature muscle cells show Ca^{2+} spikes (Miyazaki et al. 1972). Takahashi and his colleagues developed a sophisticated schema in which neurons express specific Na^+ currents and delayed K^+ currents to elicit sharp Na^+ spikes, muscle cells express specific Ca^{2+} currents and delayed K^+ currents for Ca^{2+} spikes, and epidermal cells do not express delayed rectifier K^+ currents and thus show a different sustained type of Ca^{2+} spikes (Hirano et al. 1984; Okado and Takahashi 1988; Simoncini et al. 1988; Takahashi and Okamura 1998; Nakajo et al. 1999). This cell-type-specific expression of membrane excitability can be seen and analyzed even in blastomeres that have been isolated manually and cleavage-arrested by cytochalasin B (Takahashi and Yoshii 1981). This allowed them to examine the differential expression of cell-type-specific membrane currents using a two-electrode voltage-clamp configuration (for a review, see Takahashi and Okamura 1998). The isolated and cleavage-arrested a-line blastomeres develop to express epidermis-type Ca^{2+} spikes, but the a-line blastomere isolated together with the A blastomere from eight-cell embryos and cleavage-arrested thereafter is differentiated to show Na^+ spikes (Okado and Takahashi 1988, 1990a, b). This is a simple model of neural induction and it was later shown that the induction signal from the A-line can be replaced by treatment with serine proteases or basic fibroblast growth factor (Okado and Takahashi 1993; Takahashi and Okamura 1998).

The detailed kinetics of ion channels have been investigated using the voltage-clamp and patch-clamp techniques (Okamura and Shidara 1987, 1990a, b; Shidara and Okamura 1991; Okamura and Takahashi 1993). Okamura et al. (1994) succeeded in molecular cloning of the cDNA of a neurally expressed, voltage-gated Na^+ channel named TuNa I from *Halocynthia roretzi* (Hr-Na_V1). They examined the expression pattern of the gene and also knocked down the gene expression to show its involvement in the induced Na^+ spikes in neural cells (see also Okada et al. 1997; Ono et al. 1999). On the other hand, establishment of Ca^{2+} spikes in the course of autonomous differentiation of muscle cells is now known on the basis of the expression of TuCa1 (Hr-Ca_V1) and its auxiliary subunit TuCaβ (Simoncini et al. 1988; Davis et al. 1995; Dallman et al. 1998; Nakajo et al. 1999; Dallman et al. 2000; Okagaki et al. 2001; Ohtsuka and Okamura 2007). K^+ channels are also important for shaping APs in neurons and muscle cells, and expression patterns and functions of some of the K^+ channel genes have also been studied in relation to the differentiation of excitabilities (Ono et al. 1999; Murata et al. 2001; Nakajo et al. 2003; Hill et al. 2008).

While this fruitful experimental system is now in danger because of a decrease in the number of researchers using electrophysiological techniques, the system can still provide an ideal research pipeline for understanding the developmental processes of differential membrane excitabilities. We now have a catalog of ion channel genes in *Ciona* (Okamura et al. 2005), and methods for the primary culture of neural cells and electrophysiological recording from the cultured cells have also been reported (Zanetti et al. 2007). It is expected that new fruits will be obtained by combining these research systems with genomic resources and with transgenic techniques.

16.5 Transgenic Techniques for Studying Larval Nervous and Muscular Systems

Transient transgenesis was first reported in ascidians in 1992, with recombinant plasmids being introduced into the fertilized eggs of *Ciona savignyi* and *Halocynthia roretzi* by microinjection (Hikosaka et al. 1992). In 1997, the introduction of transgenes into the fertilized eggs of *Ciona intestinalis*, in which microinjection has been considered relatively difficult, became possible by means of electroporation, at the same time as the use of the green fluorescent protein (GFP) gene in ascidians (Corbo et al. 1997). The use of GFP allowed living cells to be labeled, resulting in a great accumulation of information on neuron morphology and the structural basis of whole nervous systems. Okada and others isolated a *synaptotagmin* gene promotor/enhancer sequence from *Halocynthia*, and the construct of enhanced GFP (EGFP) placed downstream of this *cis*-regulatory sequence has been utilized for random labeling of neurons with chemical presynapses (Fig. 16.1) (Okada et al. 2001, 2002; Katsuyama et al. 2002; Imai and Meinertzhagen 2007a, b; Nishino et al. 2011). They also tried forced expression of a dominant-negative form of a voltage-gated K^+ channel (DN-TuK$_V$2) under control of the *synaptotagmin cis*-regulatory sequence by microinjecting this construct into A5.2 to change the swimming pattern of the transformants (Ono et al. 1999; Okada et al. 2002). This pioneering trial proved that transgenic manipulation of ion channel functions can affect locomotor patterns of the transgenic larva.

Live imaging based on GFP has made much progress and several fluorescent fusion proteins are now also used in ascidians, including fusions with nucleus localization signal (NLS), histone H2B (localizing with chromosomes), MAP7 (microtubule-associated protein 7, ensconsin), LifeAct (a short peptide from *Saccharomyces* that binds to F-actin), pleckstrin homology (PH) domains (binding phosphatidylinositol lipids in cell membranes), EB3 (end binding 3 to label the plus end of microtubules), or Geminin and Cdt1 (specific cell-cycle regulators) to visualize subcellular structures, cytoskeletons, and phases in cell cycles (e.g., Ogura et al. 2011; Negishi and Yasuo 2015). Stolfi and Levine (2011) utilized a UNC-76::GFP fusion protein derived from GFP and a factor involved in axon extension in *C. elegans*, which improved visualization of neurites of zebrafish olfactory neurons (see Dynes and Ngai 1998). I myself tried EGFP fused with *Ciona* synaptobrevin for visualizing presynaptic vesicular accumulation, but the effect was not clear (Fig. 16.2) (Nishino et al. 2011). On the other hand, our trial to express fluorescent protein tagged *Ciona* nAChR subunits and rapsyn in larval muscle succeeded in visualizing postsynaptic aggregates of the receptors (Fig. 16.2) (Nishino et al. 2011).

Transgenesis by electroporation established by Corbo et al. (1997) in *C. intestinalis* has pushed this ascidian species forward as an ideal model organism. This technique allows us to introduce transgenes into living cells without special skills. The basic procedure for electroporation of transgenes, however, requires dechorionation of eggs. Dechorionation is known to affect the development of embryos, especially the characteristics of left–right asymmetry. Nervous systems of the larva have many asymmetric traits, and dechorionation does indeed change them (Shimeld and Levin 2006; Nishide et al. 2012; Oonuma et al. 2016; Ryan et al. 2016). It is also known that tail elongation becomes distorted and that larvae that have developed without the chorion cannot swim normally (Nishide et al. 2012). To minimize these effects, transgenesis by microinjection without dechorionation was performed in eggs of *C. intestinalis* (Nishino et al. 2011; Oonuma et al. 2016).

Combining gene knock-down or knock-out and transgenic expression of the gene under consideration allows us to rescue the phenotypes to confirm the specificity of the knock-down/knock-out. Rescuing by an artificial fusion protein gene or a gene with an artificial mutation enables us to assess the effect(s) of the fusion or mutation respectively. We knocked down an indispensable subunit of muscle nAChR called BGDE3, and rescued the paralytic phenotype using wild-type BGDE3 and its mutant, which alters the electrophysiological properties of muscle nAChR (Nishino et al. 2011). In that experiment, the knocked-down phenotype caused by microinjection of a BGDE3-morpholino oligonucleotide was rescued by wild-type BGDE3 synthetic mRNA, the mutant mRNA, or the mutant gene expressed under the control of a muscle actin promoter. The larvae rescued by wild-type BGDE3 appeared to swim normally, whereas those rescued by the mutant showed altered patterns of movement (Nishino et al. 2011). This impairment is accounted for by a single mutation in the gene, and this kind of analysis can naturally be applied for assessing the various functions of paralogs or orthologs. We can now use tissue-/cell-type-specific genome editing (Sasaki et al. 2014; Stolfi et al. 2014; Treen et al. 2014) in addition to tissue-/cell-type-specific expression of a rescuing construct by means of well-designed transgenic constructs. Such approaches will lead to more reliable examination of a working hypothesis.

16.6 Neural Control of Metamorphosis

Attachment to the substrate via adhesive papillae transforms the swimming larva into a sessile adult. Internal signaling to trigger metamorphosis involves neural and/or neuroendocrine systems (Cloney 1982; Nakayama-Ishimura et al. 2009; Kamiya et al. 2014; Karaiskou et al. 2015; Matsunobu and Sasakura 2015). Although many substances that can induce metamorphosis have been described so far, the point of action of each remains unclear. The *C. intestinalis* larva ceases movement and regresses the tail in response to a hypothetical signal that emerges only after occurrence of competence for tail regression, approximately 12 h after hatching and after experience of continuous attachment via papillae for 28 min or more (Matsunobu and Sasakura 2015). This cessation of movement, also in addition to the emergence of competence and the memory of 28-min attachment, would reflect some unknown alterations engraved in the nervous or muscular systems in the larva. This would be one of research foci in which new transgenic approaches, including genome editing, would facilitate further cellular-/molecular-based resolution of this complicated series of phenomena in metamorphosis.

Metamorphosis, on the other hand, changes the entire nervous and muscular systems. The larval nervous and muscular systems degenerate, and those of the adult are rebuilt from undifferentiated cells that have been set aside in the larval body (Willey 1894; Hirano and Nishida 1997; Stolfi et al. 2010; Horie et al. 2011; Razy-Krajka et al. 2014). Horie et al. (2011) fully showed the ability of transgenic fluorescent proteins. Their work began by establishing a stable line, having a photoconvertible fluorescent protein, Kaede, in the nervous systems under the control of a *β2-tubulin* gene *cis*-regulatory sequence (Horie et al. 2011). Photoconversion of Kaede from green to red before metamorphosis enabled cells in the larval nervous system to be distinguished from the neural cells that differentiated after metamorphosis. They utilized several types of cell-type-specific *cis*-regulatory sequences to express Kaede in some neuron subtypes and in ependymal cells. Their analyses revealed that ependymal cells in the BV and neck region of the larval CNS are the cellular origin of the adult CNS, cerebral ganglion (CG), and that antero-posterior cellular organization of the adult CNS corresponds well with that of the larval CNS (Horie et al. 2011).

16.7 Structural and Functional Analyses of Adult Nervous Systems

The organization and function of the adult CNS, cerebral ganglion (CG), and the adult PNS have been elucidated in several species by morphological and physiological analyses (Arkett 1987; Arkett et al. 1989; Koyama and Kusunoki 1993; Tsutsui and Oka 2000; Mackie and Burighel 2005). However, compared with those in the larva, our understanding of the juvenile/adult nervous systems has not been sufficient.

The adult nervous system consists of the CNS and PNS. The CG located between the two, oral and atrial, siphons of ascidians represents the CNS, associated with several branches of nerve tracts, a dorsal strand (dorsal cord), and a neural gland. Based on its relative position, the neural gland complex composed of the ciliated duct, neural gland, and dorsal strand had been discussed as being homologous to the vertebrate adenohypophysis, but recent research has suggested that it might not be appropriate to refer to the neural gland as a homologue of the adenohypophysis (for reviews, see Goodbody 1974; Gorbman 1995). There has been an accumulation of expression data on the neural gland complex of several kinds of genes for peptides and their receptor-like G protein-coupled receptors (Deyts et al. 2006; Matsubara et al. 2016). Another aspect of interest is that this CG and neural gland complex is wholly regenerative (Dahlberg et al. 2009, and see references therein). Dahlberg et al. (2009) utilized an enhancer-trap transgenic line called E15 that expresses GFP in neural cells (Awazu et al. 2007) to show fine cellular views of CG regeneration.

The CG is oval- or rod-shaped and has anterior–posterior (oral–atrial) polarity (Mackie and Burighel 2005; Hozumi et al. 2015). One pair, or more than one pair, of nerve tracts, the number being variable among species, extends from the anterior and posterior ends of the CG (Arkett et al. 1989; Koyama and Kusunoki 1993; Mackie and Burighel 2005). Some thin nerves protrude from the left and right lateral sides of the CG in *Botryllus* (Zaniolo et al. 2002). Ascidian adults possess specialized peripheral sensory apparatuses such as the coronal organ and cupular organ, which are known as hydrodynamic sensors. The coronal organ includes secondary sensory neurons that have stereovilli and a sensory cilium resembling vertebrate inner ear hair cells, and it is located in the velum and tentacles of the oral siphon (Mackie and Burighel 2005). The cupular organ has a gelatinous cupula and primary sensory cells that look like a lateral line neuromast of vertebrates, many of which are distributed over the surface of the atrium (Mackie and Burighel 2005; Ohta et al. 2010). In contrast, chemical sensation in ascidian adults is not well understood.

Anterior and posterior tracts of the CG include both sensory and motor nerves, whereas sensory nerves projecting to the oral siphon are from the anterior tracts and those projecting to the atrial siphon are from the posterior tracts (Hozumi et al. 2015). Most of the nerves innervating ciliated cells on stigmata, gill apertures, are within the posterior tract(s), called the visceral nerve (Arkett 1987; Arkett et al. 1989; Mackie and Burighel 2005). Most of the neurons in the CG are thought to be interneurons to relay and process sensory inputs, and others are motor neurons to innervate adult muscle cells and stigmatal ciliated cells (Mackie and Burighel 2005; Dufour et al. 2006; Hozumi et al. 2015). On the other hand, there are some sensory cells in the CG that respond to light (Tsutsui and Oka 2000). A considerable number of neurons (approximately 25% of cells in the CG of *C. savignyi*) are responsive to light on or light off stimuli with membrane depolarization or hyperpolarization, although it is not clear whether the recorded responses occurred directly on the cell or via the synapses (Tsutsui and Oka 2000).

Hozumi et al. (2015) effectively utilized stable transgenic lines and transiently transgenic animals of *C. intestinalis*, and revealed the neuronal organization of the juvenile CG. The juvenile nervous systems are mostly reorganized during metamorphosis and are distinguished from those in the larva, but they revealed that the organization in juveniles is comparable with that in the larva. Peripheral neurons express the

VGluT gene, and central neurons expressing the same gene are mainly located in the anterior portion of the CG. Neurons expressing *VIAAT* are found only in the CG and their neurites are also within the CG. *VAChT*-positive neurons are mostly within the CG and extend long axons peripherally to innervate siphons, body-wall muscles, and stigmatal cells (Hozumi et al. 2015 and see also Dufour et al. 2006). Hozumi et al. (2015) further optogenetically confirmed the cholinergic regulation of the siphon and body-wall muscles and the cilia on stigmatal cells. They expressed channelrhodopsin-2 (Nagel et al. 2003; Zhang et al. 2007) fused in frame with yellow fluorescent protein (ChR2::YFP) mosaically in cholinergic neurons in the CG under the control of a *VAChT* gene *cis*-regulatory sequence. Some juveniles expressed ChR2::YFP in the CG neurons innervating the oral siphon, atrial siphon or body-wall muscle, and they contracted each of the targets in response to blue-light irradiation; others expressing ChR2::YFP in those innervating stigma evoked ciliary arrest upon irradiation (Hozumi et al. 2015). This indicated the possibility of using optogenetic techniques in this transparent organism that will enable us to examine, or manipulate, the function of genetically identified neurons and other excitable cells.

16.8 Future Direction

Research into ascidian nervous systems has made much progress. Recent ssEM studies have revealed the precise number, types, and connections of neurons in a *Ciona* larva (Ryan et al. 2016, 2017). Techniques for transient and stable transgenesis, genome editing, and fluorescence imaging have been established (see other chapters in this book, and Sasakura et al. 2012). Optogenetics is also now being applied and has enabled manipulation of the physiological activity of living ascidians (Hozumi et al. 2015; Horie et al., personal communication). These resources and techniques represent a reliable map and vehicles to solve the remaining mysteries of the nervous systems in this model species.

Many mysteries do indeed remain. We do not yet know the possible variability of neuron networks among individuals. We do not know what happens in the network during the aging of larvae, while we do know for instance that light sensitivity of the *Ciona* larva and correspondingly the neurites of photoreceptors gradually develop after hatching (Tsuda et al. 2003b; Horie et al. 2005) and that DA in the BV increases as the larvae age (Moret et al. 2005b). We also do not know how the networks change in response to the irreversible activation of larval papillae (Matsunobu and Sasakura 2015), or to what degree the view of the CG structure in juveniles proposed by Hozumi et al. (2015) is applicable to that in mature adults (Dufour et al. 2006). We are not yet convinced of the full operational logics of larval neurons and muscles to facilitate effective swimming performance that can be modulated in response to sensory stimuli.

To answer these questions, we would be able to collect *cis*-regulatory sequences specific to each subtype for making it possible to drive specific transgenes in specific neuron subtypes. It would also be useful to know the neurotransmitter(s), the receptors, and voltage-gated ion channels that each subtype utilizes to function in the network. Furthermore, we would be able to uncover the mostly unknown developmental mechanisms for each neuron subtype to make proper connections to other subtypes, i.e., the mechanisms for axon pathfinding. Changing the properties and connections of specific neuron subtypes by utilizing DNA resources and transgenics that we have at present and in the future will provide a promising approach to solving operational logics of neuron networks. This is indeed possible in *Ciona*. The effectiveness of this approach would be greatly increased if we could overcome the remaining technical obstacles by establishing effective methods for managing living adults, including stable lines, controlling their repetitive reproduction, and by establishing methods to avoid artifacts caused by dechorionation, etc.

There are many ascidian and other tunicate species on earth. They have major or minor structural and functional differences in their sensory and motor systems, reflecting their lifestyles and patterns of adaptation to different environments. Berrill (1948) described intense variation in the presence and absence of the otolith, ocellus, and their combinatorial apparatus, photolith, in larvae of stolidobranch ascidians, for instance, but we do not yet know what the developmental causes are and to what degree such differences in the sensory apparatus affect the central neural network or their output; namely, the impacts on larval behavior and ecology. We are now in the era in which morphology can be integrated with any other fields, including developmental genetics, evolutionary biology, ecology, and physiology.

Acknowledgements I am grateful for helpful comments by the editor, Dr. Yasunori Sasakura, which greatly improved the manuscript.

References

Abitua PB, Wagner E, Navarrete IA, Levine M (2012) Identification of a rudimentary neural crest in a non-vertebrate chordate. Nature 492:104–107

Arkett SA (1987) Ciliary arrest controlled by identified central neurons in a urochordate (Ascidiacea). J Comp Physiol A 161:837–847

Arkett SA, Mackie GO, Singla CL (1989) Neuronal organization of the ascidian (Urochordata) branchial basket revealed by cholinesterase activity. Cell Tissue Res 257:285–294

Awazu S, Matsuoka T, Inaba K, Satoh N, Sasakura Y (2007) High throughput enhancer trap by remobilization of transposon Minos in *Ciona intestinalis*. Genesis 45:307–317

Berrill NJ (1948) The nature of the ascidian tadpole, with reference to *Boltenia echinata*. J Morphol 82:269–285

Bone Q (1992) On the locomotion of ascidian tadpole larvae. J Mar Biol Assoc U K 72:161–186

Brown ER, Nishino A, Bone Q, Meinertzhagen IA, Okamura Y (2005) GABAergic synaptic transmission modulate swimming in the ascidian larva. Eur J Neurosci 22:2541–2548

Cloney RA (1982) Ascidian larvae and the events of metamorphosis. Am Zool 22:817–826

Cole AG, Meinertzhagen IA (2004) The central nervous system of the ascidian larva: mitotic history of cells forming the neural tube in late embryonic *Ciona intestinalis*. Dev Biol 271:239–262

Corbo JC, Levine M, Zeller RW (1997) Characterization of a notochord-specific enhancer from the *Brachyury* promoter region of the ascidian, *Ciona intestinalis*. Development 124:589–602

Coric T, Passamaneck YJ, Zhang P, Gregorio AD, Canessa CM (2008) Simple chordates exhibit a proton-independent function of acid-sensing ion channels. FASEB J 22:1914–1923

Crisp DJ, Ghobashy AFAA (1971) Responses of the larvae of *Diplosoma listerianum* to light and gravity. In: Crisp DJ (ed) Fourth European marine biology symposium. Cambridge Press, Cambridge, pp 443–465

Crowther RJ, Whittaker JR (1994) Serial repetition of cilia pairs along the tail surface of an ascidian larva. J Exp Zool 268:9–16

Dahlberg C, Auger H, Dupont S, Sasakura Y, Thorndyke M, Joly J-S (2009) Refining the *Ciona intestinalis* model of central nervous system regeneration. PLoS One 4:e4458

Dallman JE, Davis AK, Moody WJ (1998) Spontaneous activity regulates calcium-dependent K^+ current expression in developing ascidian muscle. J Physiol 511:683–693

Dallman JE, Dorman JB, Moody WJ (2000) Action potential waveform voltage clamp shows significance of different Ca^{2+} channel types in developing ascidian muscle. J Physiol 524:375–386

Davis AK, Greaves AA, Dallman JE, Moody WJ (1995) Comparison of ionic currents expressed in immature and mature muscle cells of an ascidian larva. J Neurosci 15:4875–4884

Deyts C, Casane D, Vernier P, Bourrat F, Joly J-S (2006) Morphological and gene expression similarities suggest that the ascidian neural gland may be osmoregulatory and homologous to vertebrate peri-ventricular organs. Eur J Neurosci 24:2299–2308

Dufour HD, Chettouh Z, Deyts C, de Rosa R, Goridis C, Joly J-S, Brunet J-F (2006) Precraniate origin of cranial motoneurons. Proc Natl Acad Sci U S A 103:8727–8732

Dynes J, Ngai J (1998) Pathfinding of olfactory neuron axons to stereotyped glomerular targets revealed by dynamic imaging in living zebrafish embryos. Neuron 20:1081–1091

Eakin RM, Kuda A (1971) Ultrastructure of sensory receptors in ascidian tadpoles. Z Zellforsch 112:287–312

Eaton RC, Lee RKK, Foreman MB (2001) The Mauthner cell and other identified neurons of the brainstem escape network of fish. Prog Neurobiol 63:467–485

Esposito R, Yasuo H, Sirour C, Palladino A, Spagnuolo A, Hudson C (2017) Patterning of brain precursors in ascidian embryos. Development 144:258–264

Fetcho JR (1991) Spinal network of the Mauthner cell. Brain Behav Evol 37:298–316

Goodbody I (1974) The physiology of ascidians. Adv Mar Biol 12:1–149

Gorbman A (1995) Olfactory origins and evolution of the brain pituitary endocrine system: facts and speculation. Gen Comp Endocrinol 97:171–178

Grave C (1920) *Amaroucium pellucidum* (Leidy) form *constellatum* (Verrill) I. The activities and reactions of the tadpole larva. J Exp Zool 30:239–257

Hikosaka A, Kusakabe T, Satoh N, Makabe KW (1992) Introduction and expression of recombinant genes in ascidian embryos. Develop Growth Differ 34:627–634
Hill AS, Nishino A, Nakajo K, Zhang G, Fineman JR, Selzer ME, Okamura Y, Cooper EC (2008) Ion channel clustering at the axon initial segment and node of Ranvier evolved sequentially in early chordates. PLoS Genet 4:e1000317
Hirai S, Hotta K, Kubo Y, Nishino A, Okabe S, Okamura Y, Okado H (2017) AMPA glutamate receptors are required for sensory-organ formation and morphogenesis in the basal chordate. Proc Natl Acad Sci U S A 114:3939–3944
Hirano T, Nishida H (1997) Developmental fates of larval tissues after metamorphosis in ascidian *Halocynthia roretzi*. I. Origin of mesodermal tissues of the juvenile. Dev Biol 192:199–210
Hirano T, Takahashi K, Yamashita N (1984) Determination of excitability types in blastomeres of the cleavage-arrested but differentiated embryos of an ascidian. J Physiol 347:301–325
Horie T, Orii H, Nakagawa M (2005) Structure of ocellus photoreceptors in the ascidian *Ciona intestinalis* Larva as revealed by an anti-arrestin antibody. J Neurobiol 65:241–250
Horie T, Kusakabe T, Tsuda M (2008a) Glutamatergic networks in the *Ciona intestinalis* Larva. J Comp Neurol 508:249–263
Horie T, Sakurai D, Ohtsuki H, Terakita A, Shichida Y, Usukura J, Kusakabe T, Tsuda M (2008b) Pigmented and nonpigmented ocelli in the brain vesicle of the ascidian larva. J Comp Neurol 509:88–102
Horie T, Nakagawa M, Sasakura Y, Kusakabe TG, Tsuda M (2010) Simple motor system of the ascidian larva: neuronal complex comprising putative cholinergic and GABAergic/glycinergic neurons. Zool Sci 27:181–190
Horie T, Shinki R, Ogura Y, Kusakabe TG, Satoh N, Sasakura Y (2011) Ependymal cells of chordate larvae are stem-like cells that form the adult nervous system. Nature 469:525–528
Hozumi A, Horie T, Sasakura Y (2015) Neuronal map reveals the highly regionalized pattern of the juvenile central nervous system of the ascidian *Ciona intestinalis*. Dev Dyn 244:1375–1393
Hudson C (2016) The central nervous system of ascidian larvae. Wiley Interdiscip Rev Dev Biol 5:538–561
Imai JH, Meinertzhagen IA (2007a) Neurons of the ascidian larval nervous system in *Ciona intestinalis*: I. Central nervous system. J Comp Neurol 501:316–334
Imai JH, Meinertzhagen IA (2007b) Neurons of the ascidian larval nervous system in *Ciona intestinalis*: II. Peripheral nervous system. J Comp Neurol 501:335–352
Imai KS, Stolfi A, Levine M, Satou Y (2009) Gene regulatory networks underlying the compartmentalization of the *Ciona* central nervous system. Development 136:285–395
Jeffery WR, Swalla BJ (1992) Evolution of alternate modes of development in ascidians. BioEssays 14:219–226
Kajiwara S, Yoshida M (1985) Changes in behavior and ocellar structure during the larval life of solitary ascidians. Biol Bull 169:565–577
Kamiya C, Ohta N, Ogura Y, Yoshida K, Horie T, Kusakabe TG, Satake H, Sasakura Y (2014) Nonreproductive role of gonadotropin-releasing hormone in the control of ascidian metamorphosis. Dev Dyn 243:1524–1535
Karaiskou A, Swalla BJ, Sasakura Y, Chambon J-P (2015) Metamorphosis in solitary ascidians. Genesis 53:34–47
Katsuyama Y, Matsumoto J, Okada T, Ohtsuka Y, Chen L, Okado H, Okamura Y (2002) Regulation of synaptotagmin gene expression during ascidian embryogenesis. Dev Biol 244:293–304
Katz MJ (1983) Comparative anatomy of the tunicate tadpole, *Ciona intestinalis*. Biol Bull 164:1–27
Koyama H, Kusunoki T (1993) Organization of the cerebral ganglion of the colonial ascidian *Polyandrocarpa misakiensis*. J Comp Neurol 338:549–559
Mackie GO, Burighel P (2005) The nervous system in adult tunicates: current research directions. Can J Zool 83:151–183
Mast SO (1921) Reactions to light in the larvae of the ascidians, *Amarocium constellatum* and *Amarocium pellucidum* with special reference to their photic orientation. J Exp Zool 34:149–187
Matsubara S, Kawada T, Sakai T, Aoyama M, Osugi T, Shiraishi A, Satake H (2016) The significance of *Ciona intestinalis* as a stem organism in integrative studies of functional evolution of the chordate endocrine, neuroendocrine, and nervous systems. Gen Comp Endocrinol 227:101–108
Matsunobu S, Sasakura Y (2015) Time course for tail regression during metamorphosis of the ascidian *Ciona intestinalis*. Dev Biol 405:71–81
McHenry MJ, Strother JA (2003) The kinematics of phototaxis in larvae of the ascidian *Aplidium constellatum*. Mar Biol 142:173–184
Meinertzhagen IA, Okamura Y (2001) The larval ascidian nervous system: the chordate brain from its small beginnings. Trends Neurosci 24:401–410
Miyazaki S-I, Takahashi K, Tsuda K (1972) Calcium and sodium contributions to regenerative responses in the embryonic excitable cell membrane. Science 176:1441–1443
Moody WJ, Okamura Y (2013) Neural development in simpler embryos: a retrospective of Dr. Kunitaro Takahashi's work. Neurosci Res 75:167–170
Moret F, Christiaen L, Deyts C, Blin M, Vernier P, Joly J-S (2005a) Regulatory gene expressions in the ascidian ventral sensory vesicle: evolutionary relationships with the vertebrate hypothalamus. Dev Biol 277:567–579
Moret F, Christiaen L, Deyts C, Blin M, Joly J-S, Vernier P (2005b) The dopamine-synthesizing cells in the

swimming larva of the tunicate *Ciona intestinalis* are located only in the hypothalamus-related domain of the sensory vesicle. Eur J Neurosci 21:3043–3055

Murata Y, Okado H, Katsuyama Y, Okamura Y, Kubo Y (2001) Primary structure, developmental expression and functional properties of an inward rectifier K^+ channel of the tunicate. Receptors Channels 7:387–399

Nagel G, Szellas T, Huhn W, Kateriya S, Adeishvili N, Berthold P, Ollig D, Hegemann P, Bamberg E (2003) Channelrhodopsin-2, a directly light-gated cation-selective membrane channel. Proc Natl Acad Sci U S A 100:13940–13945

Nakagawa M, Miyamoto T, Ohkuma M, Tsuda M (1999) Action spectrum for the photophobic response of *Ciona intestinalis* (Ascidiacea, Urochordata) larvae implicates retinal protein. Photochem Photobiol 70:359–362

Nakajo K, Chen L, Okamura Y (1999) Cross-coupling between voltage-dependent Ca^{2+} channels and ryanodine receptors in developing ascidian muscle blastomeres. J Physiol 515:695–710

Nakajo K, Katsuyama Y, Ono F, Ohtsuka Y, Okamura Y (2003) Primary structure, functional characterization and developmental expression of the ascidian K_V4-class potassium channel. Neurosci Res 45:59–70

Nakayama-Ishimura A, Chambon J-P, Horie T, Satoh N, Sasakura Y (2009) Delineating metamorphic pathways in the ascidian *Ciona intestinalis*. Dev Biol 326:357–367

Negishi T, Yasuo H (2015) Distinct modes of mitotic spindle orientation align cells in the dorsal midline of ascidian embryos. Dev Biol 408:66–78

Nicol D, Meinertzhagen IA (1988a) Development of the central nervous system of the larva of the ascidian, *Ciona intestinalis* L. I. The early lineages of the neural plate. Dev Biol 130:721–736

Nicol D, Meinertzhagen IA (1988b) Development of the central nervous system of the larva of the ascidian, *Ciona intestinalis* L. II. Neural plate morphogenesis and cell lineages during neurulation. Dev Biol 130:737–766

Nicol D, Meinertzhagen IA (1991) Cell counts and maps in the larval central nervous system of the ascidian *Ciona intestinalis* (L.) J Comp Neurol 309:415–429

Nishida H, Satoh N (1985) Cell lineage analysis in ascidian embryos by intracellular injection of a tracer enzyme. II. The 16- and 32-cell stages. Dev Biol 110:440–454

Nishide K, Mugitani M, Kumano G, Nishida H (2012) Neurula rotation determines left-right asymmetry in ascidian tadpole larvae. Development 139:1467–1475

Nishino A, Okamura Y, Poscopo S, Brown ER (2010) A glycine receptor is involved in the organization of swimming movements in an invertebrate chordate. BMC Neurosci 11:6

Nishino A, Baba SA, Okamura Y (2011) A mechanism for graded motor control encoded in the channel properties of the muscle ACh receptor. Proc Natl Acad Sci U S A 108:2599–2604

Nishitsuji K, Horie T, Ichinose A, Sasakura Y, Yasuo H, Kusakabe TG (2012) Cell lineage and cis-regulation for a unique GABAergic/glycinergic neuron type in the larval nerve cord of the ascidian *Ciona intestinalis*. Develop Growth Differ 54:177–186

Ogura Y, Sakaue-Sawano A, Nakagawa M, Satoh N, Miyawaki A, Sasakura Y (2011) Coordination of mitosis and morphogenesis: role of a prolonged G2 phase during chordate neurulation. Development 138:577–587

Ohmori H, Sasaki S (1977) Development of neuromuscular transmission in a larval tunicate. J Physiol 269:221–254

Ohta N, Horie T, Satoh N, Sasakura Y (2010) Transposon-mediated enhancer detection reveals the location, morphology and development of the cupular organs, which are putative hydrodynamic sensors, in the ascidian *Ciona intestinalis*. Zool Sci 27:842–850

Ohtsuka Y, Okamura Y (2007) Voltage-dependent calcium influx mediates maturation of myofibril arrangement in ascidian larval muscle. Dev Biol 301:361–373

Ohtsuka Y, Okamura Y, Obinata T (2001) Changes in gelsolin expression during ascidian metamorphosis. Dev Genes Evol 211:252–256

Ohtsuka Y, Matsumoto J, Katsuyama Y, Okamura Y (2014) Nodal signaling regulates specification of ascidian peripheral neurons through control of the BMP signal. Development 141:3889–3899

Okada T, Hirano H, Takahashi K, Okamura Y (1997) Distinct neuronal lineages of the ascidian embryo revealed by expression of a sodium channel gene. Dev Biol 190:257–272

Okada T, MacIsaac SS, Katsuyama Y, Okamura Y, Meinertzhagen IA (2001) Neuronal form in the central nervous system of the tadpole larva of the ascidian *Ciona intestinalis*. Biol Bull 200:252–256

Okada T, Katsuyama Y, Ono F, Okamura Y (2002) The development of three identified motor neurons in the larva of an ascidian, *Halocynthia roretzi*. Dev Biol 244:278–292

Okado H, Takahashi K (1988) A simple "neural induction" model with two interacting cleavage-arrested ascidian blastomeres. Proc Natl Acad Sci U S A 85:6197–6201

Okado H, Takahashi K (1990a) Differentiation of membrane excitability in isolated cleavage-arrested blastomeres from early ascidian embryos. J Physiol 427:583–602

Okado H, Takahashi K (1990b) Induced neural-type differentiation in the cleavage-arrested blastomere isolated from early ascidian embryos. J Physiol 427:603–623

Okado H, Takahashi K (1993) Neural differentiation in cleavage-arrested ascidian blastomeres induced by a proteolytic enzyme. J Physiol 463:269–290

Okagaki R, Izumi H, Okada T, Nagahora H, Nakajo K, Okamura Y (2001) The maternal transcript for truncated voltage-dependent Ca^{2+} channels in the ascidian embryo: a potential suppressive role in Ca^{2+} channel expression. Dev Biol 230:258–277

Okamura Y, Shidara M (1987) Kinetic differences between Na channels in the egg and in the neurally differentiated blastomere in the tunicate. Proc Natl Acad Sci U S A 84:8702–8706

Okamura Y, Shidara M (1990a) Changes in sodium channels during neural differentiation in the isolated blastomere of the ascidian embryo. J Physiol 431:39–74

Okamura Y, Shidara M (1990b) Inactivation kinetics of the sodium channel in the egg and the isolated, neurally differentiated blastomere of the ascidian. J Physiol 431:75–102

Okamura Y, Takahashi K (1993) Neural induction suppresses early expression of the inward-rectifier K^+ channel in the ascidian blastomere. J Physiol 463:245–268

Okamura Y, Ono F, Okagaki R, Chong JA, Mandel G (1994) Neural expression of a sodium channel gene requires cell-specific interactions. Neuron 13:937–948

Okamura Y, Nishino A, Murata Y, Nakajo K, Iwasaki H, Ohtsuka Y, Tanaka-Kunishima M, Takahashi N, Hara Y, Yoshida T, Nishida M, Okado H, Watari H, Meinertzhagen IA, Satoh N, Takahashi K, Satou Y, Okada Y, Mori Y (2005) Comprehensive analysis of the ascidian genome reveals novel insights into the molecular evolution of ion channel genes. Physiol Genomics 22:269–282

Ono F, Katsuyama Y, Nakajo K, Okamura Y (1999) Subfamily-specific posttranscriptional mechanism underlies K^+ channel expression in a developing neuronal blastomere. J Neurosci 19:6874–6886

Oonuma K, Tanaka M, Nishitsuji K, Kato Y, Shimai K, Kusakabe TG (2016) Revised lineage of larval photoreceptor cells in *Ciona* reveals archetypal collaboration between neural tube and neural crest in sensory organ formation. Dev Biol 420:178–185

Pasini A, Amiel A, Rothbacher RA, Lemaire P, Darras S (2006) Formation of the ascidian epidermal sensory neurons: insights into the origin of the chordate peripheral nervous system. PLoS Biol 4:e225

Pasini A, Manenti R, Rothbächer LP (2012) Antagonizing retinoic acid and FGF/MAPK pathway control posterior body patterning in the invertebrate chordate *Ciona intestinalis*. PLoS One 7:e46193

Razy-Krajka F, Brown ER, Horie T, Callebert J, Sasakura Y, Joly J-S, Kusakabe TG, Vernier P (2012) Monoaminergic modulation of photoreception in ascidian: evidence for a proto-hypothalamo-retinal territory. BMC Biol 10:45

Razy-Krajka F, Lam K, Wang W, Stolfi A, Joly M, Bonneau R, Christiaen L (2014) Collier/OLF/EBF-dependent transcriptional dynamics control pharyngeal muscle specification from primed cardiopharyngeal progenitors. Dev Cell 29:263–276

Ryan K, Lu Z, Meinertzhagen IA (2016) The CNS connectome of a tadpole larva of *Ciona intestinalis* (L.) highlights sidedness in the brain of a chordate sibling. Elife 5:e16962

Ryan K, Lu Z, Meinertzhagen IA (2017) Circuit homology between decussating pathways in the *Ciona* larval CNS and the vertebrate startle-response pathway. Curr Biol 27:721–728

Sakurai D, Gada M, Kohmura Y, Horie T, Iwamoto H, Ohtsuki H, Tsuda M (2004) The role of pigment cells in the brain of ascidian larva. J Comp Neurol 475:70–82

Sasaki H, Yoshida K, Hozumi A, Sasakura Y (2014) CRISPR/Cas9-mediated gene knockout in the ascidian *Ciona intestinalis*. Develop Growth Differ 56:499–510

Sasakura Y, Mita K, Ogura Y, Horie T (2012) Ascidians as excellent chordate models for studying the development of the nervous system during embryogenesis and metamorphosis. Develop Growth Differ 54:420–437

Satoh N (2014) Developmental genomics of ascidians. Wiley-Blackwell

Satoh N (2016) Chordate origins and evolution. Academic

Shidara M, Okamura Y (1991) Developmental changes in delayed rectifier K^+ currents in the muscular- and neural-type blastomere of ascidian embryos. J Physiol 443:277–305

Shimeld SM, Levin M (2006) Evidence for the regulation of left-right asymmetry in *Ciona intestinalis* by ion flux. Dev Dyn 235:1543–1553

Simoncini L, Block ML, Moody WJ (1988) Lineage-specific development of calcium currents during embryogenesis. Science 242:1572–1575

Stolfi A, Levine M (2011) Neuronal subtype specification in the spinal cord of a protovertebrate. Development 138:995–1004

Stolfi A, Gainous B, Young JJ, Mori A, Levine M, Christiaen L (2010) Early chordate origins of the vertebrate second heart field. Science 329:565–568

Stolfi A, Gandhi S, Salek F, Christiaen L (2014) Tissue-specific genome editing in *Ciona* embryos by CRISPR/Cas9. Development 141:4115–4120

Stolfi A, Ryan K, Meinertzhagen IA, Christiaen L (2015) Migratory neuronal progenitors arise from the neural plate borders in tunicates. Nature 527:371–374

Sulston JE, Schierenberg E, White JG, Thomson JN (1983) The embryonic cell lineage of the nematode *Caenorhabditis elegans*. Dev Biol 100:64–119

Svane I, Young CM (1989) The ecology and behavior of ascidian larvae. Oceanogr Mar Biol Rev 27:45–90

Takahashi K, Okamura Y (1998) Ion channels and early development of neural cells. Physiol Rev 78:307–337

Takahashi K, Yoshii M (1981) Development of sodium, calcium and potassium channels in the cleavage-arrested embryo of an ascidian. J Physiol 315:515–529

Takahashi K, Miyazaki S, Kidokoro Y (1971) Development of excitability in embryonic muscle cell membranes in certain tunicates. Science 171:415–417

Takamura K (1998) Nervous network in larvae of the ascidian *Ciona intestinalis*. Dev Genes Evol 208:1–8

Takamura K, Egawa T, Ohnishi S, Okada T, Fukuoka T (2002) Developmental expression of ascidian neurotransmitter synthesis genes. I. Choline acetyltransferase and acetylcholine transporter genes. Dev Genes Evol 212:50–53

Takamura K, Minamida N, Okabe S (2010) Neural map of the larval central nervous system in the ascidian *Ciona intestinalis*. Zool Sci 27:191–203

Taniguchi K, Nishida H (2004) Tracing cell fate in brain formation during embryogenesis of the ascidian *Halocynthia roretzi*. Develop Growth Differ 46:163–180

Terakubo HQ, Nakajima Y, Sasakura Y, Horie T, Konno A, Takahashi H, Inaba K, Hotta K, Oka K (2010) Network structure of projections extending from peripheral neurons in the tunic of ascidian larva. Dev Dyn 239:2278–2287

Torrence SA (1986) Sensory endings of the ascidian static organ (Chordata, Ascidiacea). Zoomorphology 106:61–66

Torrence SA, Cloney RA (1982) Nervous system of ascidian larvae: caudal primary sensory neurons. Zoomorphology 99:103–115

Torrence SA, Cloney RA (1983) Ascidian larval nervous system: primary sensory neurons in adhesive papillae. Zoomorphology 102:111–123

Treen N, Yoshida K, Sakuma T, Sasaki H, Kawai N, Yamamoto T, Sasakura Y (2014) Tissue-specific and ubiquitous gene knockouts by TALEN electroporation provide new approaches to investigating gene function in *Ciona*. Development 141:481–487

Tsuda M, Sakurai D, Goda M (2003a) Direct evidence for the role of pigment cells in the brain of ascidian larvae by laser ablation. J Exp Biol 206:1409–1417

Tsuda M, Kawakami I, Shiraishi S (2003b) Sensitization and habituation of the swimming behavior in ascidian larvae to light. Zool Sci 20:13–22

Tsutsui H, Oka Y (2000) Light-sensitive voltage responses in the neurons of the cerebral ganglion of *Ciona savignyi* (Chordata: Ascidiacea). Biol Bull 198:26–28

White JG, Southgate E, Thomson JN, Brenner S (1986) The structure of the nervous system of the nematode *Caenorhabditis elegans*. Philos Trans R Soc Lond B 314:1–340

Willey A (1894) Amphioxus and the ancestry of the vertebrates. Macmillan, New York

Yokoyama TD, Hotta K, Oka K (2014) Comprehensive morphological analysis of individual peripheral neuron dendritic arbors in ascidian larvae using the photoconvertible protein Kaede. Dev Dyn 243:1362–1373

Zanetti L, Ristoratore F, Francone M, Piscopo S, Brown ER (2007) Primary cultures of nervous system cells from the larva of the ascidian *Ciona intestinalis*. J Neurosci Methods 165:191–197

Zaniolo G, Lane NJ, Burighel P, Manni L (2002) Development of the motor nervous system in ascidians. J Comp Neurol 443:124–135

Zega G, Thorndyke MC, Brown ER (2006) Development of swimming behaviour in the larva of the ascidian *Ciona intestinalis*. J Exp Biol 209:3405–3412

Zega G, Biggiogero M, Groppelli S, Candiani S, Oliveri D, Parodi M, Pestarino M, De Bernardi F, Pennati R (2008) Developmental expression of glutamic acid decarboxylase and of γ-aminobutyric acid type B receptors in the ascidian *Ciona intestinalis*. J Comp Neurol 506:489–505

Zhang F, Aravanis AM, Adamantidis A, de Lacea L, Deisseroth K (2007) Circuit-breakers: optical technologies for probing neural signals and systems. Nat Rev Neurosci 8:577–581

Index

Y. Sasakura (ed.), *Transgenic Ascidians*, Advances in Experimental Medicine and Biology 1029,
https://doi.org/10.1007/978-981-10-7545-2

MIX
Papier aus verantwortungsvollen Quellen
Paper from responsible sources
FSC® C105338

If you have any concerns about our products,
you can contact us on
ProductSafety@springernature.com

In case Publisher is established outside the EU,
the EU authorized representative is:
Springer Nature Customer Service Center GmbH
Europaplatz 3, 69115 Heidelberg, Germany

Printed by Libri Plureos GmbH
in Hamburg, Germany